T0136965

Emerging Technology and Architecture for Big-data Analytics

Anupam Chattopadhyay • Chip Hong Chang
Hao Yu

Editors

Emerging Technology and Architecture for Big-data Analytics

 Springer

Editors
Anupam Chattopadhyay
School of Computer Science
    and Engineering, School of Physical
    and Mathematical Sciences
Nanyang Technological University
Singapore

Chip Hong Chang
School of Electrical and Electronic
    Engineering
Nanyang Technological University
Singapore

Hao Yu
School of Electrical and Electronic
    Engineering
Nanyang Technological University
Singapore

ISBN 978-3-319-85497-7   ISBN 978-3-319-54840-1   (eBook)
DOI 10.1007/978-3-319-54840-1

Printed on acid-free paper

This Springer imprint is published by Springer Nature
The registered company is Springer International Publishing AG
The registered company address is: Gewerbestrasse 11, 6330 Cham, Switzerland

Preface

Everyone loves to talk about big data, of course for various reasons. We got into that discussion when it seemed that there is a serious problem that big data is throwing down to the system, architecture, circuit and even device specialists. The problem is of scale, of which everyday computing experts were not really aware of. The last big wave of computing is driven by embedded systems and all the infotainment riding on top of that. Suddenly, it seemed that people loved to push the envelope of data and it does not stop growing at all.

According to a recent estimate done by Cisco® Visual Networking Index (VNI), global IP traffic crossed the zettabyte threshold in 2016 and grows at a compound annual growth rate of 22%. Now, zettabyte is 10^{18} bytes, which is something that might not be easily appreciated. To give an everyday comparison, take this estimate. The amount of data that is created and stored somewhere in the Internet is 70 times that of the world's largest library—Library of Congress in Washington DC, USA. Big data is, therefore, an inevitable outcome of the technological progress of human civilization. What lies beneath that humongous amount of information is, of course, knowledge that could very much make or break business houses. No wonder that we are now rolling out course curriculum to train data scientists, who are gearing more than ever to look for a needle in the haystack, literally. The task is difficult, and here enters the new breed of system designers, who might help to downsize the problem.

The designers' perspectives that are trickling down from the big data received considerable attention from top researchers across the world. Upfront, it is the storage problem that had to be taken care of. Denser and faster memories are very much needed, as ever. However, big data analytics cannot work on idle data. Naturally, the next vision is to reexamine the existing hardware platform that can support intensive data-oriented computing. At the same time, the analysis of such a huge volume of data needs a scalable hardware solution for both big data storage and processing, which is beyond the capability of pure software-based data analytic solutions. The main bottleneck that appeared here is the same one, known in computer architecture community for a while—memory wall. There is a growing mismatch between the access speed and processing speed for data. This disparity no doubt will affect the big data analytics the hardest. As such, one needs to redesign an energy-efficient hardware platform for future big data-driven computing. Fortunately, there are novel and promising researches that appeared in this direction.

A big data-driven application also requires high bandwidth with maintained low-power density. For example, Web-searching application involves crawling, comparing, ranking, and paging of billions of Web pages or images with extensive memory access. The microprocessor needs to process the stored data with intensive memory access. The present data storage and processing hardware have well-known bandwidth wall due to limited accessing bandwidth at I/Os, but also power wall due to large leakage power in advanced CMOS technology when holding data by charge. As such, a design of scalable energy-efficient big data analytic hardware is a highly challenging problem. It reinforces well-known issues, like memory and power wall that affects the smooth downscaling of current technology nodes. As a result, big data analytics will have to look beyond the current solutions—across architectures, circuits, and technologies—to address all the issues satisfactorily.

In this book, we attempt to give a glimpse of the things to come. A range of solutions are appearing that will help a scalable hardware solution based on the emerging technology (such as nonvolatile memory device) and architecture (such as in-memory computing) with the correspondingly well-tuned data analytics algorithm (such as machine learning). To provide a comprehensive overview in this book, we divided the contents into three main parts as follows:

Part I: State-of-the-Art Architectures and Automation for Data Analytics
Part II: New Approaches and Applications for Data Analytics
Part III: Emerging Technology, Circuits, and Systems for Data Analytics

As such, this book aims to provide an insight of hardware designs that capture the most advanced technological solutions to keep pace with the growing data and support the major developments of big data analytics in the real world. Through this book, we tried our best to justify different perspectives in the growing research domain. Naturally, it would not be possible without the hard work from our excellent contributors, who are well-established researchers in their respective domains. Their chapters, containing state-of-the-art research, provide a wonderful perspective of how the research is evolving and what practical results are to be expected in future.

Singapore   Anupam Chattopadhyay
                                                                                      Chip Hong Chang
                                                                                              Hao Yu

Contents

About the Editors

Anupam Chattopadhyay received his BE degree from Jadavpur University, India, in 2000. He received his MSc from ALaRI, Switzerland, and PhD from RWTH Aachen in 2002 and 2008, respectively. From 2008 to 2009, he worked as a member of consulting staff in CoWare R&D, Noida, India. From 2010 to 2014, he led the MPSoC Architectures Research Group in UMIC Research Cluster at RWTH Aachen, Germany, as a junior professor. Since September 2014, he has been appointed as an assistant professor in the School of Computer Science and Engineering (SCSE), NTU, Singapore. He also holds adjunct appointment at the School of Physical and Mathematical Sciences, NTU, Singapore.

During his PhD, he worked on automatic RTL generation from the architecture description language LISA, which was commercialized later by a leading EDA vendor. He developed several high-level optimizations and verification flow for embedded processors. In his doctoral thesis, he proposed a language-based modeling, exploration, and implementation framework for partially reconfigurable processors, for which he received outstanding dissertation award from RWTH Aachen, Germany.

Since 2010, Anupam has mentored more than ten PhD students and numerous master's/bachelor's thesis students and several short-term internship projects. Together with his doctoral students, he proposed domain-specific high-level synthesis for cryptography, high-level reliability estimation flows, generalization of classic linear algebra kernels, and a novel multilayered coarse-grained reconfigurable architecture. In these areas, he published as a (co)author over 100 conference/journal papers, several book chapters for leading press, e.g., Springer, CRC, and Morgan Kaufmann, and a book with Springer. Anupam served in several TPCs of top conferences like ACM/IEEE DATE, ASP-DAC, VLSI, VLSI-SoC, and ASAP. He regularly reviews journal/conference articles for ACM/IEEE DAC, ICCAD, IEEE TVLSI, IEEE TCAD, IEEE TC, ACM JETC, and ACM TEC; he also reviewed book proposal from Elsevier and presented multiple invited seminars/tutorials in prestigious venues. He is a member of ACM and a senior member of IEEE.

Chip Hong Chang received his BEng (Hons) degree from the National University of Singapore in 1989 and his MEng and PhD degrees from Nanyang Technological University (NTU) of Singapore, in 1993 and 1998, respectively. He served as a technical consultant in the industry prior to joining the School of Electrical and Electronic Engineering (EEE), NTU, in 1999, where he is currently a tenure associate professor. He holds joint appointments with the university as assistant chair of School of EEE from June 2008 to May 2014, deputy director of the 100-strong Center for High Performance Embedded Systems from February 2000 to December 2011, and program director of the Center for Integrated Circuits and Systems from April 2003 to December 2009. He has coedited four books, published 10 book chapters, 87 international journal papers (of which 54 are published in the IEEE Transactions), and 158 refereed international conference papers. He has been well recognized for his research contributions in hardware security and trustable computing, low-power and fault-tolerant computing, residue number systems, and digital filter design. He mentored more than 20 PhD students, more than 10 MEng and MSc research students, and numerous undergraduate student projects.

Dr. Chang had been an associate editor for the IEEE Transactions on Circuits and Systems I from January 2010 to December 2012 and has served IEEE Transactions on Very Large Scale Integration (VLSI) Systems since 2011, IEEE Access since March 2013, IEEE Transactions on Computer-Aided Design of Integrated Circuits and Systems since 2016, IEEE Transactions on Information Forensic and Security since 2016, Springer Journal of Hardware and System Security since 2016, and Microelectronics Journal since May 2014. He had been an editorial advisory board member of the Open Electrical and Electronic Engineering Journal since 2007 and an editorial board member of the Journal of Electrical and Computer Engineering since 2008. He also served Integration, the VLSI Journal from 2013 to 2015. He also guest-edited several journal special issues and served in more than 50 international conferences (mostly IEEE) as adviser, general chair, general vice chair, and technical program cochair and as member of technical program committee. He is a member of the IEEE Circuits and Systems Society VLSI Systems and Applications Technical Committee, a senior member of the IEEE, and a fellow of the IET.

Dr. Hao Yu obtained his BS degree from Fudan University (Shanghai China) in 1999, with 4-year first-prize Guanghua scholarship (top 2) and 1-year Samsung scholarship for the outstanding student in science and engineering (top 1). After being selected by mini-CUSPEA program, he spent some time in New York University and obtained MS/PhD degrees both from electrical engineering department at UCLA in 2007, with major in integrated circuit and embedded computing. He has been a senior research staff at Berkeley Design Automation (BDA) since 2006, one of top 100 start-ups selected by Red Herring at Silicon Valley. Since October 2009, he has been an assistant professor at the School of Electrical and Electronic Engineering and also an area director of VIRTUS/VALENS Centre of Excellence, Nanyang Technological University (NTU), Singapore.

Dr. Yu has 165 peer-reviewed and referred publications [conference (112) and journal (53)], 4 books, 5 book chapters, 1 best paper award in ACM Transactions on Design Automation of Electronic Systems (TODAES), 3 best paper award nominations (DAC'06, ICCAD'06, ASP-DAC'12), 3 student paper competition finalists (SiRF'13, RFIC'13, IMS'15), 1 keynote paper, 1 inventor award from semiconductor research cooperation (SRC), and 7 patent applications in pending. He is the associate editor of Journal of Low Power Electronics; reviewer of IEEE TMTT, TNANO, TCAD, TCAS-I/II, TVLSI, ACM-TODAEs, and VLSI Integration; and a technical program committee member of several conferences (DAC'15, ICCAD'10-12, ISLPED'13-15, A-SSCC'13-15, ICCD'11-13, ASP-DAC'11-13'15, ISCAS'10-13, IWS'13-15, NANOARCH'12-14, ISQED'09). His main research interest is about the emerging technology and architecture for big data computing and communication such as 3D-IC, THz communication, and nonvolatile memory with multimillion government and industry funding. His industry work at BDA is also recognized with an EDN magazine innovation award and multimillion venture capital funding. He is a senior member of IEEE and member of ACM.

Part I
State-of-the-Art Architectures and Automation for Data-Analytics

Chapter 1
Scaling the Java Virtual Machine on a Many-Core System

Karthik Ganesan, Yao-Min Chen, and Xiaochen Pan

1.1 Introduction

Today, many big data applications use the Java SE platform [13], also called Java Virtual Machine (JVM), as the run-time environment. Examples of such applications include Hadoop Map Reduce [1], Apache Spark [3], and several graph processing platforms [2, 11]. In this chapter, we call these applications the *JVM applications*. Such applications can benefit from modern multicore servers with large memory capacity and the memory bandwidth needed to access it. However, with the enormous amount of data to process, it is still a challenging mission for the JVM platform to scale well with respect to the needs of big data applications. Since the JVM is a multithreaded application, one needs to ensure that the JVM performance can scale well with the number of threads. Therefore, it is important to understand and improve performance and scalability of JVM applications on these multicore systems.

To be able to scale JVM applications most efficiently, the JVM and the various libraries must be scalable across multiple cores/processors and be capable of handling heap sizes that can potentially run into a few hundred gigabytes for some applications. While such scaling can be achieved by scaling-out (multiple JVMs) or scaling-up (single JVM), each approach has its own advantages, disadvantages, and performance implications. Scaling-up, also known as vertical scaling, can be very challenging compared to scaling-out (also known as horizontal scaling), but also has a great potential to be resource efficient and opens up the possibility

K. Ganesan
Oracle Corporation, 5300 Riata Park Court Building A, Austin, TX 78727, USA
e-mail: karthik.ganesan@oracle.com

Y.-M. Chen (✉) • X. Pan
Oracle Corporation, 4180 Network Circle, Santa Clara, CA 95054, USA
e-mail: yaomin.chen@oracle.com; deb.pan@oracle.com

© Springer International Publishing AG 2017
A. Chattopadhyay et al. (eds.), *Emerging Technology and Architecture for Big-data Analytics*, DOI 10.1007/978-3-319-54840-1_1

for features like multi-tenancy. If done correctly, scaling-up usually can achieve higher CPU utilization, putting the servers operating in a more resource and energy efficient state. In this work, we restrict ourselves to the challenges of scaling-up on enterprise-grade systems to provide a focused scope. We elaborate on the various performance bottlenecks that ensue when we try to scale up a single JVM to multiple cores/processors, discuss the potential performance degradation that can come out of these bottlenecks, provide solutions to alleviate these bottlenecks, and evaluate their effectiveness using a representative Java workload.

To facilitate our performance study we have chosen a business analytics workload written in the Java language because Java is one of the most popular programming languages with many existing applications built on it. Optimizing JVM for a representative Java workload would benefit many JVM applications running on the same platform. Towards this purpose, we have selected the LArge Memory Business Data Analytics (LAMBDA) workload. It is derived from the SPECjbb2013 benchmark,[1,2] developed by Standard Performance Evaluation Corporation (SPEC) to measure Java server performance based on the latest features of Java [15]. It is a server side benchmark that models a world-wide supermarket company with multiple point-of-sale stations, multiple suppliers, and a headquarter office which manages customer data. The workload stores all its retail business data in memory (Java heap) without interacting with an external database that stores data on disks. For our study we modify the benchmark in such a way as to scale to very large Java heaps (hundreds of GBs). We condition its run parameter setting so that it will not suffer from an abnormal scaling issue due to inventory depletion.

As an example, Fig. 1.1 shows the throughput performance scaling on our workload as we increase the number of SPARC T5 CPU cores from one to 16.[3] By

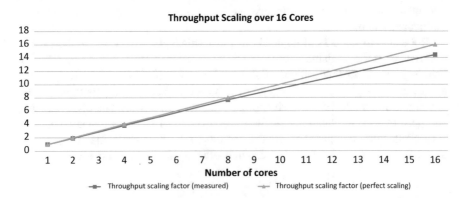

Fig. 1.1 Single JVM scaling on a SPARC T5 server, running the LAMBDA workload

---

[1]The use of SPECjbb2013 benchmark conforms to SPEC Fair Use Rule [16] for research use.

[2]The SPECjbb2013 benchmark has been retired by SPEC.

[3]Experimental setup for this study is described in Sect. 1.2.3.

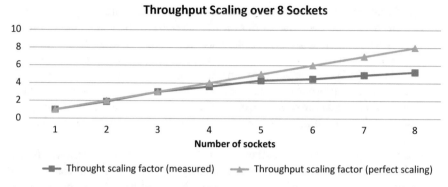

Fig. 1.2 Single JVM scaling on a SPARC M6 server with JDK8 Build 95

contrast, the top ("perfect scaling") curve shows the ideal case where the throughput increases linearly with the number of cores. In reality, there is likely certain system level, OS, Java VM, or application bottleneck to prevent the applications from scaling linearly. And quite often it is a combination of multiple factors that causes the scaling to be non-linear. The main goal of the work described in this chapter is to facilitate application scaling to be as close to linear as possible.

As an example of sub-optimal scaling, Fig. 1.2 shows the throughput performance scaling on our workload as we increase the number of SPARC M6 CPU nsockets from one to eight.[4] There are eight processors ("sockets") on an M6-8 server, and we can run the workload subject to using only the first N sockets. By contrast, the top ("perfect scaling") curve shows the ideal case where the throughput increases linearly with the number of sockets. Below, we discuss briefly the common factors that lead to sub-optimal scaling. We will expand on the key ideas later in this chapter.

1. *Sharing of data objects.* When shared objects that are rarely written to are cached locally, they have the potential to reduce space requirements and increase efficiency. But, the same shared objects can become a bottleneck when being frequently written to, incurring remote memory access latency in the order of hundreds of CPU cycles. Here, a remote memory access can mean accessing the memory not affined to the local CPU, as in a Non-Uniform Memory Access (NUMA) system [5], or accessing a cache that is not affined to the local core, in both cases resulting in a migratory data access pattern [8]. Localized implementations of such shared data objects have proven to be very helpful in improving scalability. A case study that we use to explain this is the concurrent hash map initialization that uses a shared random seed to randomize the layout of hash maps. This shared random seed object causes major synchronization overhead when scaling an application like LAMBDA which creates many transient hash maps.

---

[4]Experimental setup for this study is described in Sect. 1.2.3.

2. *Application and system software locks.* On large systems with many cores, locks in both user code and system libraries for serialized implementations can be equally lethal in disrupting application scaling. Even standard system calls like malloc in *libc* library tend to have serial portions which are protected by per-process locks. When the same system call is invoked concurrently by multiple threads of same process on a many-core system, these locks around serial portions of implementation become a critical bottleneck. Special implementations of memory allocator libraries like MT hot allocators [18] are available to alleviate such bottlenecks.

3. *Concurrency framework.* Another major challenge involved in scaling is due to inefficient implementations of concurrency frameworks and collection data structures (e.g., concurrent hash maps) using low level Java concurrency control constructs. Utilizing concurrency utilities like JSR166 [10] that provide high quality scalable implementations of concurrent collections and frameworks has a significant potential to improve scalability of applications. One such example is performance improvement of 57% for a workload like LAMBDA derived out of a standard benchmark when using JSR166.

4. *Garbage collection.* As a many-core system is often provisioned with a proportionally large amount of memory, another major challenge in scaling a single JVM on a large enterprise system involves efficiently scaling the Garbage Collection (GC) algorithm to handle huge heap sizes. From our experience, garbage collection pause times (stop-the-world young generation collections) can have a significant effect on the response time of application transactions. These pause times typically tend to be proportional to the nursery size of the Java heap. To reduce the pause times, one solution is to eliminate serial portions of GC phases, parallelizing them to remove such bottlenecks. One such case study includes improvements to the G1 GC [6] to handle large heaps and a parallelized implementation of "*Free Cset*" phase of G1, which has the potential to improve the throughput and response time on a large SPARC system.

5. *NUMA.* The time spent collecting garbage can be compounded due to remote memory accesses on a NUMA based system if the GC algorithm is oblivious to the NUMA characteristics of the system. Within a processor, some cache memories closest to the core can have lower memory access latencies compared to others and similarly across processors of a large enterprise system, some memory banks that are closest to the processor can have lower access latencies compared to remote memory banks. Thus, incorporating the NUMA awareness into the GC algorithm can potentially improve scalability. Most of the scaling bottlenecks that arise out of locks on a large system also tend to become worse on NUMA systems as most of the memory accesses to lock variables end up being remote memory accesses.

The different scalability optimizations discussed in this chapter are accomplished by improving the system software like the Operating System or the Java Virtual Machine instead of changing the application code. The rest of the chapter is organized as follows: Sect. 1.2 provides the background including the methodologies and tools used in the study and the experimental setup. Section 1.3 addresses the sharing of data objects. Section 1.4 describes the scaling of memory allocators. Section 1.5 expounds on the effective usage of concurrency API. Section 1.6 elaborates on scalable Garbage Collection. Section 1.7 discusses scalability issues in NUMA systems and Sect. 1.8 concludes with future directions.

1.2 Background

The scaling study is often an iterative process as shown in Fig. 1.3. Each iteration consists of four phases: workload characterization, bottleneck identification, performance optimization, and performance evaluation. The goal of each iteration is to remove one or more performance bottlenecks to improve performance. It is an iterative process because a bottleneck may hide other performance issues. When the bottleneck is removed, performance scaling may still be limited by another bottleneck or improvement opportunities which were previously overshadowed by the removed bottleneck.

1. *Workload characterization.* Each iteration starts with characterization using a representative workload. Section 1.2.1 describes selecting a representative workload for this purpose. During workload characterization, performance tools are used in monitoring and capturing key run-time status information and statistics. Performance tools will be described in more detail in Sect. 1.2.2. The result of the characterization is a collection of profiles that can be used in the bottleneck identification phase.
2. *Bottleneck identification.* This phase typically involves modeling, hypothesis testing, and empirical analysis. Here, a bottleneck refers to the cause, or limiting factor, for sub-optimal scaling. The bottleneck often points to, but is not limited to, inefficient process, thread or task synchronization, an inferior algorithm or sub-optimal design and code implementation.
3. *Performance optimization.* Once a bottleneck is identified in the previous phase, in the current phase we try to work out an alternative design or implementation to alleviate the bottleneck. Several possible implementations may be proposed and a comparative study can be conducted to select the best alternative. This phase itself can be an iterative process where several alternatives are evaluated either through analysis or through actual prototyping and subsequent testing.

Fig. 1.3 Iterative process for performance scaling: (1) workload characterization, (2) bottleneck identification, (3) performance optimization, and (4) performance evaluation

4. *Performance evaluation.* With the implementation from the performance optimization work in the previous phase, we evaluate whether the performance scaling goal is achieved. If the goal is not yet reached even with the current optimization, we go back to the workload characterization phase and start another iteration.

At each iteration, Amdahl's law [9] is put to practice in the following sense. The goal of many-core scaling is to minimize the serial portion of the execution and maximize the degree of parallelism (DOP) whenever parallel execution is possible. For applications running on enterprise servers, the problem can be solved by resolving issues in the hardware and the software levels. At the hardware level, multiple hardware threads can share an execution pipeline and when a thread is stalled from loading data from memory, other threads can proceed with useful instruction execution in the pipeline. Similarly, at the software level, multiple software threads are mapped to these hardware threads by the operating system in a time-shared fashion. To achieve maximum efficiency, sufficient number of software threads or processes are needed to keep feeding sequences of instructions to ensure that the processing pipelines are busy. A software thread or process being blocked (such as when waiting for a lock) can lead to reduction in parallelism. Similarly, shared hardware resources can potentially reduce parallelism in execution due to hardware constraints. While the problem, as defined above, consists of software-level and hardware-level issues, in this chapter we focus on the software-level issues and consider the hardware micro-architecture as a given constraint to our solution space.

The iterative process continues until the performance scaling goal is reached or adjusted to reflect what is actually feasible.

1.2.1 Workload Selection

In order to expose effectively the scaling bottlenecks of Java libraries and the JVM, one needs to use a Java workload that can scale to multiple processors and large heap sizes from within a single JVM without any inherent scaling problems in the application design. It is also desirable to use a workload that is sensitive to GC pause times as the garbage collector is one of the components that is most difficult to scale when it comes to using large heap sizes and multiple processors. We have found the LAMBDA workload quite suitable for this investigation. The workload implements a usage model based on a world-wide supermarket company with an IT infrastructure that handles a mix of point-of-sale requests, online purchases, and data-mining operations. It exercises modern Java features and other important performance elements, including the latest data formats (XML), communication using compression, and messaging with security. It utilizes features such as the fork-join pool framework and concurrent hash maps, and is very effective in exercising JVM components such as Garbage Collector by tracking response times as small as 10 ms in granularity. It also provides support for virtualization and cloud environments.

The workload is designed to be inherently scalable, both horizontally and vertically using the run modes called *multi-JVM* and *composite* modes respectively. It contains various aspects of e-commerce software, yet no database system is used. As a result, the benchmark is very easy to install and use. The workload produces two final performance metrics: maximum throughput (operations per second) and weighted throughput (operations per second) under response time constraint. Maximum throughput is defined as the maximum achievable injection rate on the System under Test (SUT) until it becomes unsettled. Similarly weighted throughput is defined as the geometric mean of maximum achievable Injection Rates (IR) for a set of response time Service Level Agreements (SLAs) of 10, 50, 100, 200, and 500 ms using the 99th percentile data. The maximum throughput metric is a good measurement of maximum processing capacity, while the weighted throughput gives good indication of the responsiveness of the application running on a server.

1.2.2   Performance Analysis Tools

To study application performance scaling, performance observability tools are needed to illustrate what happens inside a system when running a workload. The performance tools used for our study include Java GC logs, Solaris operating system utilities including cpustat, prstat, mpstat, lockstat, and the Solaris Studio Performance Analyzer.

1. *GC logs.* The logs are very vital in understanding the time spent in garbage collection, allowing us to specify correctly JVM settings targeting the most efficient way to run the workload achieving the least overhead from GC pauses when scaling to multiple cores/processors. An example segment is shown in Fig. 1.4, for the G1 GC [6]. There, we see the breakdown of a stop-the-world (STW) GC event that lasts 0.369 s. The total pause time is divided into four parts: Parallel Time, Code Root Fixup, Clear, and Other. The parallel time represents the time spent in the parallel processing by the 25 GC worker threads. The other parts comprise the serial phase of the STW pause. As seen in the example, Parallel Time and Other are further divided into subcomponents, for which statistics are reported. At the end of the log, we also see the heap occupancy changes from 50.2 GB to 3223 MB. The last line describes that the total user time spent by all GC threads consists of 8.10 s in user land and 0.01 s in the system (kernel), while the elapsed real time is 0.37 s.

2. *cpustat.* The Solaris cpustat [12] utility on SPARC uses hardware counters to provide hardware level profiling information such as cache miss rates, accesses to local/remote memory, and memory bandwidth used. These statistics are invaluable in identifying bottlenecks in the system and ensure that we use the system to the fullest potential. Cpustat provides critical information such as system utilization in terms of cycles per instruction (CPI) and its reciprocal instructions per cycle (IPC) statistics, instruction mix, branch prediction related

```
2016-05-18T16:53:58.019-0700: [GC pause (G1 Evacuation Pause) (young), 0.3690834 secs]
   [Parallel Time: 317.8 ms, GC Workers: 25]
      [GC Worker Start (ms): Min: 333072.4, Avg: 333072.7, Max: 333073.0, Diff: 0.6]
      [Ext Root Scanning (ms): Min: 1.1, Avg: 1.7, Max: 4.4, Diff: 3.3, Sum: 43.1]
      [Update RS (ms): Min: 1.7, Avg: 4.7, Max: 6.7, Diff: 5.0, Sum: 116.8]
         [Processed Buffers: Min: 4, Avg: 14.2, Max: 29, Diff: 25, Sum: 355]
      [Scan RS (ms): Min: 61.3, Avg: 63.5, Max: 64.1, Diff: 2.8, Sum: 1587.1]
      [Object Copy (ms): Min: 246.5, Avg: 246.9, Max: 247.7, Diff: 1.2, Sum: 6172.8]
      [Termination (ms): Min: 0.0, Avg: 0.0, Max: 0.0, Diff: 0.0, Sum: 0.2]
      [GC Worker Other (ms): Min: 0.1, Avg: 0.4, Max: 0.7, Diff: 0.6, Sum: 9.9]
      [GC Worker Total (ms): Min: 316.6, Avg: 317.2, Max: 317.8, Diff: 1.2, Sum: 7929.7]
      [GC Worker End (ms): Min: 333389.6, Avg: 333389.9, Max: 333390.2, Diff: 0.6]
   [Code Root Fixup: 0.0 ms]
   [Clear CT: 7.2 ms]
   [Other: 44.0 ms]
      [Choose CSet: 0.1 ms]
      [Ref Proc: 1.6 ms]
      [Ref Enq: 0.1 ms]
      [Free CSet: 33.6 ms]
   [Eden: 47.0G(47.0G)->0.0B(47.1G) Survivors: 992.0M->960.0M Heap: 50.2G(60.0G)-
>3223.2M(60.0G)]
   [Times: user=8.10 sys=0.01, real=0.37 secs]
```

Fig. 1.4 Example of a segment in the Garbage Collector (GC) log showing (1) total GC pause time; (2) time spent in the parallel phase and the number GC worker threads; (3) amounts of time spent in the Code Root Fixup and Clear CT, respectively; (4) amount of time spent in the other part of serial phase; and (5) reduction in heap occupancy due to the GC

```
Section: System Utilization
--------------------
Stat                        Total

CPI per-core (avg.)         0.64
CPI per-thread (avg.)       5.115
CPI per-core (MIPS est.)    0.73
IPC per-core (avg)          1.56
IPC per-socket (avg)        50.00
Core Util (@select)         79.2%
Core Util (@Instr Cnt)      67.6%
```

Fig. 1.5 An example of cpustat output that shows utilization related statistics. In the figure, we only show the System Utilization section, where CPI, IPC, and Core Utilization are reported

statistics, cache and TLB miss rates, and other memory hierarchy related statistics. Figure 1.5 shows a partial cpustat output that provides system utilization related statistics.

3. *prstat and mpstat.* Solaris prstat and mpstat utilities [12] provide resource utilization and context switch information dynamically to identify phase behavior and time spent in system calls in the workload. This information is very useful in finding bottlenecks in the operating system. Figures 1.6 and 1.7 are examples of a prstat and mpstat output, respectively. The prstat utility looks at resource usage from the process point of view. In Fig. 1.6, it shows that at time instant 2:13:11 the JVM process, with process ID 1472, uses 63 GB of memory, 90% of CPU, and 799 threads while running the workload. However, at time 2:24:33,

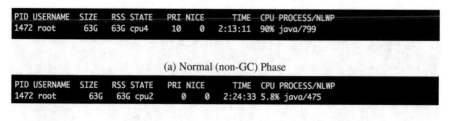

(a) Normal (non-GC) Phase

PID	USERNAME	SIZE	RSS	STATE	PRI	NICE	TIME	CPU	PROCESS/NLWP
1472	root	63G	63G	cpu2	0	0	2:24:33	5.8%	java/475

(b) GC Phase

Fig. 1.6 An example of prstat output that shows dynamic process resource usage information. In (**a**), the JVM process (PID 1472) is on cpu4 and uses 90% of the CPU. By contrast, in (**b**) the process goes into GC and uses 5.8% of cpu2

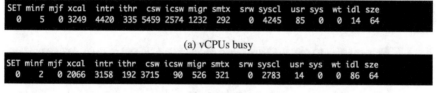

(a) vCPUs busy

SET	minf	mjf	xcal	intr	ithr	csw	icsw	migr	smtx	srw	syscl	usr	sys	wt	idl	sze
0	2	0	2066	3158	192	3715	90	526	321	0	2783	14	0	0	86	64

(b) vCPUs idle

Fig. 1.7 An example of mpstat output. In (**a**) we show the dynamic system activities when the processor set (ID 0) is busy. In (**b**) we show the activities when the processor set is fairly idle

the same process has gone into the garbage collection phase, resulting in CPU usage dropped to 5.8% and the number of threads reduced to 475. By contrast, rather than looking at a process, mpstat takes the view from a vCPU (hardware thread) or a set of vCPUs. In Fig. 1.7 the dynamic resource utilization and system activities of a "processor set" is shown. The processor set, with ID 0, consists of 64 vCPUs. The statistics are taken during a sampling interval, typically one second or 5 s. One can contrast the difference in system activities and resource usage taken during a normal running phase (Fig. 1.7a) and during a GC phase (Fig. 1.7b).

4. *lockstat and plockstat.* Lockstat [12] helps us to identify the time spent spinning on system locks and plockstat [12] provides the same information regarding user locks enabling us to understand the scaling overhead that is coming out of spinning on locks. The plockstat utility provides information in three categories: *mutex block*, *mutex spin*, and *mutex unsuccessful spin*. For each category it lists the time (in nanoseconds) in descending order of the locks. Therefore, on the top of the list is the lock that consumes the most time. Figure 1.8 shows an example of plockstat output, where we only extract the lock on the top from each category. For the mutex block category, the lock at address 0x10015ef00 was called 19 times during the capturing interval (1 s for this example). It was

```
Mutex block

Count     nsec Lock                         Caller
-----------------------------------------------------------------------------
   19    66258 0x10015ef00                  libumem.so.1`umem_cache_alloc+0x50
   31    16040 0x100194e80                  libumem.so.1`umem_cache_alloc+0x50

Mutex spin

Count     nsec Lock                         Caller
-----------------------------------------------------------------------------
  116    38090 0x10015ed00                  libumem.so.1`umem_cache_free+0x68
   92    37586 0x10015ed40                  libumem.so.1`umem_cache_free+0x68

Mutex unsuccessful spin

Count     nsec Lock                         Caller
-----------------------------------------------------------------------------
   49    97353 0x100152e80                  libumem.so.1`umem_cache_alloc+0x50
   42    97960 0x100152d40                  libumem.so.1`umem_cache_alloc+0x50
```

Fig. 1.8 An example of plockstat output, where we show the statistics from three types of locks

called by "libumem.so.1'umem_cache_alloc+0x50" and consumed 66258 ns of
CPU time. The locks in the other categories, mutex spin and mutex unsuccessful
spin, can be understood similarly.

5. *Solaris studio performance analyzer.* Lastly, Solaris Studio Performance Analyzer [14] provides insights into program execution by showing the most
   frequently executed functions, caller-callee information along with a timeline
   view of the dynamic events in the execution. This information about the code
   is also augmented with hardware counter based profiling information helping
   to identify bottlenecks in the code. In Fig. 1.9, we show a profile taken while
   running the LAMBDA workload. From the profile we can identify hot methods
   that use a lot of CPU time. The hot methods can be further analyzed using the
   call tree graph, such as the example shown in Fig. 1.10.

1.2.3 Experimental Setup

Two hardware platforms are used in our study. The first is a two-socket system
based on the SPARC T5 [7] processor (Fig. 1.11), the fifth generation multicore
microprocessor of Oracle's SPARC T-Series family. The processor has a clock
frequency of 3.6 GHz, 8 MB of shared last level (L3) cache, and 16 cores where
each core has eight hardware threads, providing a total of 128 hardware threads,
also known as virtual CPUs (vCPUs), per processor. The SPARC T5-2 system used
in our study has two SPARC T5 processors, giving a total of 256 vCPUs available
for application use. The SPARC T5-2 server runs Solaris 11 as its operating system.
Solaris provides a configuration utility ("psrset") to condition an application to use

Excl. Total CPU		Incl. Total CPU		Excl. User Lock		Incl. User Lock		Name
(sec.)	(%)	(sec.)	(%)	(sec.)	(%)	(sec.)	(%)	
36 164.598	100.00	36 164.598	100.00	116 402.805	100.00	116 402.805	100.00	<Total>
2 404.662	6.65	3 631.450	10.04	8.566	0.01	9.627	0.01	java.util.concurrent.ConcurrentHashMap.get(java.lang.Object)
1 075.052	2.97	3 725.256	10.30	0.	0.	508.506	0.44	JVM_LatestUserDefinedLoader
810.997	2.24	812.648	2.25	0.	0.	0.	0.	java.lang.Long.equals(java.lang.Object)
627.709	1.74	770.179	2.13	0.	0.	0.	0.	java.io.ObjectInputStream$BlockDataInputStream.readUTFSpan(java.lang.StringBuilder, long)
590.493	1.63	757.610	2.09	5.054	0.00	5.054	0.00	java.util.HashMap.getNode(int, java.lang.Object)
554.578	1.53	554.588	1.53	35.525	0.03	35.525	0.03	java.util.HashMap.resize()
504.913	1.40	504.913	1.40	158.991	0.14	158.991	0.14	com.oracle.security.ucrypto.NativeCipher.nativeInit(int, boolean, byte[], byte[], int, byte[])
502.552	1.39	1 296.197	3.58	0.	0.	0.	0.	org.spec.jbb.core.purchase.CustomerSpecificPurchaseAgent.getNextProducts(org.spec.jbb.hq.entity.CustomerProfile)
471.300	1.30	471.300	1.30	0.	0.	0.	0.	java.util.HashMap$HashIterator.nextNode()
434.894	1.20	589.833	1.63	0.	0.	36.946	0.03	ParNewGeneration::copy_to_survivor_space_avoiding_promotion_undo(ParScanThreadState*, oopDesc*, unsigned long, markOopDesc*)
433.894	1.20	433.894	1.20	0.	0.	0.	0.	java.io.ByteArrayInputStream.read(byte[], int, int)
429.911	1.19	11 681.431	32.30	0.	0.	1 364.144	1.17	java.io.ObjectInputStream.readNonProxyDesc(boolean)
391.034	1.08	1 325.847	3.67	0.	0.	0.	0.	frame::sender(RegisterMap*)const
361.453	1.00	773.011	2.14	0.	0.	0.	0.	vframeStreamCommon::skip_reflection_related_frames()
341.369	0.94	341.369	0.94	0.	0.	0.	0.	Klass::is_subclass_of(const Klass*)const
341.199	0.94	377.474	1.04	0.	0.	0.010	0.00	java.util.concurrent.locks.AbstractQueuedSynchronizer.release(int)
340.248	0.94	432.463	1.20	16.532	0.01	18.653	0.02	org.spec.jbb.core.comm.Interconnect.mergeHints(java.util.Collection)
316.762	0.88	316.762	0.88	30.211	0.03	30.211	0.03	com.oracle.security.ucrypto.NativeCipher.nativeFinal(long, boolean, byte[], int)
309.386	0.86	934.754	2.58	0.	0.	0.	0.	frame::frame(long*,long*,bool)
301.891	0.83	301.891	0.83	0.	0.	0.	0.	java.io.ByteArrayInputStream.read()
296.567	0.82	932.903	2.58	14.190	0.01	49.715	0.04	java.util.HashMap.putVal(int, java.lang.Object, java.lang.Object, boolean, boolean)
293.856	0.81	524.357	1.45	0.	0.	0.	0.	VerifyFixClassname
291.364	0.81	620.654	1.72	66.757	0.06	66.757	0.06	java.util.AbstractCollection.toArray()
272.310	0.75	272.310	0.75	76.363	0.07	76.363	0.07	com.oracle.security.ucrypto.NativeCipher.nativeUpdate(long, boolean, byte[], int, int, byte[], int)
271.960	0.75	1 662.554	4.60	0.	0.	2.843	0.00	org.spec.jbb.sm.tx.AbstractSMTransaction.resolveProduct(java.lang.Long)
270.309	0.75	270.309	0.75	13.089	0.01	13.089	0.01	java.lang.Long.valueOf(long)
268.288	0.74	1 535.034	4.24	0.	0.	0.	0.	java.io.ObjectInputStream$BlockDataInputStream.readUTFBody(long)
267.427	0.74	981.887	2.72	17.282	0.01	17.842	0.02	org.spec.jbb.sm.inventory.LockedRandomBarcodeInventory$QuantityInfo.reserve(int)
258.891	0.72	573.681	1.59	0.	0.	0.	0.	java.util.ArrayList.ensureCapacityInternal(int)
250.946	0.69	784.509	2.17	2.722	0.00	567.037	0.49	java.lang.String.intern()
248.394	0.69	1 768.357	4.89	31.032	0.03	153.988	0.13	org.spec.jbb.hq.tx.SupermarketAudit.summarizeReceipt(org.spec.jbb.hq.entity.Receipt)
242.079	0.67	242.079	0.67	0.	0.	0.	0.	java.util.Arrays.copyOf(java.lang.Object[], int)
241.259	0.67	271.610	0.75	0.	0.	0.	0.	org.spec.jbb.core.collections.CollectionUtils$2.compare(java.lang.Object, java.lang.Object)
238.187	0.66	238.187	0.66	0.	0.	0.	0.	next_utf2unicode
227.459	0.63	227.459	0.63	0.	0.	0.	0.	UNICODE::as_utf8(unsigned short*,int,char*,int)
226.589	0.63	226.589	0.63	0.	0.	0.	0.	CodeHeap.find_start(void*)const
225.428	0.62	225.428	0.62	1.001	0.00	1.001	0.00	java.util.Arrays.copyOf(char[], int)
223.346	0.62	258.551	0.71	1.091	0.00	1.091	0.00	java.io.ObjectOutputStream$BlockDataOutputStream.getUTFLength(java.lang.String)
206.985	0.57	206.985	0.57	0.	0.	0.	0.	java.lang.String.hashCode()
206.414	0.57	206.414	0.57	0.	0.	0.	0.	java.lang.Long.hashCode(long)
203.913	0.56	364.995	1.01	0.	0.	0.	0.	skip_over_fieldname

Fig. 1.9 An example of Oracle Solaris Studio Performer Analyzer profile, where we show the methods ranked by exclusive cpu time

```
Call Tree FUNCTIONS  Filtered view  Threshold 1%  Sort by metric  Metric Attributed Total CPU Time
590.493  (100%)  <Total>
  590.473  (100%)  java.util.concurrent.ForkJoinWorkerThread.run()
    590.473  (100%)  java.util.concurrent.ForkJoinPool.runWorker(java.util.concurrent.ForkJoinPool$WorkQueue)
      590.473  (100%)  java.util.concurrent.ForkJoinPool.scan(java.util.concurrent.ForkJoinPool$WorkQueue, int)
        590.473  (100%)  java.util.concurrent.ForkJoinPool$WorkQueue.runTask(java.util.concurrent.ForkJoinTask)
          590.473  (100%)  java.util.concurrent.ForkJoinTask.doExec()
            590.473  (100%)  org.spec.jbb.core.threadpools.ForkJoinBatchTask.exec()
              590.473  (100%)  org.spec.jbb.core.threadpools.AbstractPool.processOne(int, org.spec.jbb.core.ExecutionHandler, org.spec.jbb.core.comm.Incoming)
                519.363  (88%)  org.spec.jbb.hq.HQ.execute(org.spec.jbb.core.comm.Incoming)
                  519.353  (88%)  org.spec.jbb.core.tx.SimpleTransactionExecutor.execute(org.spec.jbb.core.tx.Transaction)
                    342.149  (58%)  org.spec.jbb.hq.tx.BusinessReportTransaction.execute()
                      342.149  (58%)  org.spec.jbb.hq.tx.SupermarketAudit.generate()
                        342.149  (58%)  org.spec.jbb.hq.tx.SupermarketAudit.collect()
                          341.669  (58%)  org.spec.jbb.hq.tx.SupermarketAudit.summarizeReceipt(org.spec.jbb.hq.entity.Receipt)
                            311.218  (53%)  org.spec.jbb.core.collections.AbstractMultiSet.add(java.lang.Object, int)
                              311.218  (53%)  org.spec.jbb.core.collections.HashMultiSet.update(java.lang.Object, int)
                                311.218  (53%)  java.util.HashMap.get(java.lang.Object)
                                  311.218  (53%)  java.util.HashMap.getNode(int, java.lang.Object)
                            26.989  (5%)  org.spec.jbb.core.collections.UpdateHashMap.update(java.lang.Object, java.lang.Object, org.spec.jbb.core.collections.BinaryFunction)
                            2.512  (0%)  org.spec.jbb.core.collections.AbstractMultiMap$1.iterator()
                            0.951  (0%)  org.spec.jbb.core.collections.AbstractMultiMap$1$1.next()
                          0.460  (0%)  org.spec.jbb.hq.tx.SupermarketAudit.summarizeInvoice(org.spec.jbb.sp.Invoice)
                      0.020  (0%)  org.spec.jbb.hq.tx.SupermarketAudit.analyze()
                    118.933  (20%)  org.spec.jbb.hq.tx.AssociativityOfProductTransaction.execute()
                      118.743  (20%)  java.util.HashSet.contains(java.lang.Object)
                        118.743  (20%)  java.util.HashMap.containsKey(java.lang.Object)
                          118.743  (20%)  java.util.HashMap.getNode(int, java.lang.Object)
                      0.150  (0%)  org.spec.jbb.core.collections.AbstractMultiSet.addAll(java.util.Collection)
                      0.040  (0%)  org.spec.jbb.util.DbProperties.getDataMiningHottestCount(java.lang.String)
                    48.424  (8%)  org.spec.jbb.hq.tx.AssociativityOfCategoryTransaction.execute()
                    6.264  (1%)  org.spec.jbb.hq.tx.PurchaseReceiptTransaction.execute()
                    3.262  (1%)  org.spec.jbb.hq.tx.ProcessHQInvoicesTransaction.execute()
                    0.180  (0%)  org.spec.jbb.hq.tx.AdjustSupermarketInventoryTransaction.execute()
                    0.040  (0%)  org.spec.jbb.hq.tx.CustomerBuyingBehaviorTransaction.execute()
                    0.040  (0%)  org.spec.jbb.hq.tx.SupplierReceiptTransaction.execute()
                  0.010  (0%)  org.spec.jbb.hq.tx.request.AssociativityOfCategoryRequest.getTransaction(org.spec.jbb.hq.HQ, org.spec.jbb.core.tx.TransactionContext)
                62.143  (11%)  org.spec.jbb.sm.SM.execute(int, org.spec.jbb.core.comm.Incoming)
                8.966  (2%)  org.spec.jbb.sp.SP.execute(int, org.spec.jbb.core.comm.Incoming)
  0.020  (0%)  java.lang.Thread.run()
```

Fig. 1.10 An example of Oracle Solaris Studio Performer Analyzer call tree graph

only a subset of vCPUs. Our experimental setup includes running the LAMBDA workload on configurations of 1 core (8 vCPUs), 2 cores (16 vCPUs), 4 cores (32 vCPUs), 8 cores (64 vCPUs), 1 socket (16 cores/128 vCPUs), and 2 sockets (32 cores/256 vCPUs).

The second hardware platform is an eight-socket SPARC M6-8 system that is based on the SPARC M6 [17] processor (Fig. 1.12). The SPARC M6 processor has a clock frequency of 3.6 GHz, 48 MB of L3 cache, and 12 cores. Same as SPARC T5, each M6 core has eight hardware threads. This gives a total of 96 vCPUs per

Fig. 1.11 SPARC T5
processor [7]

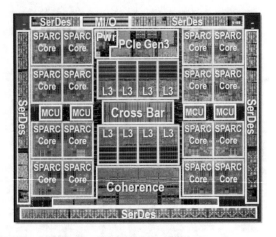

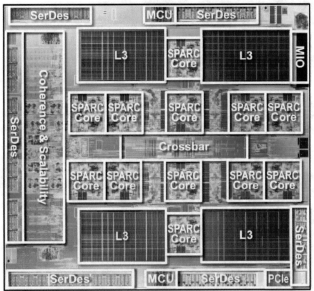

Fig. 1.12 SPARC M6 processor [17]

processor socket, for a total of 768 vCPUs for the full M6-8 system. The SPARC
M6-8 server runs Solaris 11. Our setup includes running the LAMBDA workload on
configurations of 1 socket (12 cores/96 vCPUs), 2 sockets (24 cores/192 vCPUs), 4
sockets (48 cores/384 vCPUs), and 8 sockets (96 cores/384 vCPUs).

Several JDK versions have been used in the study. We will call out the specific
versions in the sections to follow.

1.3 Thread-Local Data Objects

A globally shared data object when protected by locks on the critical path of application leads to the serial part of Amdahl's law. This causes less than perfect scaling. To improve degree of parallelism, the strategy is to "unshare" such data objects that cannot be efficiently shared. Whenever possible, we try to use data objects that are local to the thread, and not shared with other threads. This can be more subtle than it sounds, as the following case study demonstrates.

Hash map is a frequently used data structure in Java programming. To minimize the probability of collision in hashing, JDK 7u6 introduced an alternative hash map implementation that adds randomness in the initiation of each *HashMap* object. More precisely, the alternative hashing introduced in JDK 7u6 includes a feature to randomize the layout of individual map instances. This is accomplished by generating a random mask value per hash map. However, the implementation in JDK 7u6 uses a *shared* random seed to randomize the layout of hash maps. This shared random seed object causes significant synchronization overhead when scaling an application like LAMBDA which creates many transient hash maps during the run. Using Solaris Studio Analyzer profiles, we observed that for an experiment run with 48 cores of M6, CPUs were saturated and 97% of CPU time was spent in the *java.util.Random.nextInt()* function achieving less than 15% of the system's projected performance. The problem came out of *java.util.Random.nextInt()* updating global state, causing synchronization overhead as shown in Fig. 1.13.

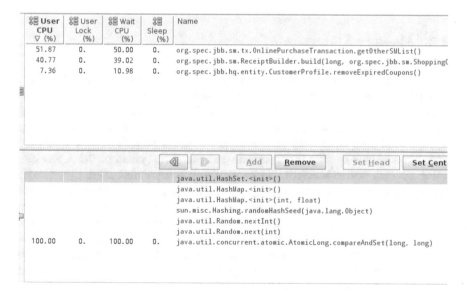

Fig. 1.13 Scaling bottleneck due to *java.util.Random.nextInt*

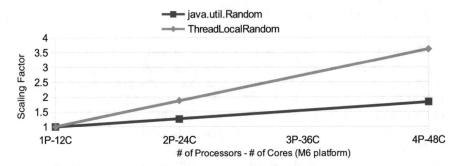

Fig. 1.14 LAMBDA Scaling with *ThreadLocalRandom* on M6 platform

The OpenJDK bug JDK-8006593 tracks the aforementioned issue and uses a thread-local random number generator, *ThreadLocalRandom* to resolve the problem, thereby eliminating the synchronization overhead and improving performance of the LAMBDA workload significantly. When using the *ThreadLocalRandom* class, a generated random number is isolated to the current thread. In particular, the random number generator is initialized with an internally generated seed. In Fig. 1.14, we can see that the 1-to-4 processor scaling improved significantly from a scaling factor of 1.83 (when using *java.util.Random*) to 3.61 (when using *java.util.concurrent.ThreadLocalRandom*). The same performance fix improves the performance of a 96-core 8-processor large M6 system by 4.26 times.

1.4   Memory Allocators

Many in-memory business data analytics applications allocate and deallocate memory frequently. While Java uses an internal heap and most of the allocations happen within this heap, there are components of applications that end up allocating outside the Java heap using native memory allocators provided by the operating system. One such commonly seen component would be native code, which are code parts written specific to a hardware and operating system platform accessed using the Java Native Interface. Native code uses system *malloc()* to dynamically allocate memory. Many business analytics applications use crypto functionality for security purposes and most of the implementations for crypto functions are hand optimized native code which allocates memory outside the Java heap. Similarly, network I/O components are also frequently implemented to allocate and access memory outside the Java heap. In business analytics applications, we see many such crypto and network I/O functions used regularly resulting in calls to the OS system call *malloc()* from within the JVM.

Most modern operating systems, like Solaris, have a heap segment, which allows for dynamic allocation of space during run time using system calls such as *malloc()*. When such a previously allocated object is deallocated, the space used by the object can be reused. For the most efficient allocation and reuse of space, the solution is to maintain a heap inventory (alloc/free list) stored in a set of data structures in the process address space. In this way, calling *free()* does not return the memory back to the system; it is put in the free-list. The traditional implementation (such as the default memory allocator in libc) protects the entire inventory using a single per-process lock. Calls to memory allocation and de-allocation routines manipulate this set of data structures while holding the lock. This single lock causes a potential performance bottleneck when we scale a single JVM to a large number of cores and the target Java application has *malloc()* calls from components like network I/O or crypto. When we profiled the LAMBDA workload using Solaris Studio Analyzer, we found that the *malloc()* calls were showing higher than expected CPU time. A further investigation using the lockstat and plockstat tools revealed a highly contended lock called the depot lock. The depot lock protects the heap inventory of free pages. This motivated us to explore scalable implementations of memory allocators.

A set of newer memory allocators, called Multi-Thread (MT) Hot allocators [18], partition the inventory and the associated locks into arrays to reduce the contention on the inventory. A value derived from the caller's CPU ID is used as an index into the array. It is worth noting that a slight side effect of this approach is that it can cause more memory usage. This happens because instead of a single free-list of memory, we now have a disjoint set of free-lists. This tends to require more space since we will have to ensure each free-list has sufficient memory to avoid run-time allocation failures.

The *libumem* [4] memory allocator is an MT-Hot allocator included in Solaris. To evaluate the improvement from this allocator, we use the LD_PRELOAD environment variable to preload this library, there by *malloc()* implementation in this library is used over the default implementation in the libc library. The improvement in performance seen when using libumem over libc is shown in Fig. 1.15. With the MT-hot allocator, the performance in terms of throughput increases by 106%, 213%, and 478% for 8-core (half processor), 16-core (1 processor), and 32-core

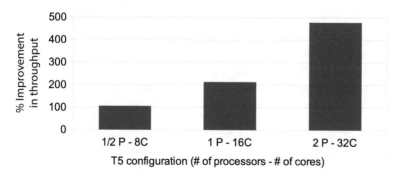

Fig. 1.15 LAMBDA workload throughput improvement with MT-hot alloc over libc *malloc()* on T5-2

(2 processors) configurations, respectively, on T5-2 in comparison to *malloc()* in libc. Note that while JVM uses *mmap()*, instead of *malloc()*, for allocation of its garbage-collectable heap region, the JNI part of JVM does use *malloc()*, especially for the crypto and security related processing. The workload LAMBDA has a significant part of operation in crypto and security, so the effect of MT Hot allocator is quite significant. After switching to an MT-Hot allocator, the hottest observed lock "depot lock" in the memory allocator disappeared and reduced the time spent in locks by a factor of 21. This confirmed the necessity of an MT-Hot memory allocator for successful scaling.

1.5   Java Concurrency API

Ever since JDK 1.2, Java has included a standard set of collection classes called the Java *collections* framework. A collection is an object that represents a group of objects. Some of the fundamental and popularly used collections are dynamic arrays, linked lists, trees, queues, and hashtables. The collections framework is a unified architecture that enables storage and manipulation of the collections in a standard way, independent of underlying implementation details. Some of the benefits of the collections framework include reduced programming effort by providing data structures and algorithms for programmers to use, increased quality from high performance implementation and enabling reusability and interoperability. The collection framework is used extensively in almost every Java program these days. While these pre-implemented collections make the job of writing single threaded application so much easier, writing concurrent multithreaded programs is still a difficult job. Java provided low level threading primitives such as synchronized blocks, Object.wait and Object.notify, but these were too fine grained facilities forcing programmers to implement high level concurrency primitives, which are tediously hard to implement correctly and often were non-performant.

Later, a concurrency package, comprising several concurrency primitives and many collection-related classes, as part of the JSR 166 [10] library, was developed. The library was aimed at providing high quality implementation of classes to include atomic variables, special-purpose locks, barriers, semaphores, high performant threading utilities like thread pools and various core collections like queues and hashmaps designed and optimized for multithreaded programming. The concurrency APIs developed by the JSR 166 working group were included as part of the JDK 5.0. Since then both Java SE 6 and Java SE 7 releases introduced updated versions of the JSR 166 APIs as well as several new additional APIs. Availability of this library relieves the programmer from redundantly crafting these utilities by hand, similar to what the collections framework did for data structures. Our early evaluation of Java SE 7 found a major challenge in scaling from the implementations of concurrent collection data structures (such as concurrent hash maps) using low level Java concurrency control constructs. We explored utilizing concurrency utilities from JSR 166, leveraging the scalable implementations of concurrent collections and frameworks and saw very significant improvement in the scalability of applications. Specifically, the LAMBDA workload code uses the Java class *java.util.concurrent.ConcurrentHashMap*. The efficiency of its underlying implementation affects performance quite significantly. For example, comparing the *ConcurrentHashMap* implementation of JDK8 over JDK7, there is an improvement of about 57% in throughput due to the improved JSR 166 implementation.

1.6   Garbage Collection

Automatic Garbage Collection (GC) is the cornerstone of memory management in Java enabling developers to allocate new objects without worrying about deallocation. The Garbage Collector reclaims memory for reuse ensuring that there are no memory leaks and also provides security from vulnerabilities in terms of memory safety. But, automatic garbage collection comes at a small performance cost for resolving these memory management issues. It is an important aspect of real world enterprise application performance, as GC pause times translate into unresponsiveness of an application. Shorter GC pauses will help the applications to meet more stringent response time requirements. When heap sizes run into a few hundred gigabytes on contemporary many-core servers, achieving low pause times require the GC algorithm to scale efficiently with the number of cores. Even when an application and the various dependent libraries are ensured to scale well without any bottlenecks, it is important that the GC algorithm also scales well to achieve scalable performance.

It may be intuitive to think that the garbage collector will identify and eliminate dead objects. But, in reality it is more appropriate to say that the garbage collector rather tracks the various live objects and copies them out, so that the remaining space can be reclaimed. The reason that such an implementation is preferred in the modern collectors is that, most of the objects die young and it is much faster to copy the fewer remaining live objects out than tracking and reclaiming the space of each of the dead objects. This will also give us a chance to compact the remaining live objects ensuring a defragmented memory. Modern garbage collectors have a generational approach to this problem, maintaining two or more allocation regions (generations) with objects grouped into these regions based on their age. For example, the G1 GC [6] reduces heap fragmentation by incremental parallel copying of live objects from one or more sets of regions (called Collection Set or CSet in short) into different new region(s) to achieve compaction. The G1 GC [6] tracks references into regions using independent Remembered Sets (RSets). These RSets enable parallel and independent collection of these regions because each region's RSet can be scanned independently for references into that region as opposed to scanning the entire heap. The G1 GC has a multiphase complex algorithm that has both parallel and serial code components contributing to Stop The World (STW) evacuation pauses and concurrent collection cycles.

With respect to the LAMBDA workload, pauses due to GC directly affect the response time metric monitored by the benchmark. If the GC algorithm does not scale well, long pauses will exceed the latency requirements of the benchmark resulting in lower throughput. In our experiments with monitoring the LAMBDA workload on an M6 server, we had some interesting observations. While at the regular throughput phase of the benchmark run, the system CPUs were fully utilized almost at 100%. By contrast, there was much more CPU headroom (75%) during a GC phase, hinting at possible serial bottlenecks in Garbage Collection. By collecting and analyzing code profiles using Solaris Studio Analyzer, the time the worker threads of the LAMBDA workload spend waiting on conditional variables increase from 3%, for a 12-core (single-processor) run, to to 16%, for a 96-core (8-processor) run on M6. This time was mostly spent in lwp_cond_wait() waiting for the young generation stop-the-world garbage collection, observed to be in sync with the GC events based on a visual timeline review of Studio Analyzer profiles. Further the call stack of the worker threads consists of the *SafepointSynchronize::block()* function consuming 72% of time clearly pointing at the scalability issue in garbage collection.

G1 GC [6] provides a breakdown of the time spent in various phases to the user via verbose GC logs. Analyzing these logs pointed to a major serial component "*Free Cset*," for which the processing time was proportional to the size of the heap (mainly the nursery component responsible for the storage of the young objects). This particular phase of the GC algorithm was not parallelized and some of the considerations included the cost involved in thread creation for parallel execution. While thread creation may be a major overhead and an overkill for small heaps, such a cost can be amortized if the heap size is large and running into hundreds of gigabytes. A parallelized implementation of the "*Free Cset*" phase was created for testing purposes as part of the JDK bug JDK-8034842. We noticed that this parallelized implementation for the "*Free Cset*" phase of G1 GC provided major reduction in pause times for this phase for the LAMBDA workload. The pause times for this phase went down from 230 ms to 37 ms for scaled runs on 8 processors (96 cores) of M6. The ongoing work in fully parallelizing the *FreeCset* phase is tracked in the JDK bug report JDK-8034842. Also, we observed that a major part of the scaling overhead that came out of garbage collection on large many-core systems was from accesses to remote memory banks in a Non-Uniform Memory Access (NUMA) system. We examine this impact further in the following subsection.

1.7   Non-uniform Memory Access (NUMA)

Most of the modern many-core systems are shared memory systems that have Non-Uniform Memory Access (NUMA) latencies. Modern operating systems like Solaris have memory (DRAM, cache) banks and CPUs classified into a hierarchy of locality groups (lgroup). Each lgroup includes a set of CPU and memory banks, where the leaf lgroups include the CPUs and memory banks that are closest to each other in

Fig. 1.16 Machine with
single latency is represented
by only one lgroup

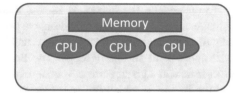

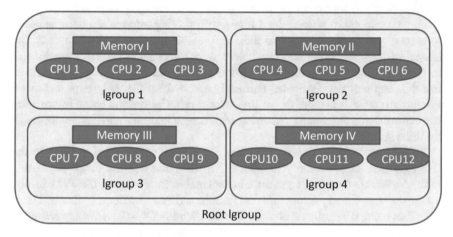

Fig. 1.17 Machine with multiple latency is represented by multiple lgroups

terms of access latency, with the hierarchy being organized similarly up to the root. Figure 1.16 shows a typical system with a single memory latency, represented by one lgroup. Figure 1.17 shows a system with multiple memory latencies, represented by multiple lgroups. In this organization, the CPUs 1–3 belong to lgroup1 and will have the least latency to access Memory I. Similarly, CPUs 4–6 to Memory II, CPUs 7–9 to Memory III, and CPUs 10–12 to Memory IV will have the least local access latencies. When a CPU accesses a memory location that is outside its local lgroup, a longer remote memory access latency will be incurred.

In systems with multiple lgroups, it would be most desirable to have the data that is being accessed by the CPUs in their nearest lgroups, thus incurring shortest access latencies. Due to high remote memory access latency, it is very important that the operating system be aware of the NUMA characteristics of the underlying hardware. Additionally, it is a major value add if the Garbage Collector in the Java Virtual Machine is also engineered to take these characteristics into account. For example, the initial allocation of space for each thread can be made so that it is in the same lgroup as that of the CPU on which the thread is running. Secondly, the GC algorithm can also make sure that when data is compacted or copied from one generation to another, some preference can be given to ensure that the data is not copied to a remote lgroup with respect to the thread that is most frequently accessing the data. This will enable easier scaling across multiple cores and multiple processors of large enterprise systems.

To understand the impact of remote memory accesses on the performance of garbage collector and the application, we profiled the LAMBDA workload with the help of pmap and Solaris tools cpustat and busstat, breaking down the distribution of heap/stack to various lgroups. The Solaris tool pmap provides a snapshot of process data at a given point of time in terms of the number of pages, size of pages, and the lgroup in which the pages are resident. This can be used to get a spatial breakdown of the Java heap to various lgroups. The utility cpustat on SPARC uses hardware counters to provide hardware level profiling information such as cache miss rates and access latencies to local and remote memory banks. Similarly, the busstat utility provides memory bandwidth usage information, again broken down at memory bank/lgroup granularity. Our initial set of observations using pmap showed that the heap was not distributed uniformly across the different lgroups and that a few lgroups were used more frequently than the rest. Cpustat and bustat information corroborated this observation, showing high access latencies and bandwidth usage for these stressed set of lgroups.

To alleviate this, we tried using key JVM flags which provide hints to the GC algorithm about memory locality. First, we found that the usage of the flag -*XX:+UseNUMAInterleaving* can be indispensable in hinting to the JVM to distribute the heap equally across different lgroups and avoid bottlenecks that will arise from data being concentrated on a few lgroups. While -*XX:+UseNUMAInterleaving* will only avoid concentration of data in particular banks, flags like -*XX:+UseNUMA* when used with Parallel Old Garbage Collector have the potential to tailor the algorithm to be aware of NUMA characteristics and increase locality. Further, operating system flags like *lpg_alloc_prefer* in Solaris 11 and *lgrp_mem_pset_aware* in Solaris 12, when set to true, hint to the OS to allocate large pages in the local lgroup rather than allocating them in a remote lgroup. This can be very effective in improving memory locality in scaled runs. The *lpg_alloc_prefer* flag, when set to true can increase the throughput of the LAMBDA workload by about 65% on the M6 platform, showing the importance of data locality. While ParallelOld is an effective stop-the-world collector, concurrent garbage collectors like CMS and G1 GC [6] are most useful in real world response time critical application deployments. The enhancement requests that track the implementation of NUMA awareness into G1 GC and CMS GC are JDK-7005859 and JDK-6468290.

1.8   Conclusion and Future Directions

We present an iterative process for performance scaling JVM applications on many-core enterprise servers. This process consists of workload characterization, bottleneck identification, performance optimization, and performance evaluation in each iteration. As part of workload characterization, we first provide an overview of the various tools that are provided as part of modern operating systems most useful to profile the execution of workloads. We use a data analytics workload, LAMBDA as an example to explain the process of performance scaling. We identify various bottlenecks in scaling this application such as synchronization overhead due to shared objects, serial resource bottleneck in memory allocation, lack of usage of high level concurrency primitives, serial implementations of Garbage Collection phases, and uneven distribution of heap on a NUMA machine oblivious to the NUMA characteristics by using the profiled data. We further discuss in depth the root cause of each bottleneck and present solutions to address them. These solutions include unsharing of shared objects, usage of multicore friendly allocators such as MT-Hot allocators, high performance concurrency constructs as in JSR166, parallelized implementation of Garbage Collection phases, and NUMA aware garbage collection. Taken together, the overall improvement for the proposed solutions is more than 16 times on an M6-8 server for the LAMBDA workload in terms of maximum throughput.

Future directions include hardware accelerations to address scaling bottlenecks, increased emphasis on the response time metric where GC performance and scalability will be a key factor, and horizontal scaling aspects of big data analytics where disk and network I/O will play crucial roles.

Acknowledgements We would like to thank Jan-Lung Sung, Pallab Bhattacharya, Staffan Friberg, and other anonymous reviewers for their valuable feedback to improve the chapter.

References

1. Apache, Apache Hadoop (2017). Available: https://hadoop.apache.org
2. Apache Software Foundation, Apache Giraph (2016). Available https://giraph.apache.org
3. Apache Spark (2017). Available https://spark.apache.org
4. Oracle, Analyzing Memory Leaks Using the libumem Library [online]. https://docs.oracle.com/cd/E19626-01/820-2496/geogv/index.html
5. W. Bolosky, R. Fitzgerald, M. Scott, Simple but effective techniques for numa memory management. SIGOPS Oper. Syst. Rev. **23**(5), 19–31 (1989)
6. D. Detlefs, C. Flood, S. Heller, T. Printezis, Garbage-first garbage collection, in *Proceedings of the 4th International Symposium on Memory Management* (2004), pp. 37–48
7. J. Feehrer, S. Jairath, P. Loewenstein, R. Sivaramakrishnan, D. Smentek, S. Turullols, A. Vahidsafa, The Oracle Sparc T5 16-core processor scales to eight sockets. IEEE Micro **33**(2), 48–57 (2013)
8. K. Ganesan, L.K. John, Automatic generation of miniaturized synthetic proxies for target applications to efficiently design multicore processors. IEEE Trans. Comput. **63**(4), 833–846 (2014)
9. M.D. Hill, M.R. Marty, Amdahl's law in the multicore era. Computer **41**(07), 33–38 (2008)
10. D. Lea, Concurrency JSR-166 interest site (2014). http://gee.cs.oswego.edu/dl/concurrency-interest/
11. Neo4j, Neo4j graph database (2017). Available https://neo4j.com
12. Oracle, Man pages section 1M: system Administration Commands (2016). [Online]. Available http://www.oracle.com
13. Oracle Corporation, Java SE platform (2017). Available http://www.oracle.com/technetwork/java/javase/overview/index.html
14. Oracle solaris studio performance analyzer (2014). http://docs.oracle.com/cd/E18659_01/html/821\discretionary-1379/

15. C. Pogue, A. Kumar, D. Tollefson, S. Realmuto, Specjbb2013 1.0: an overview, in *Proceedings of the 5th ACM/SPEC International Conference on Performance Engineering* (2014), pp. 231–232
16. Standard Performance Evaluation Corporation, Spec fair use rule. academic/research usage (2015). [Online]. Available http://www.spec.org/fairuse.html#Academic
17. A. Vahidsafa, S. Bhutani, SPARC M6 oracle's next generation processor for enterprise systems, in *Hotchips 25* (2013). [Online]. Available http://www.hotchips.org/wp-content/uploads/hc_archives/hc25/HC25.90-Processors3-epub/HC25.27.920-SPARC-M6-Vahidsafa-Oracle.pdf
18. R.C. Weisner, How memory allocation affects performance in multithreaded programs (2012). http://www.oracle.com/technetwork/articles/servers-storage-dev/mem\discretionary-alloc\discretionary-1557798.html

Chapter 2
Accelerating Data Analytics Kernels with Heterogeneous Computing

Guanwen Zhong, Alok Prakash*, and Tulika Mitra

2.1 Introduction

The past decade has witnessed an unprecedented and exponential growth in the amount of data being produced, stored, transported, processed, and displayed. The journey of zettabyte of data from the myriad of end-user devices in the form of PCs, tablets, smart phones through the ubiquitous wired/wireless communication infrastructure to the enormous data centers forms the backbone of computing today. Efficient processing of this huge amount of data is of paramount importance. The underlying computing platform architecture plays a critical role in enabling efficient data analytics solutions.

Computing systems made an irreversible transition towards multi-core architectures in early 2000. As of now, homogeneous multi-cores are prevalent in all computing systems starting from smart phones to PCs to enterprise servers. Unfortunately, homogeneous multi-cores cannot provide the desired performance and energy-efficiency for diverse application domains. A promising alternative design is heterogeneous multi-core architecture where cores with different functional characteristics (CPU, GPU, FPGA, etc.) and/or performance-energy characteristics (simple versus complex micro-architecture) co-exist on the same die or in the same

---

*Alok completed this project while working at SoC, NUS

G. Zhong • T. Mitra (✉)
School of Computing, National University of Singapore, Singapore, Singapore
e-mail: guanwen@comp.nus.edu.sg; tulika@comp.nus.edu.sg

A. Prakash
School of Computer Science and Engineering, Nanyang Technological University, Singapore, Singapore
e-mail: alok@ntu.edu.sg

© Springer International Publishing AG 2017
A. Chattopadhyay et al. (eds.), *Emerging Technology and Architecture for Big-data Analytics*, DOI 10.1007/978-3-319-54840-1_2

system. Given an application, only the cores that best fit the application can be exploited leading to faster and power-efficient computing.

Another reason behind the emergence of the heterogeneous computing is the thermal design power constraint [14, 24, 25, 28, 31–33]. While the number of cores on die continues to increase due to Moore's Law [23], the failure of Dennard scaling [11] has led to rising power density that forces a significant fraction of the cores to be kept powered down at any point in time. This phenomenon, known as the "Dark Silicon" [12], provides opportunities for heterogeneous computing as only the appropriate cores need to switch on for efficient processing under thermal constraints.

Heterogeneous computing architectures can be broadly classified into two categories: *performance heterogeneity* and *functional heterogeneity*. Performance heterogeneous multi-core architectures consist of cores with different power-performance characteristics but all sharing the same instruction-set architecture. The difference stems from distinct micro-architectural features such as in-order core versus out-of-order core. The complex cores can provide better performance at the cost of higher power consumption, while the simpler cores exhibit low-power behavior alongside lower performance. This is also known as single-ISA heterogeneous multi-core architecture [18] or asymmetric multi-core architecture. The advantage of this approach is that the same binary executable can run on all different core types depending on the context and no additional programming effort is required. Examples of commercial performance heterogeneous multi-cores include ARM big.LITTLE [13] integrating high-performance out-of-order cores with low-power in-order cores, nVidia Kal-El (brand name Tegra3) [26] consisting of four high-performance cores with one low-power core, and more recently Wearable Processing Unit (WPU) from Ineda consisting of cores with varying power-performance characteristics [16]. An instance of the ARM big.LITTLE architecture integrating quad-core ARM Cortex-A15 (big core) and quad-core ARM Cortex-A7 (small core) appears in the Samsung Exynos 5 Octa SoC driving high-end Samsung Galaxy S4 and S5 smart phones.

As mentioned earlier, a large class of heterogeneous multi-cores comprise of cores with different functionality. This is fairly common in the embedded space where a multiprocessor system-on-chip (MPSoC) consists of general-purpose processor cores, GPU, DSP, and various hardware accelerators (e.g., video encoder/decoder). The heterogeneity is introduced here to meet the performance demand under stringent power budget. For example, 3G mobile phone receiver requires 35–40 giga operations per second (GOPS) at 1W budget, which is impossible to achieve without custom designed ASIC accelerator [10]. Similarly, embedded GPUs are ubiquitous today in mobile platforms to enable not only mobile 3D gaming but also general-purpose computing on GPU for data-parallel (DLP) compute-intensive tasks such as voice recognition, speech processing, image processing, gesture recognition, and so on.

Heterogeneous computing systems, however, present a number of unique challenges. For heterogeneous multi-cores where the cores have the same instruction-set architecture (ISA) but different micro-architecture [18], the issue is to identify at runtime the core that best matches the computation in the current context. For heterogeneous multi-cores consisting of cores with different functionality, for example CPU, GPU, and FPGAs, the difficulty lies in porting computational kernels of data analytics applications to the different computing elements. While high-level programming languages such as C, C++, Java are ubiquitous for CPUs, they are not sufficient to expose the large-scale parallelism required for GPUs and FPGAs. However, improving productivity demands fast implementation of computational kernels from high-level programming languages to heterogeneous computing elements. In this chapter, we will focus on acceleration of data analytics kernels on field programmable gate arrays (FPGAs).

With the advantages of reconfigurability, customization, and energy efficiency, FPGAs are widely used in embedded domains such as automotive, wireless communications, etc. that demand high performance with low energy consumption. As the capacity keeps increasing together with better power efficiency (e.g., 16 nm UltraScale+ from Xilinx and 14 nm Stratix 10 from Altera), FPGAs become an attractive solution to high-performance computing domains such as data-centers [35]. However, complex hardware programming model (Verilog or VHDL) hinders its acceptance to average developers and it makes FPGA development a time-consuming process even as the time-to-market constraints continue to tighten.

To improve FPGA productivity and abstract hardware development using complex programming models, both academia [3, 7] and industry [2, 40, 43] have spent efforts on developing high-level synthesis (HLS) tools that enable automated translation of applications written in high-level specifications (e.g., C/C++, SystemC) to register-transfer level (RTL). Via various optimizations in the form of pragmas/directives (for example, loop unrolling, pipelining, array partitioning), HLS tools have the ability to explore diverse hardware architectures. However, this makes it non-trivial to select appropriate options to generate a high-quality hardware design on an FPGA due to the large optimization design space and non-negligible HLS runtime.

Therefore, several works [1, 22, 29, 34, 37, 39, 45] have been proposed using compiler-assisted static analysis approaches, similar to the HLS tools, to predict accelerator performance and explore the large design space. However, the static analysis approach suffers from its inherently conservative dependence analysis [3, 7, 38]. It might lead to false dependences between operations and limit the exploitable parallelism on accelerators, ultimately introducing inaccuracies in the predicted performance. Moreover, some works rely on HLS tools to improve the prediction accuracy by obtaining performance for a few design points and extrapolating for the rest. The time spent by their methods ranges from minutes to hours and is affected by design space, and number of design points to be synthesized with HLS tools.

In this work, we predict accelerator performance by leveraging a dynamic analysis approach and exploit run-time information to detect true dependences between operations. As our approach obviates the invocation of HLS tools, it enables rapid design space exploration (DSE). In particular, our contributions are two-fold:

- We propose Lin-Analyzer, a high-level analysis tool, to predict FPGA performance accurately according to different optimizations (loop unrolling, loop pipelining, and array partitioning) and perform rapid DSE. As Lin-Analyzer does not generate any RTL implementations, its prediction and DSE are fast.
- Lin-Analyzer has the potential to identify bottlenecks of hardware architectures with different optimizations enabled. It can facilitate hardware development with HLS tools and designers can better understand where the performance impact comes from when applying diverse optimizations.

The goal of Lin-Analyzer is to explore a large design space at an early stage and suggest the best suited optimization pragma combination for an application mapping on FPGAs. With the recommended pragma combination, a HLS tool should be invoked to generate the final synthesized accelerator. Experimental evaluation with different computational kernels from the data analytics applications confirms that Lin-Analyzer returns the optimal recommendation and its runtime varies from seconds to minutes with complex design spaces. This provides an easy translation path towards acceleration of data analytics kernels on heterogeneous computing systems featuring FPGAs.

2.2   Motivation

As the complexity of accelerator designs continues to rise, the traditional time-consuming manual RTL design flow is unable to satisfy the increasingly strict time-to-market constraints. Hence, design flows based on HLS tools such as Xilinx Vivado HLS [43] that start from high-level specifications (e.g., C/C++/SystemC) and automatically convert them to RTL implementations become an attractive solution to designers.

The HLS tools typically provide optimization options in the form of pragmas/directives to generate hardware architectures with different performance/area trade-offs. Pragma options like loop unrolling, loop pipelining, and array partitioning have the most significant impact on hardware performance and area [8, 21, 44]. Loop unrolling is a technique to exploit instruction-level parallelism inside loop iterations, while loop pipelining enables different loop iterations to run in parallel. Array partitioning is used to alleviate memory bandwidth constraints by allowing multiple data reads or writes to be completed in one cycle.

However, this diverse set of pragma options necessitate designers to explore a large design space to select the appropriate set of pragma settings that meets performance and area constraints in the system. The large design space created by the multitude of available pragma settings makes the design space exploration a significantly time-consuming work, especially due to the non-negligible runtime of HLS tools using the DSE step. We highlight the time complexity of this step by using the example of Convolution3D kernel, typically used in big data domain.

Listing 2.1 Convolution3D kernel

```
...
/* Constant values of a window filter: {c11 ,... ,c21 ,... ,c33} */
loop_1: for (i = 1; i < N−1; i++) {
  loop_2: for (j = 1; j < M−1; j++) {
    loop_3: for (k = 1; k < K−1; k++) {
      b[i][j][k]=c11*a[i −1][j −1][k−1]+
                 c13*a[i +1][j −1][k−1]+c21*a[i −1][j −1][k−1]+
                 c23*a[i +1][j −1][k−1]+c31*a[i −1][j −1][k−1]+
                 c33*a[i +1][j −1][k−1]+c12*a[i  ][j −1][k  ]+
                 c22*a[i  ][j  ][k  ]+c32*a[i  ][j +1][k  ]+
                 c11*a[i −1][j −1][k+1]+c13*a[i +1][j −1][k+1]+
                 c21*a[i −1][j  ][k+1]+c23*a[i +1][j  ][k+1]+
                 c31*a[i −1][j +1][k+1]+c33*a[i +1][j +1][k+1];
    }
  }
}
```

Table 2.1 HLS runtime of Convolution3D

Input size	Loop pipelining	Loop unrolling	Array partitioning	HLS runtime
32*32*32	Disabled	loop_3 factor:30	a, cyclic, 2 b, cyclic, 2	44.25 s
	loop_3, yes	loop_3 factor:15	a, cyclic, 16 b, cyclic, 16	1.78 h
	loop_3, yes	loop_3 factor:16	a, cyclic, 16 b, cyclic, 16	3.25 h

Table 2.2 Exploration time of convolution 3D: exhausted vs. Lin-Analyzer

		Exploration time	
Input size	Design space	Exhaustive HLS-based DSE	Lin-Analyzer
32*32*32	120	10 days[a]	29.30 s

[a]For few design points with complex pragmas, the HLS tool takes a long time and thus we stop the program after 10 days

Listing 2.1 shows the Convolution3D kernel. We use a commercial HLS tool, Xilinx Vivado HLS [43], to generate an FPGA-based accelerator for this kernel with different pragma combinations and observe the runtime for this step, as shown in Table 2.1. It is noteworthy that the runtime varies from seconds to hours for different choices of pragmas. As the internal workings of the Vivado HLS tool is not available publicly, we do not know the exact reasons behind this highly variable synthesis time. Other techniques proposed in the existing literature, such as [29], that depend on automatic HLS-based design space exploration are also limited by this long HLS runtime.

Next, we perform an extensive design space exploration for this kernel using the Vivado HLS tool by trying the exhaustive combination of pragma settings. Table 2.2 shows the runtime for this step. It can be observed that even for a relatively smaller input size of $(32 * 32 * 32)$, HLS-based DSE takes more than 10 days.

However, in order to find good-quality hardware accelerator designs, it is imperative to perform the DSE step rapidly and reliably. This provides designers with important information about the accelerators, such as FPGA performance/area/power at an early design stage. For these reasons, we develop Lin-Analyzer, a pre-RTL, high-level analysis tool for FPGA-based accelerators. The proposed tool can rapidly and reliably predict the effect of various pragma settings and combinations on the resulting accelerator's performance and area. As shown in the last column of Table 2.2, Lin-Analyzer can perform the same DSE as the HLS-based DSE, but in the order of seconds versus days. In the next section, we describe the framework of our proposed tool.

2.3 Automated Design Space Exploration Flow

The automated design space exploration flow leverages the high-level FPGA-based performance analysis tool, Lin-Analyzer [46], to correlate FPGA performance with given optimization pragmas for a target kernel in the form of nested loops. With the chosen pragma that leads to the best predicted FPGA performance within resource constraints returned by Lin-Analyzer, the automated process invokes HLS tools to generate an FPGA implementation with good quality. The overall framework is shown in Fig. 2.1. The following subsections describe more details in Lin-Analyzer.

2.3.1 The Lin-Analyzer Framework

Lin-Analyzer is a high-level performance analysis tool for FPGA-based accelerators without register-transfer-level (RTL) implementations. It leverages dynamic analysis method and performs prediction on dynamic data dependence graphs (DDDGs) generated from program traces. The definition of DDDG is given below.

Definition 1 A **DDDG** is a directed, acyclic graph $G(V_G, E_G)$, where $V_G = V_{\mathrm{op}}$ and $E_G = E_r \cup E_m$. V_{op} is the set containing all operation nodes in G. Edges in E_r represent data dependences between register nodes, while edges in E_m denote data dependences between memory load/store nodes.

As the DDDG is generated from a trace, basic blocks of the trace have been merged. If we apply any scheduling algorithms on DDDG, operations can be scheduled across basic blocks. The inherent feature of using dynamic execution

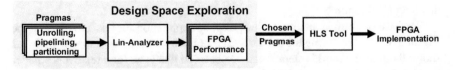

Fig. 2.1 The proposed automated design space exploration flow

trace is that it automatically enables global code motion optimization. In contrast, almost all of the current state-of-the-art HLS tools use static analysis and therefore, need to leverage advanced scheduling algorithms such as System of Difference Constraints (SDC) scheduling [4, 7, 43, 44] to perform global code optimization. However, the inherent feature of the dynamic trace coupled with the dataflow nature of accelerators makes DDDG a good candidate for modeling hardware behavior [38].

With the DDDG, Lin-Analyzer mimics HLS tools and estimate performance of FPGA-based accelerators directly from algorithms in high-level specifications such as C/C++ without generating RTL implementations.

2.3.2   Framework Overview

Figure 2.2 shows the Lin-Analyzer framework. As we can see, Lin-Analyzer consists of three stages: *Instrumentation, DDDG Generation & Pre-optimizationi* and *DDDG Scheduling*. It starts from high-level specifications (C/C++) of an

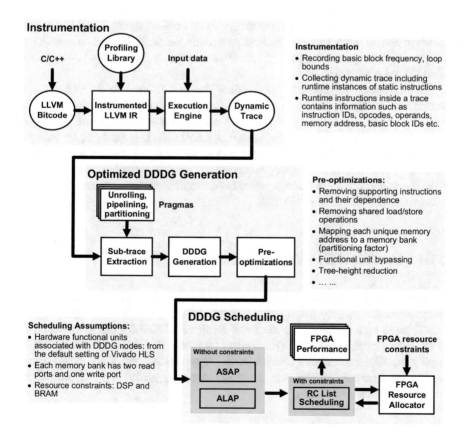

Fig. 2.2 The Lin-Analyzer framework

algorithm with changes and optimizations. By inserting profiling functions into the original codes, Lin-Analyzer collects dynamic trace in *Instrumentation* stage. According to pragmas provided by users, Lin-Analyzer extracts a *sub-trace* from the dynamic trace and builds a dynamic data dependence graph (DDDG) to represent hardware accelerators. As initial DDDG usually contains unnecessary information and needs to be optimized, Lin-Analyzer performs pre-optimizations on it and creates a new DDDG. With the optimized DDDG, it schedules the nodes with resource constraints and estimates performance of the FPGA-based accelerator for the given algorithm. As the analysis is based on DDDG of the relevant *sub-trace* and utilizes fast scheduling algorithm, runtime of Lin-Analyzer is small even for kernels with relative large data size and complex pragma combination such as complete loop unrolling, large array partitioning factors, and loop pipelining.

2.3.3 Instrumentation

A program trace of the kernel containing dynamic instance of static instructions is required for DDDG generation. In this work, we utilize the Low-Level Virtual Machine (LLVM) [19] to instrument programs and collect traces. LLVM leverages passes to perform code analysis, optimization, and modification based on a machine-independent intermediate representation (IR), which is a Static Single Assignment (SSA) based representation.

Lin-Analyzer first converts an application in C/C++ into LLVM IR and instrument the IR by inserting profiling functions. The profiling functions are implemented in the *Profiling Library* and used to record basic block frequency and trace information. With the instrumented LLVM IR, Lin-Analyzer invokes the embedded *Execution Engine*, an LLVM Just-in-Time (JIT) compiler, to run the IR with input data if available. After execution, a run-time trace is dumped into the disk. The dynamic trace includes runtime instances of static instructions and detailed information can be found in Fig. 2.2.

2.3.4 Optimized DDDG Generation

To perform analysis on whole dynamic trace is inefficient and slow, as trace typically contains million or even billion of instruction instances. Therefore, Lin-Analyzer only focuses on a subset of the trace and creates a DDDG for the *sub-trace*. Size of the *sub-trace* is based on pragmas given by users. The initial generated DDDG usually includes unnecessary operations and dependences, and is not good enough to represent hardware accelerators. Thus, Lin-Analyzer performs pre-optimizations on DDDGs before scheduling.

2.3.4.1   Sub-trace Extraction

Size of a *sub-trace* is related to loop unrolling and loop pipelining pragmas. A kernel in the form of a nested (perfectly or non-perfectly) loop can be represented by $L = \{L_1, .., L_i, .., L_K\}$ with K loop levels and the innermost loop level is L_K. Users can apply loop unrolling pragma at any loop levels in L. Assume a given unrolling factor tuple is $\{U_1, .., U_i, .., U_K\}$, where U_i is the factor of i-th loop level. Lin-Analyzer extracts U_i iterations of Loop L_i as the *sub-trace* if its inner loops ($\{L_{i+1}, L_{i+2}, \ldots, L_K\}$) are completely unrolled; otherwise, the *sub-trace* only includes U_K iterations of Loop L_K.

According to Vivado HLS [43], the HLS tool only considers loop pipelining when the pipelining pragma is applied at one loop level L_i in L and all its inner loops ($L' = \{L_{i+1}, L_{i+2}, \ldots, L_K\}$) are forced to be completely unrolled irrespective of their unrolling factors. In this case, the *sub-trace* contains all instruction instances of the inner loops L'. If L_i is the innermost loop level ($i = K$), Lin-Analyzer extracts U_K iterations of L_K as the *sub-trace*.

2.3.4.2   DDDG Generation & Pre-optimizations

Once the *sub-trace* is ready, Lin-Analyzer generates a dynamic data dependence graph (DDDG) to represent the hardware accelerator.In our implementation, a node in the DDDG represents a dynamic instance of an LLVM IR instruction, while an edge represents register- or memory-dependence between nodes. We only consider true dependences. Anti- or output-dependences are not included, as they could be potentially eliminated by optimizations. As we work with dynamic traces, control dependences are not considered.

The initial generated DDDG normally contains supporting instructions and dependences between loop index variables, which cannot model hardware accelerators properly [38]. Therefore, we perform several optimizations before scheduling.

- *Removing supporting instructions and their dependences*: Some of the instructions in a nested loop are related to computation directly, while others are supporting instructions that are used to keep computation in the correct sequence such as instructions related to loop indices, instructions used to obtain memory address of a pointer or based address of an array, etc. Those instructions might potentially introduce true dependences that are not relevant to actual computation, for example, dependence between loop index variables. To remove those information in DDDG, Lin-Analyzer assigns zero latency to those nodes.
- *Removing redundant load/store operations*: A program might potentially contain redundant memory accesses (load or store). This redundancy increases memory (BRAM) bandwidth requirement of a hardware accelerator. To save memory bandwidth, Lin-Analyzer removes redundant memory access operations.

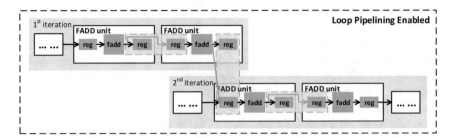

Fig. 2.3 Functional unit bypassing used in Vivado HLS when loop pipelining enabled

- *Associating memory banks with memory addresses*: In our implementation, we assume a memory bank only allows one write and two reads at the same cycle. To restrict the DDDG with the memory constraint, Lin-Analyzer maps each unique address of load/store instructions to a memory bank index. The number of memory banks supported is related to array partitioning factors provided by users.
- *Tree height reduction*: An application might sometimes contain long expression chains. To expose potential parallelism and reduce height of the chains, we employ tree height reduction similar to Shao's work [38].
- *Functional unit bypassing*: When applying loop pipelining and unrolling pragmas at the innermost loop level, we observe (through RTL simulation) that HLS tool (Vivado HLS [43]) enables functional unit bypassing optimization, which is shown in Fig. 2.3. The optimization bypasses output registers of pipelined functional units and directly sends results to the next units connected. Lin-Analyzer also enables similar optimization when users apply loop pipelining and unrolling pragmas at the innermost loop level.

2.3.5 DDDG Scheduling

Lin-Analyzer leverages Resource-Constrained List Scheduling (RCLS) algorithm to schedule nodes on a DDDG. The algorithm takes the optimized DDDG generated by the previous stage and a priority list as inputs. The priority list is obtained from As-Soon-As-Possible (ASAP) and As-Late-As-Possible (ALAP) scheduling policies. The RCLS algorithm works with the following assumptions:

1. Nodes in the DDDG are associated with hardware functional units. Configurations of those units follow the default setting of Vivado HLS such as functional types, latencies, and resource consumption;
2. Data is stored into memory banks (FPGA BRAM) and each bank supports one write and two reads at the same cycle;

3. Nodes that are supporting instructions are removed by assigning zero latency;
4. As most of the accelerator designs are restricted by BRAM and DSP resource, in current implementations, we only consider these two resource constraints.

Based on the above assumptions, Lin-Analyzer finds the minimum latency of the DDDG utilizing ASAP policy, which only schedules a node with the condition that all predecessors of the node are completed. ALAP policy schedules a node as late as possible when its successors are all finished. RCLS scheduling takes timestamps (a priority list) of nodes returned by ALAP scheduling as an input. Both ASAP and ALAP have no resource limitation, which is infeasible. Therefore, Lin-Analyzer leverages resource-constrained list scheduling policy to obtain a feasible schedule of minimum latency within FPGA resource constraints.

The RCLS policy schedules a DDDG node with the following conditions:

- All predecessors of the node have been scheduled and completed;
- Among the unscheduled ready nodes, the node has the highest priority;
- There are sufficient FPGA resources for allocating the node.

The resource management (allocation and release) is provided by FPGA Resource Allocator (FRA). To schedule a type T node, FRA checks if there exists an allocated T functional unit available. If all T units are occupied and there are still sufficient resources, FRA allocates a new T functional unit for the node and records its occupied status; otherwise, the node is assigned with an available allocated T functional unit. Functional units consist of pipelined and non-pipelined designs. In this work, we utilize pipelined functional units for floating-point operations and the rest uses non-pipelined units. For pipelined units, if a node using this kind of unit is scheduled, the occupied pipelined unit will be released in the next cycle by FRA. For non-pipelined unit, an occupied functional unit will be released only if the associated node finishes.

When RCLS policy finishes scheduling all nodes in the DDDG, Lin-Analyzer obtains the final schedule and execution latency of the DDDG. With the loop bounds and latency of *sub-trace*, Lin-Analyzer predicts execution cycles of an FPGA-based design for the kernel.

2.3.6 Enabling Design Space Exploration

Designers can use HLS tools to develop diverse hardware implementations by inserting various optimization pragmas. The three prominent pragmas, loop unrolling, loop pipelining, and array partitioning, have significant impact on FPGA performance and resource consumption [9, 29]. Therefore, the three pragmas are supported in this work and Lin-Analyzer enables rapid design space exploration with this feature.

Loop Unrolling With this optimization, HLS tools can schedule instructions of multiple loop iterations and exploit more instruction-level parallelism. To mimic the optimization on HLS tools, Lin-Analyzer properly selects size of *sub-trace* according to unrolling factors as explained in Sect. 2.3.4.1 and predicts performance on the optimized DDDG generated. Assume that designers provide loop unrolling optimization with factor u, Lin-Analyzer extracts a *sub-trace* containing dynamic instruction instances of u loop iterations and generates an initial un-optimized DDDG. After pre-optimization in Sect. 2.3.4.2, Lin-Analyzer schedules nodes in the new generated optimized DDDG with RCLS policy and obtains latency *IL* of the *sub-trace* according to loop unrolling configuration. With loop bounds, unrolling factor u and latency *IL* of the *sub-trace*, Lin-Analyzer predicts performance of the FPGA-based accelerator.

Loop Pipelining Operations in a loop iteration i are executed in sequence. The next iteration $i + 1$ of the loop can only start execution when all operations inside the current loop iteration i are complete. Loop pipelining optimization enables operations in the next loop iteration $i + 1$ begin execution without waiting for the current loop iteration i to be finished. This concurrent execution manner significantly improves performance of hardware accelerators. With pipelining optimization enabled, performance of an accelerator is determined by an initiation interval (*II*) of the loop. *II* is a constant clock cycle period required between the start of two consecutive loop iterations. To predict performance of accelerators with loop pipelining enabled, Lin-Analyzer does not perform scheduling and calculates the minimum initiation interval (*MII*) to approximate the *II* instead. This can reduce the size of *sub-trace* and help to reduce Lin-Analyzer's runtime. The calculation of *MII* is done by the following Eqs. (2.1)–(2.4),

$$MII = \max(RecMII, ResMII) \tag{2.1}$$

$$ResMII = \max(ResMII_{\text{mem}}, ResMII_{\text{op}}) \tag{2.2}$$

$$ResMII_{\text{mem}} = \max_{m} \left(\left\lceil \frac{R_m}{RPorts_m} \right\rceil, \left\lceil \frac{W_m}{WPorts_m} \right\rceil \right) \tag{2.3}$$

$$ResMII_{\text{op}} = \max_{n} \left(\left\lceil \frac{Fop\_Par_n}{Fop\_used_n} \right\rceil \right) \tag{2.4}$$

where *RecMII* is the recurrence-constrained *MII* and *ResMII* is the resource-constrained *MII*. $ResMII_{\text{mem}}$ is used to analyze *MII* that is restricted by memory bandwidth, while $ResMII_{\text{op}}$ is limited by number of floating-point hardware units. The number of memory read and write operations of array m within a pipelined stage are represented by R_m and W_m, respectively. The number of read and write ports of array m depends on number of memory banks associated, which is related with array partitioning factors. The available number of read and write ports of array m are denoted by $RPorts_m$ and $WPorts_m$, respectively. Fop_Par_n and Fop_used_n are the number of floating-point functional unit of type n returned by ALAP scheduling and RCLS policy, respectively. Fop_Par_n denotes the maximum number of functional units that can run simultaneously without resource constraints.

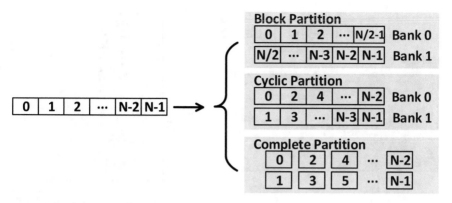

Fig. 2.4 Array partitioning example for three strategies with factor 2 [43]. For simplicity, the partitioning factor used here is two

In this work, we use latency *IL* of the selected *sub-trace* as its pipeline depth. With loop bounds, pipeline depth *IL*, and the estimated *MII*, Lin-Analyzer predicts performance of the FPGA-based accelerator using the equation in [20].

Array Partitioning Data in FPGA-based accelerator is stored into one or multiple memory banks which are composed of FPGA BRAM resource. As memory ports per bank are limited, the number of read/write through the same bank at the same cycle is restricted (we assume two-read and one-write ports per memory bank). Accelerators might suffer from this memory bandwidth bottleneck. In Vivado HLS [43], it supports array partitioning pragma to split data into multiple memory banks to improve the bandwidth. The partitioning strategies include three types, *block*, *cyclic*, and *complete* as shown in Fig. 2.4. To simulate array partitioning optimization, Lin-Analyzer first maps addresses of load and store operations in the DDDG to memory banks, and leverages FRA to keep track of read/write ports used each bank and prevent RCLS scheduling from violating memory port constraints. Memory bank number $Bank\_N_m$ related to array partitioning factor is calculated as below,

$$Bank\_N_m = \begin{cases} (addr_m)/(\lceil size_m/pf \rceil) & \text{if block} \\ (addr_m) \text{ modulo } (pf) & \text{if cyclic} \end{cases} \tag{2.5}$$

where $addr_m$ represents a memory address of array m, $size_m$ denotes array size of m, and pf describes the partition factor. Memory-port constraint is released for *complete* array partitioning, as the whole array is implemented with registers.

An Example Figures 2.5 and 2.6 show two examples to describe how Lin-Analyzer estimates FPGA performance when given different pragmas. In the examples, the *fadd* functional unit has 5-cycle latency and it is a pipelined design. Memory operations (load and store) have 1-cycle latency. These FPGA node latencies follow the default setting of Vivado HLS.

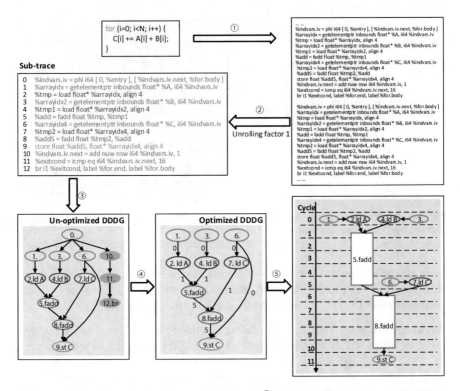

Fig. 2.5 An example without optimization pragma: ① Lin-Analyzer instruments the source code and generates dynamic trace; ② Given loop unrolling factor *uf* (*uf* = 1, which means no optimization), Lin-Analyzer extracts dynamic instruction instances of one loop iteration as a *sub-trace*; ③ With the *sub-trace*, our tool generates an un-optimized DDDG to represent the hardware accelerator; ④ Lin-Analyzer performs pre-optimizations on the un-optimized DDDG; ⑤ Lin-Analyzer performs RCLS scheduling on the optimized DDDG. Latency *IL* of the *sub-trace* returned from the scheduling graph is 12 cycles and the total FPGA execution cycle of the loop is (12 ∗ *N*) cycles, where *N* is the loop bound

Figure 2.5 shows the example without optimization, which means that loop unrolling *uf* and array partitioning factors *pf* are equal to 1 and no loop pipelining is enabled. In Fig. 2.5, the instructions in the *sub-trace* highlighted in green are *supporting* instructions, which are used to keep computation being carried out in the correct manner. Lin-Analyzer removes the *supporting* instructions by assigning zero-latency as their edge weights. As we can see that, there is a true dependence between Instruction 0 and 10, which are related to loop indices. This kind of dependence is removed after performing optimization on an initial DDDG. With the optimizations mentioned in Sect. 2.3.4.2, Lin-Analyzer schedules the DDDG leveraging RCLS policy. The final scheduling graph is shown in Fig. 2.5. Array *A*, *B*, and *C* consume only one memory bank (BRAM consumption depends on their size) because of *pf* = 1 and can support two-read and one-write operations simultaneously. Based on the scheduling, we know the latency of *uf*

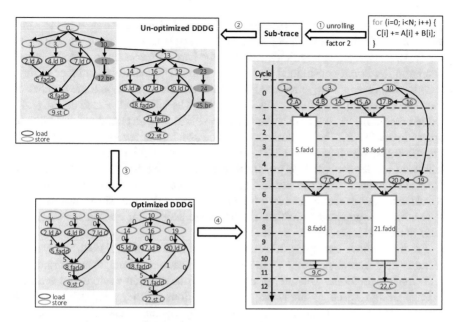

Fig. 2.6 An example with loop unrolling pragma ($uf = 2$): With a new optimization pragma ①, Lin-Analyzer extracts a new *sub-trace*, creates an initial DDDG accordingly ②, performs pre-optimization to generate the optimized DDDG ③, and schedules nodes on the DDDG. Latency *IL* of the *sub-trace* returned from the scheduling graph is 13 cycles in this case and the total FPGA execution cycle of the loop with unrolling factor $uf = 2$ is $(13 * N/2)$ cycles

loop iterations and Lin-Analyzer predicts its FPGA performance for this kernel without optimization. In this example, the hardware accelerator uses one 32-bit *fadd* functional unit, which consumes 2 DSPs.

Figure 2.6 shows the same example with loop unrolling enabled ($uf = 2$). As the time spent on collecting the whole trace is a one-time cost, Lin-Analyzer reuses the whole trace from the previous example and extracts a new *sub-trace* according to loop unrolling pragma provided. With the new *sub-trace*, Lin-Analyzer follows similar steps in the previous example and predicts performance of the hardware accelerator with loop unrolling enabled without generating any RTL implementations. In Fig. 2.6, Instruction 9 and 22 in blue are used to store data in Array C. As we do not enable array partitioning pragma ($pf = 1$), Array C only allows one write operation per cycle due to memory bank constraint and thus Lin-Analyzer spends two cycles on Instruction 9 and 22. In this example, the hardware accelerator shares one 32-bit *fadd* functional unit, which consumes 2 DSPs.

When we apply loop pipelining pragma on the example in Fig. 2.5, we follow the same steps in the figure and calculate the initiation interval *II* with Eqs. (2.1)–(2.4). In the example, there is no recurrence loop dependence and thus *RecMII* is 0. The number of memory read and write operations (R_A and W_A) of Array A within a pipelined stage from the figure is 1 for both. The available number of read and write ports of Array A are 2 and 1, respectively. With Eq. (2.3) for Array A, we have

$\lceil \frac{R_A}{RPorts_A} \rceil = \lceil \frac{1}{2} \rceil = 1$, while $\lceil \frac{W_A}{WPorts_A} \rceil = \lceil \frac{1}{1} \rceil = 1$. Data of Array B is the same with that of A. For Array C, R_C is 0, as there is no memory read. According to Eq. (2.4), $ResMII_{mem}$ is 1, which means that the hardware accelerator with the given configuration is not constrained by memory ports. The maximum number of *fadd* functional unit Fop_Par_{fadd} returned by ALAP scheduling is 1, and the number of fadd functional unit Fop_used_{fadd} used in RCLS scheduling is 1. Based on Eq. (2.4), we get $ResMII_{op} = 1$. We calculate $MII = 1$ with Eq. (2.1) and use its value to approximate the II. The pipeline depth of the accelerator leverages the latency of the *sub-trace*, which is $IL = 12$. Therefore, the total FPGA execution cycle of the hardware accelerator with loop pipelining enabled is $(II*(N-1)+IL = N-1+12 = N + 11)$ cycles. As we can see in Fig. 2.5, the two *fadd* instructions, *5.fadd* and *8.fadd*, can start execution at every $6 * i$ ($i \in [1, 2, \ldots]$) cycles simultaneously and thus with loop pipelining enabled, the hardware accelerator consumes two 32-bit *fadd* functional units, which uses 4 DSPs.

From the above examples, Lin-Analyzer can explore different hardware architectures of a kernel rapidly by changing combinations of pragmas without any RTL implementations. This ability makes Lin-Analyzer can explore and evaluate a large design space of hardware implementations in the order of seconds to minutes. However, similar to other works using dynamic analysis [15, 38, 41], if different program inputs have significant impacts on behaviors of an application, Lin-Analyzer might also suffer from inaccuracy when predicting performance. In this case, selecting a representative input for generating trace is necessary and crucial. Moreover, in current implementation, as Lin-Analyzer only optimizes for FPGA performance, it tries to use available resources as much as possible if necessary. Area-performance tradeoff in accelerator design will be included inside our framework in future.

2.4 Acceleration of Data Analytics Kernels

The experiment is set up on a computer with an Intel Xeon CPU E5-2620 running at 2.10 GHz with 64 GB RAM and the OS used is Ubuntu 14.04. We leverage Xilinx Vivado HLS version 2014.4 as the HLS tool and frequency of accelerators is set to 100MHz. Our target FPGA device is Xilinx ZC702 Evaluation Kit [43]. We select four kernels related to big-data applications for evaluation.

- **GEMM**: This kernel is a generic matrix–matrix multiplication application from Polybench Benchmark Suite [30]. It is widely used in machine learning applications such as Convolutional Neural Network [36].
- **KMeans**: This kernel is a clustering algorithm, which is used extensively in data-mining. It is modified from Rodinia Benchmark Suite [6].
- **CONV2D & CONV3D**: Convolution 2D/3D can be used to implement edge detection and smoothing as a filter. It is an important computation in signal/image processing, machine learning, and elsewhere [17, 42]. The two kernels are adapted from Polybench Benchmarks Suite GPU version [30].

2.4.1   Estimation Accuracy

Given loop unrolling, loop pipelining, and array partitioning pragmas, Lin-Analyzer predicts FPGA performance for kernels in C/C++. As Vivado HLS is based on static analysis, it might conservatively add false loop-carried dependences and limit the exploitable parallelism of accelerators. To analyze estimation quality of Lin-Analyzer, we describe prediction accuracy separately for different pragma combinations.

2.4.1.1   Loop Unrolling and Loop Pipelining

Considering loop unrolling and loop pipelining pragmas, Fig. 2.7 shows the performance (execution cycle counts) comparison of Lin-Analyzer and Vivado HLS for *GEMM, KMeans, CONV2D*, and *CONV3D* kernels. The Y-axis denotes the execution cycle counts of different configurations, while the X-axis describes various configuration combinations consisting of loop unrolling and loop pipelining. As we can see from the figure, the predicted performance from Lin-Analyzer (the yellow dashed lines with triangles) matches the ones from Vivado HLS (the green solid lines with stars) very closely for all four kernels. The average difference between the execution cycle counts returned from Vivado HLS and the

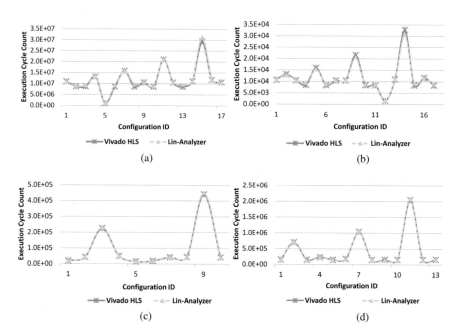

Fig. 2.7 Accuracy of Lin-Analyzer compared to Vivado HLS considering loop unrolling and loop pipelining pragmas. (**a**) GEMM. (**b**) KMeans. (**c**) CONV2D. (**d**) CONV3D

Table 2.3 Performance comparison with loop unrolling and loop pipelining enabled: Lin-Analyzer vs. Vivado HLS

Benchmark	GEMM	KMeans	CONV2D	CONV3D
Difference (%)	3.25	3.78	1.63	3.75

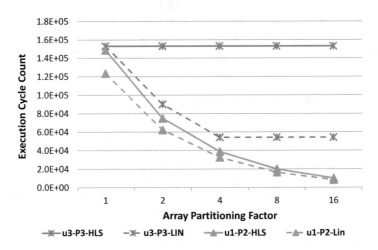

Fig. 2.8 Performance comparison with loop unrolling, loop pipelining, and array partitioning: Lin-Analyzer vs. Vivado HLS for CONV3D. Due to false loop-carried dependences, Vivado HLS generates inefficient designs, which leads to difference with predictions from Lin-Analyzer

ones estimated by Lin-Analyzer is calculated for the same configuration across all combinations and shown in Table 2.3. From the table, Lin-Analyzer can predict performance of FPGA-based accelerators with loop unrolling and loop pipelining enabled within 4.0% difference across all four kernels, which is quite accurate.

2.4.1.2 Array Partitioning

Figure 2.8 demonstrates the result comparison between Lin-Analyzer and Vivado HLS for *CONV3D* kernel with loop unrolling, loop pipelining, and array partitioning enabled. In Fig. 2.8, we fix loop unrolling and loop pipelining configuration and analyze performance when varying array partitioning factors. The Y-axis denotes the execution cycle counts of different configurations, while the X-axis describes different array partitioning factors applied varying from 1 to 16 in step of 2. (ui-Pj) represents a configuration combination consisting of loop unrolling factor i applied at the innermost loop level and loop pipelining applied at loop level j. Solid and dashed lines represent results from Vivado HLS and Lin-Analyzer, respectively.

As a memory bank on FPGAs has limited ports, which potentially hinders HLS tools to exploit more parallelism, array partitioning pragma is designed to split data into multiple memory banks and increase memory bandwidth. This pragma usually works with loop unrolling or loop pipelining. In Fig. 2.8, performance (the red solid line with stars, *u3-P3-HLS*) from Vivado HLS with configuration *u3-P3* remains constant when applying different array partitioning factors, which means that increasing memory bandwidth has no impact on hardware performance. The results from Lin-Analyzer (the red dashed line with stars, *u3-P3-LIN*) show different behavior compared to Line *u3-P3-HLS*. It demonstrates that increasing memory bandwidth can actually improve performance. The reason that leads to the performance discrepancy of Vivado HLS and Lin-Analyzer can be explained as follows. HLS tools rely on static analysis and perform conservative dependence analysis. It might potentially add *false loop-carried dependences*. In the example above, the *false loop-carried dependences* introduced by Vivado HLS leads to a high recurrence II (*RecII*) values and the *MII* in Eq. (2.1) is dominated by *RecII*. Therefore, increasing memory bandwidth in this case cannot help to exploit more parallelism. As Lin-Analyzer relies on dynamic trace and all dependences are known, Line *u3-P3-LIN* shows that increasing memory bandwidth can help to reduce execution cycles of accelerators. A hand-written RTL code or enabling dependence pragma to disable specific loop-carried dependence in HLS tools can effectively improve hardware performance as predicted by Lin-Analyzer in Line *u3-P3-LIN*. In addition, by simulating RTL codes generated by Vivado HLS, we find that for some configurations with array partitioning enabled, there exist redundant memory loads. This further deteriorates the hardware performance due to the memory inefficiency in Vivado HLS designs when compared to optimized hand-written RTL implementations.

Although results from the two might be different, Lin-Analyzer can accurately predict the hardware performance trends with array partitioning enabled. Moreover, Lin-Analyzer also can help designers to better understand design bottlenecks and generate high-quality FPGA-based accelerators with HLS tools.

2.4.2   Rapid Design Space Exploration

As mentioned in Sect. 2.3, given various pragma combinations consisting of loop unrolling, loop pipelining, and array partitioning, Lin-Analyzer can rapidly evaluate hardware performance accordingly and enable design space exploration to find the high-quality design point without generating RTL implementations. The design space we consider is shown below,

- *Loop unrolling factor*: Its range includes divisors of loop bound *N*.
- *Loop pipelining*: Its range includes *True* and *False*.
- *Array partitioning*: The factor can vary from 1 to 16 in steps of 2. The partitioning types are *cyclic*, *block*, and *complete*.

Table 2.4 demonstrates the design space exploration results with exhaustive HLS-based method and Lin-Analyzer. Kernels considered in this work are listed in Column 1. Number of loop levels and design space of each kernel are shown

Table 2.4 Design space exploration results

Benchmark	Loop levels	Design space	Configuration		Total DSE Time (s)			
			Exhaustive	Lin-Analyzer	Exhaustive	Lin-Analyzer		
						Profiling	DSE	Total
GEMM	3	85	1,1,2	1,1,2	36579.48	176.38	8.99	185.37
KMeans	2	136	(8,12,16),1,1	(8,12,16),1,1	3922.33	1.26	45.47	46.73
CONV2D	2	62	16,1,1	16,1,1	26573.13	5.48	17.28	22.76
CONV3D	3	65	16,1,2	16,1,2	21586.68	12.85	9.62	22.47

Configuration format is *(array partitioning factor, loop unrolling factor, pipeline level)*

in Column 2 and 3, respectively. To evaluate accuracy of DSE with Lin-Analyzer, we leverage Vivado HLS to perform exhaustive DSE with the same design space (shown as *Exhaustive* in Table 2.4) and record HLS exploration time and execution cycles of the generated accelerator implementations.

The optimal design points of each kernel given by *Exhaustive* DSE and Lin-Analyzer are shown in Column 4 and 5 in Table 2.4. The configuration format used here is (array partitioning factor i, loop unrolling factor j, pipeline level k). The pipeline level k means that a pipelining pragma is applied at loop level L_k. Column 4 and 5 in Table 2.4 demonstrate that the configurations, which achieves the best performance for each kernel within the design space given, recommended by *Exhaustive* method and Lin-Analyzer are exactly the same.

The exploration time of *Exhaustive* method is shown in Column 6 in Table 2.4. As Lin-Analyzer relies on dynamic trace, its exploration time consists of two parts: *Profiling* and *DSE*. The *Profiling* part is the time spent on collecting dynamic trace, which is a one-time overhead and can be amortized. The total exploration time of DSE with Lin-Analyzer is shown in Column 9 in Table 2.4. Comparing Column 9 with 6, we can see that the exploration time needed by Lin-Analyzer is only a fraction of the time using *Exhaustive* method while recommending the correct configuration combinations. Exploration time speedup of each kernel normalized to *Exhaustive* method is shown in Fig. 2.9. The results in Fig. 2.9 confirm that Lin-Analyzer is capable to perform rapid architectural exploration and the average speedup achieves 617X for the four kernels.

To evaluate the quality of the best design points (within the design space considered) given by our automated DSE flow, we compare their execution time of FPGA implementations with CPU-based performance for all the kernels. CPU-based performance is obtained by running single-thread C implementations of the same kernels on one Intel Xeon CPU E5-2620 at 2.1 GHz and one ARM Cortex-A15 core (from Odroid-XU3 [27]) at 2.0 GHz. We utilize '-O3' as the GCC optimization option. Besides, we also run OpenCL implementations for the four kernels using 4 Cortex-A15 from Odroid-XU3 [27]. The OpenCL implementations are obtained from Polybench Benchmark Suite GPU version [30] and Rodinia Benchmark Suite [6].

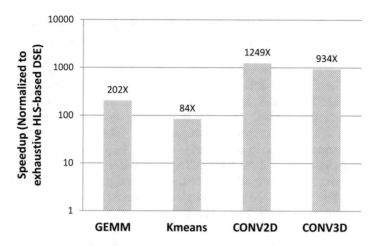

Fig. 2.9 Exploration time speedup compared to exhaustive HLS-based DSE

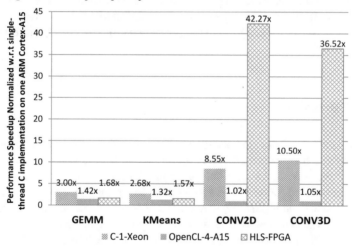

Fig. 2.10 Speedup of different implementations normalized to single-thread C design on one ARM Cortex-A15 core. FPGA implementations, *HLS-FPGA*, leverage the best design points returned by the proposed automated DSE framework. *C-1-Xeon* denotes single-thread C implementation on one Intel Xeon core, while *OpenCL-4-A15* represents OpenCL implementations using four ARM Cortex-A15 cores

The results are shown in Fig. 2.10. In the figure, we use *C-1-Xeon*, *C-1-A15*, *OpenCL-4-A15*, and *HLS-FPGA* to denote the corresponding implementations. For *GEMM* and *KMeans*, performance of *HLS-FPGA* can achieve around 1.6x speedup compared to that of *C-1-A15* and better than *OpenCL-4-A15*. Compared to implementations on the high-end CPU, *C-1-Xeon*, performance of *HLS-FPGA* is slightly slower. The reason for *GEMM* is that its *II* is dominated by memory bandwidth. Due to the limited FPGA BRAM resource, we cannot leverage large array partitioning factors to increase memory bandwidth. For *KMeans*, as it can be easily vectorized and has less computation operations compared to other kernels, CPU and OpenCL implementations can be well optimized to match FPGA performance. *CONV2D* and *CONV3D* have extensive memory reads and computations and there is no dependence among different output data. However, they are memory-bound kernels on the CPU, as the ratio between their arithmetic operations and memory accesses is low [5]. In Fig. 2.10, due to the memory-bound problem, *OpenCL-4-A15* can not achieve good speedup compared to *C-1-A15*. However, *HLS-FPGA* achieves around 40x speedup compared to *C-1-A15* implementations and roughly 5x speedup compared with *C-1-Xeon*. The reason is that with array partitioning enabled, Vivado HLS can exploit more instruction-level parallelism and utilize deep pipelining. Thus, their FPGA implementations can instantiate lots of functional units for computation and occupy up to 92% DSP resource. This demonstrates that the design points returned by our automated DSE framework have high quality.

As Lin-Analyzer does not rely on HLS tools or generate any RTL implementations, its runtime scales linearly with more complex configuration combinations (larger unrolling and partitioning factors, pipelining at higher loop levels, etc.). This makes Lin-Analyzer be an attractive complementary tool for HLS to perform design space exploration.

2.5   Conclusion

In this chapter, we focus on accelerating data analytics kernels on heterogeneous computing systems featuring FPGAs. In particular, we present a toolchain, called Lin-Analyzer, that allows easy but performance-efficient implementation of data analytics kernels on FPGA-based accelerators. Lin-Analyzer relies on the dynamic data dependence graph (DDDG) to avoid the false data dependences created by the static analysis techniques used in most existing techniques including commercial HLS tools. This results in an accurate performance estimation of FPGA-based accelerators without resorting to time-consuming HLS runs. The tool also helps in identifying design bottlenecks while exploring various pragmas such as loop unrolling, pipelining, and array partitioning. Lastly, Lin-Analyzer can assist HLS developers in identifying potential limitations of the HLS tool. Our experimental evaluation with a number of data analytics kernels confirms the effectiveness of Lin-Analyzer.

Acknowledgements This work was partially supported by the Singapore Ministry of Education Academic Research Fund Tier 2 MOE2015-T2-2-088.

References

1. S. Bilavarn, G. Gogniat, J.L. Philippe, L. Bossuet, Design space pruning through early estimations of area/delay tradeoffs for FPGA implementations. IEEE Trans. Comput. Aided Des. Integr. Circuits Syst. **25**. Doi:10.1109/TCAD.2005.862742
2. Cadence Inc. C-to-Silicon Compiler (2015)
3. A. Canis, J. Choi, M. Aldham et al., LegUp: high-level synthesis for FPGA-based processor/accelerator systems, in *Proceedings of the 19th ACM/SIGDA International Symposium on Field Programmable Gate Arrays (FPGA'2011)*, Monterey (2011)
4. A. Canis, D. Brown, J.H., Anderson, Modulo SDC scheduling with recurrence minimization in high-level synthesis, in *The 24th International Conference on Field Programmable Logic and Applications (FPL)*, Munich (2014)
5. S. Che, M. Boyer, J. Meng, D. Tarjan, J.W. Sheaffer, K. Skadron, A performance study of general-purpose applications on graphics processors using CUDA. J. Parallel Distrib. Comput. **68**(10), 1370–1380 (2008)
6. S. Che, J.W. Sheaffer, M. Boyer, L.G. Szafaryn, L. Wang, K. Skadron, A characterization of the Rodinia benchmark suite with comparison to contemporary CMP workloads, in *2010 IEEE International Symposium on in Workload Characterization (IISWC)* (2010), pp. 1–11
7. J. Cong, Z. Zhang, An efficient and versatile scheduling algorithm based on SDC formulation, in *The 43rd ACM/IEEE Design Automation Conference (DAC'2006)*, San Francisco (2006)
8. J. Cong, W. Jiang, B. Liu, Y. Zou, Automatic memory partitioning and scheduling for throughput and power optimization, in *IEEE/ACM International Conference on Computer-Aided Design - Digest of Technical Papers*, San Jose, CA (2009)
9. J. Cong, M. Huang, P. Pan, Y. Wang, P. Zhang, *Source-to-Source Optimization for HLS, FPGAs for Software Programmers*, chap. 8 (Springer International Publishing, Cham, 2016), pp. 137–163. Doi:http://dx.doi.org/10.1145/2209291.2209302. ISBN 978-3-319-26408-0
10. W.J. Dally, J.D. Balfour, D. Black-Schaffer, J. Chen, R.C. Harting, V. Parikh, J. Park, D. Sheffield, Efficient embedded computing. IEEE Comput. **41**(7), 27–32 (2008)
11. R.H. Dennard, F.H. Gaensslen, V.L. Rideout, E. Bassous, A.R. LeBlanc, Design of ion-implanted MOSFET's with very small physical dimensions. IEEE J. Solid State Circuits **9**(5), 256–268 (1974)
12. H. Esmaeilzadeh, E. Blem, R. St Amant, K. Sankaralingam, D. Burger, Dark silicon and the end of multicore scaling, in *2011 38th Annual International Symposium on Computer Architecture (ISCA)* (IEEE, New York, 2011), pp. 365–376
13. A.P. Greenhalgh, Big.LITTLE processing with ARM Cortex-A15 & Cortex-A7 (2011)
14. M. Guevara, B. Lubin, B.C. Lee, Navigating heterogeneous processors with market mechanisms, in *2013 IEEE 19th International Symposium on High Performance Computer Architecture (HPCA2013)* (IEEE, New York, 2013), pp. 95–106
15. J. Holewinski, R. Ramamurthi, M. Ravishankar, N. Fauzia, L.N. Pouchet, A. Rountev, P. Sadayappan, Dynamic trace-based analysis of vectorization potential of applications, in *The 33rd ACM SIGPLAN Conference on Programming Language Design and Implementation (PLDI)*, Beijing (2012)
16. Ineda Systems, Hierarchical computing (2014). [Online]
17. Y. Jia, E. Shelhamer, J. Donahue, S. Karayev, J. Long, R. Girshick, S. Guadarrama, T. Darrell, Caffe: convolutional architecture for fast feature embedding. Preprint (2014). arXiv:1408.5093
18. R. Kumar, K.I. Farkas, N.P. Jouppi, P. Ranganathan, D.M. Tullsen, Single-ISA heterogeneous multi-core architectures: the potential for processor power reduction, in *MICRO* (2003), pp. 81–92
19. C. Lattner, V. Adve, LLVM: a compilation framework for lifelong program analysis & transformation, in *Proceedings of the International Symposium on Code Generation and Optimization: Feedback-directed and Runtime Optimization (CGO)*, Palo Alto, CA (2004)
20. P. Li, P. Zhang, L.N. Pouchet, J. Cong, Resource-Aware Throughput Optimization for High-Level Synthesis, in *The 2015 ACM/SIGDA International Symposium on Field-Programmable Gate Arrays (FPGA)*, Monterey, CA (2015)

21. Y. Liang, K. Rupnow, Y. Li, D. Min, M.N. Do, D. Chen, High-level synthesis: productivity, performance, and software constraints. J. Electr. Comput. Eng. **2012** (2012). Doi:10.1155/2012/649057
22. H. Liu, L.P. Carloni, On learning-based methods for design-space exploration with high-level synthesis, in *The 50th Annual Design Automation Conference (DAC)*, Austin (2013)
23. G.E. Moore, Cramming more components onto integrated circuits. Proc. IEEE **86**(1), 82–85 (1998)
24. T.S. Muthukaruppan, M. Pricopi, V. Venkataramani, T. Mitra, S. Vishin, Hierarchical power management for asymmetric multi-core in dark silicon era, in *Proceedings of the 50th Annual Design Automation Conference* (ACM, New York, 2013), p. 174
25. T.S. Muthukaruppan, A. Pathania, T. Mitra, Price theory based power management for heterogeneous multi-cores, in *Proceedings of the 19th International Conference on Architectural Support for Programming Languages and operating systems* (ACM, New York, 2014), pp. 161–176
26. nVidia, Variable SMP—a multi-core CPU architecture for low power and high performance (2011)
27. Odroid-XU3. http://goo.gl/Nn6z3O
28. A. Pathania, Q. Jiao, A. Prakash, T. Mitra, Integrated CPU-GPU power management for 3D mobile games," in *Proceedings of the the 51st Annual Design Automation Conference on Design Automation Conference* (ACM, New York, 2014), pp. 1–6
29. N. Pham, A.K. Singh, A. Kumar, M.M.A. Khin, Exploiting loop-array dependencies to accelerate the design space exploration with high level synthesis, in *Proceedings of the 2015 Design, Automation & Test in Europe Conference & Exhibition*, San Jose, CA (2015)
30. L. Pouchet, PolyBench/C3.2 (2012)
31. M. Pricopi, T. Mitra, Bahurupi: a polymorphic heterogeneous multi-core architecture. ACM Trans. Archit. Code Optim. **8**(4), 22 (2012)
32. M. Pricopi, T. Mitra, Task scheduling on adaptive multi-core. IEEE Trans. Comput. **63**(10), 2590–2603 (2014)
33. M. Pricopi, T.S. Muthukaruppan, V. Venkataramani, T. Mitra, S. Vishin, Power-performance modeling on asymmetric multi-cores, in *2013 International Conference on Compilers, Architecture and Synthesis for Embedded Systems (CASES)* (2013), pp. 1–10
34. A. Prost-Boucle, O. Muller, F. Rousseau, A fast and autonomous HLS methodology for hardware accelerator generation under resource constraints, in *Euromicro Conference on Digital System Design (DSD)*, Los Alamitos, CA (2013)
35. A. Putnam, A.M. Caulfield, E.S. Chung, D. Chiou, K. Constantinides et al., A reconfigurable fabric for accelerating large-scale datacenter services, in *Proceeding of the 41st Annual International Symposium on Computer Architecuture* (IEEE, New York, 2014), pp. 13–24
36. J. Redmon, S. Divvala, R. Girshick, A. Farhadi, You only look once: unified, real-time object detection. Preprint (2015). arXiv:1506.02640
37. B.C. Schafer, K. Wakabayashi, Divide and conquer high-level synthesis design space exploration. ACM Trans. Des. Autom. Electron. Syst. **17**(3), Article 29 (2012), 19pp. Doi:http://dx.doi.org/10.1145/2209291.2209302
38. Y. Shao, B. Reagen, G.Y. Wei, D. Brooks, Aladdin: a pre-RTL, power-performance accelerator simulator enabling large design space exploration of customized architectures, in *The 41st Annual International Symposium on Computer Architecture (ISCA)*, Minneapolis (2014)
39. B. So, M.W. Hall, P.C. Diniz, A compiler approach to fast hardware design space exploration in FPGA-based systems, in *Proceedings of the ACM SIGPLAN 2002 Conference on Programming Language Design and Implementation*, Berlin (2002)
40. Synopsys Inc. (2015)
41. M.A. Todd, S.S. Gurindar, Dynamic dependency analysis of ordinary programs, in *The 19th Annual International Symposium on Computer Architecture*, New York (1992)
42. F.M. Vallina, C. Kohn, P. Joshi, Zynq all programmable SoC Sobel filter implementation using the Vivado HLS tool. Application Note XAPP890, Xilinx (2012)
43. Xilinx Inc. (2015)
44. Z. Zhang, B. Liu, SDC-based modulo scheduling for pipeline synthesis, in *IEEE/ACM International Conference on Computer-Aided Design (ICCAD)*, San Jose, CA (2013)

45. G. Zhong, V. Venkataramani, Y. Liang, T. Mitra, S. Niar, Design space exploration of multiple loops on FPGAs using high level synthesis, in *2014 IEEE 32nd International Conference on Computer Design (ICCD)*, Seoul (2014)
46. G. Zhong, A. Prakash, Y. Liang, T. Mitra, S. Niar, Lin-analyzer: a high-level performance analysis tool for FPGA-based accelerators, in *The 53rd Annual Design Automation Conference (DAC)*, Austin (2016)

Chapter 3
Least-squares-solver Based Machine Learning Accelerator for Real-time Data Analytics in Smart Buildings

Hantao Huang and Hao Yu

3.1 Introduction

Among various energy consumers, it is reported that over 70% electricity is consumed by more than 79 million residential buildings and 5 million commercial buildings in the USA [1]. There is an increasing need to develop cyber-physical energy management system (EMS) for modern buildings composed of both micro-grid and smart IoT hardware [2]. For smart energy management system, collecting information from IoT devices can help recognize energy consumption profile and perform accurately load forecasting with consideration of occupants behavior. As such, load balance can be achieved based on demand-response strategy for better energy efficiency [3].

One direct application of demand-response strategy in energy management system (EMS) is the real-time dynamic electricity price [4] based on the demand. An accurate load forecasting can help schedule the energy demand to reduce the electricity cost. However, energy data analytics for load forecasting is challenging since it is greatly affected by occupants behavior and environmental factors [5]. Occupants behavior is of random nature and very hard to predict [6]. Using real-time sensed data from occupation location, power meters and various sensors can capture occupants behavior for more accurate data analytics. However,uploading data to the cloud and processing backend take latency and edge device such as smart-gateway is computational resource limited. Therefore, a computationally efficient

H. Huang
Nanyang Technological University, 50 Nanyang Avenue, Block S3.2, Level B2,
Singapore 639798, Singapore

H. Yu (✉)
School of Electrical and Electronic Engineering, Nanyang Technological University,
50 Nanyang Avenue, Block S3.2, Level B2, Singapore 639798, Singapore
e-mail: haoyu@ntu.edu.sg

© Springer International Publishing AG 2017
A. Chattopadhyay et al. (eds.), *Emerging Technology and Architecture for Big-data Analytics*, DOI 10.1007/978-3-319-54840-1_3

51

data analytics (machine learning algorithm) is greatly needed for real-time smart building energy management system.

Machine learning algorithms can be broadly classified into: supervised learning, unsupervised learning, and reinforcement learning [7]. Supervised learning based neural network is widely applied for energy data analytics. Supervised learning will learn the connection between two subset of data, inputs and outputs, to build a model. Two central problems under supervised learning are classification and regression. Both problems share the same goal to build a mode to predict the output based on the input. However, the difference between two problems is the fact the dependent attribute (output) is categorical for classification and numerical for regressions [8].

In the smart building EMS, the pre-trained model from machine learning algorithms will be loaded in the embedded system to perform data analytics such as short-term load forecasting. However, previous works [3, 9, 10] have limitations in twofold. Firstly, since various factors affect load forecasting, the pre-trained model cannot be adjusted with the new arrival data. Moreover, traditional supporting vector machine and neural network based algorithms [3] consume large hardware resource to analyze energy data with poor efficiency and latency. Secondly, previous energy data analytics [9, 10] ignores the real-time occupant profile, whose distribution at different functionalized location (office, resting area, kitchen, etc.) can significantly affect the short-term energy load forecasting accuracy. As such, the energy management system of building towards comfort and energy-efficiency is still not optimized.

In this work, we present a fast machine learning accelerator for smart building data analytics. A computational efficient machine learning is developed using a regularized least-squares solver with incremental square-root-free Cholesky factorization. A scalable and parameterized hardware architecture is developed in a pipeline and parallel fashion for both regularized least-squares and matrix-vector multiplication. With the high utilization of the FPGA hardware resource, our implementation has 128-PE in parallel operated at 50-MHz. Experimental results have shown that the proposed machine-learning accelerator (on FPGA) has good forecasting accuracy with an average speed-up of $4.56\times$ and $89.05\times$, when compared to general CPU and embedded CPU. Moreover , $450.2\times$, $261.9\times$ and $98.92\times$ energy saving can be achieved comparing to general CPU, embedded CPU and GPU.

The rest of this chapter is organized as follows. The Internet of Things (IoT) based smart-grid and smart building are presented in Sect. 3.2. The machine learning algorithm based on least-squares and backwards propagation is discussed in Sect. 3.3. Then Sect. 3.4 elaborates the Cholesky decomposition based least-squares solver. In Sect. 3.5, detailed implementation on FPGA hardware is elaborated. Experimental results regarding accuracy, speed-up, and energy consumption by FPGA implementation are presented in Sect. 3.6 with conclusion drawn in Sect. 3.7.

3.2 IoT System Based Smart Building

3.2.1 Smart-Grid Architecture

The overall Internet of Things (IoT) based smart-grid and smart building system is illustrated in Fig. 3.1. The key components from smart gird are the two-directional main electricity power grid and additional renewable energy based electricity power grid. By utilizing smart-grid, customers cannot only buy electricity from main power grid but also sell electricity from renewable solar energy to generate profits with dynamic prices. Smart building is the main element of the smart-grid for power consumption. Therefore, accurately predicting the energy demand of building can support the balance between supply and demand of smart-grid.

3.2.2 Smart Gateway for Real-Time Data Analytics

Smart gateway is the major control center, harboring the ability in storage and computation. Our smart gateway will be BeagleBoard-xM. As Fig. 3.1 shows, smart building is an IoT based system with various connected sensors. Environment sensors can collect information on light intensity, humidity, and temperature and send the data to micro-controller to understand environment. Energy sensor are used to collect the current of each appliance and through smart sockets, on-off control can be performed according to save energy. Moreover, occupancy provides information about the activity of occupants and location base services such as lighting and air-con can be provided accordingly. All these smart control is operated based on the pattern defined in the micro-controller and learnt by supervised learning process. Therefore, it is important to recognize (classify) the environment and occupants behavior to respond accordingly for customized services. Moreover, accurately predicting the energy demand of next minute or hour and then adjusting the supply

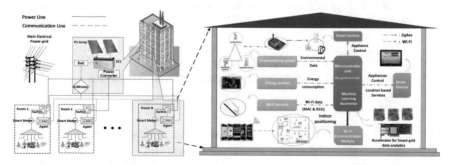

Fig. 3.1 Internet of things (IoT) based smart-grid and smart building system with renewable solar energy

are the key to achieve load balancing. However, due to the limited computation resource of smart-gateway, an FPGA based machine learning accelerator is designed to perform fast interference, model update, and re-train the machine learning model.

3.2.3 Problem Formulation for Data Analytics

In this chapter, data analytics refers to supervised machine learning, which is classification problem and regression problem. The classification problem is used for recognitions and regression problem is for prediction such as load forecasting. Details of each problem formulation are shown as below.

Objective 1: Minimize the error rate of classification.

$$\min e = \sum_{i}^{N} x_i/N \tag{3.1}$$

$$\text{s.t. } x_i = f(fe_1, fe_2, \ldots) = \{0, 1\}$$

where $f(\cdot)$ represents the trained model from training data and $fe_1, fe_2 \ldots$ represent input data for this model. $x_i = 0$ represents the accurate prediction, $x_i = 1$ represents the false prediction, and N represents the number of predictions.

Objective 2: Improve the accuracy of energy demand forecasting with time interval t.

$$\min er = \sum_{t=0}^{t=23} (y_t - f(E_t, M_t, T_t))^2 \tag{3.2}$$

where y_t is actual energy demand at time t and $f(E_t, M_t, T_t)$ is the model predicted result with input features: energy consumption data E_t, occupants motion profile M_t and environmental T_t until time t. $f(\cdot)$ is the machine learning trained model. Once the new energy consumption data is ready, the machine learning model will be re-trained with new arrival data to build up customized energy forecasting model.

3.3 Background on Neural Network Based Machine Learning

In this section, the fundamental of neural network based machine learning is introduced with comparison of two training methods.

Neural network (NN) is a family of network models inspired by biological neural network to build the link for a large number of input–output data pair. It typically has two computational phases: training phase and testing phase.

Table 3.1 A list of parameters definitions in machine learning

Parameter	Elements	Definitions
X	$[x_{11}, x_{12}, x_{13}, \ldots, x_{Nn}]$	A set of n dimension data in N training samples
T	$[t_{11}, t_{12}, t_{13}, \ldots, t_{NM}]$	A set of M target classes in N training samples
H	$[h_{11}, h_{12}, h_{13}, \ldots, h_{NL}]$	A set of L hidden nodes in N training samples
A	$[a_{11}, a_{12}, a_{13}, \ldots, a_{nL}]$	Input weight matrix between X and H
Γ	$[\gamma_{11}, \gamma_{12}, \gamma_{13}, \ldots, \gamma_{LM}]$	Output weight matrix between H and T
Y	$[y_1, y_2, y_3, \ldots, y_M]$	A set of M model outputs
β	N.A.	Learning rate set by designers

- In the training phase, the weight coefficients of the neural network model are first determined using training data by minimizing the squares of error difference between trial solution and targeted data in a so-called ℓ_2-norm method.
- In the testing phase, the neural network model with determined weight coefficients is utilized for classification or calculation given the new input of data.

Formally, the detailed descriptions of each parameter are summarized in Table 3.1. Given a neural network with n inputs and M outputs shown in Fig. 3.2, a dataset $(x_1, t_1), (x_2, t_2), \ldots, (x_N, t_N)$ is composed of paired input data **X** and training data **T** with N number of training samples, n dimensional input features and M classes. During the training, one needs to minimize the ℓ_2-norm error function with determined weights: **A** (at input layer) and $\boldsymbol{\Gamma}$ (at output layer):

$$E = ||\mathbf{T} - F(\mathbf{A}, \boldsymbol{\Gamma}, \mathbf{X})||_2 \qquad (3.3)$$

where $F(\cdot)$ is the mapping function from the input to the output of the neural network.

The output function of this neural network classifier is

$$\mathbf{Y} = \mathbf{H}F(\mathbf{A}, \boldsymbol{\Gamma}, \mathbf{X}), \quad \mathbf{Y} = \{y_1, y_2, \ldots y_m\}$$
$$\text{Label}(X) = \underset{i \in \{1,2,\ldots,m\}}{\arg\max} \; y_i \qquad (3.4)$$

where $\mathbf{Y} \in \mathbb{R}^{N \times m}$. Here N represents the number of testing samples. The index of maximum value **Y** is found and identified as the predicted class.

3.3.1 Backward Propagation for Training

The first method to minimize the error function E is the Backward Propagation (BP) method. As shown in Fig. 3.2a, the weights are firstly initially guessed for forward

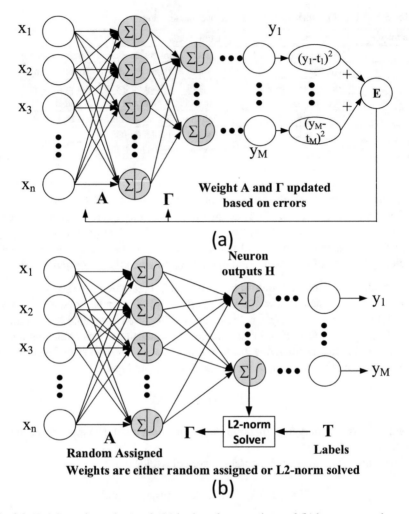

Fig. 3.2 Trainings of neural network: (**a**) backward propagation; and (**b**) least-square solver

propagation. Based on the trial error, the weights are further calculated backward with derivatives of weights calculated by

$$\nabla E = \left(\frac{\partial E}{\partial a_{11}}, \frac{\partial E}{\partial a_{12}}, \frac{\partial E}{\partial a_{13}} \cdots \frac{\partial E}{\partial a_{DL}} \right) \tag{3.5}$$

where D is the output dimension of previous layer and L is the input dimension of the next layer. For the current layer, each weight can be updated as

$$a_{dl} = a_{dl} - \beta * \frac{\partial E}{\partial a_{dl}}, \ d = 1, 2, \ldots, D, \ l = 1, 2, \ldots, L \tag{3.6}$$

where β is the learning constant that defines the step length of each iteration in the negative gradient direction. Note that the BP method requires to store the derivatives of each weight. It is expensive for hardware realization. More importantly, it may be trapped on local minimal with long converging time. Hence, the BP based training is usually performed off-line and has large latency when analyzing the real-time sensed data.

3.3.2 Least-Squares Solver for Training

One can directly solve the least-squares problem using the least-squares solvers of the ℓ_2-norm error function E [11–13]. As shown in Fig. 3.2b, the input weight $\mathbf{A}$ can be first randomly assigned and one can directly solve output weight $\boldsymbol{\Gamma}$ as follows.

We first find the relationship between the hidden neural node and input training data as

$$\mathbf{preH} = \mathbf{XA} + \mathbf{B}, \ \mathbf{H} = \frac{1}{1 + e^{-\mathbf{preH}}} \tag{3.7}$$

where $\mathbf{X} \in \mathbb{R}^{N \times n}$. $\mathbf{A} \in \mathbb{R}^{n \times L}$ and $\mathbf{B} \in \mathbb{R}^{N \times L}$ is random generated input weight and bias formed by a_{ij} and b_{ij} between $[-1, 1]$. N and n are the training size and the dimension of training data, respectively. The output weight $\boldsymbol{\Gamma}$ is computed based on pseudo-inverse $(L < N)$:

$$\boldsymbol{\Gamma} = (\mathbf{H}^T\mathbf{H})^{-1}\mathbf{H}^T\mathbf{T} \tag{3.8}$$

However, performing pseudo-inverse is also expensive for hardware realization.

The comparison of BP and least-squares solver can be summarized as follows. BP is a relative simple implementation by gradient descent objective function with good performances. However, it suffers from the long training time and may get stuck in the local optimal point. On the other hand, least-squares solver can learn very fast, but pseudo-inverse is too expensive for calculating. Therefore, solving ℓ_2-norm minimization efficiently becomes the bottleneck of the training process.

3.3.3 Feature Extraction with Behavior Cognition

Input features are very important to train an accurate machine learning model. In this chapter, occupants behavior is analyzed based on the active occupant motion in each room since it indicates the potential behavior of occupants in the room [14]. Rooms inside the same house have vastly different occupants behavior profiles due to different functionalities. Therefore, we extracted behavior profiles for different rooms, respectively. For each room i, there are four states represented by S for occupants positioning:

$$S = \begin{cases} s_1 : 0 & \text{no occupant in the room } i \\ s_2 : 0 \longrightarrow 1 & \text{occupants entering the room } i \\ s_3 : 1 & \text{occupants in the room } i \\ s_4 : 1 \longrightarrow 0 & \text{occupants leaving the room } i \end{cases} \tag{3.9}$$

where motion state S is detected by indoor positioning system via WiFi data every minute. The probability of occupants motion for room i can be expressed as:

$$M_i(t) = \frac{Ti(s_2) + Ti(s_3)}{Ti}, \quad t = 1, 2, 3, \ldots, 96 \tag{3.10}$$

where $Ti(s_j)$ represents the time duration with corresponding state s_j. $M_i(t)$ is occupant motion probability of room i in Ti time interval. Figure 3.3 presents an example of motion probability in different rooms. As a conclusion, all the features and their descriptions for data analytics are summarized in Table 3.2.

Our work differs from previous works [15, 16] such as sequential learning or recursive learning from two manifold. Firstly, the training data size in our work is fixed size with adding new arrival data and removing old data. This is preferred since environment and occupants change with several levels of seasonality [3]. Old data from months ago tend to bias the new change of load demand. Secondly, our learning algorithm focuses on tuning the size of neural network. Since training data is changed, it is more effective to re-train the model than using sequential learning method to update the out-dated model.

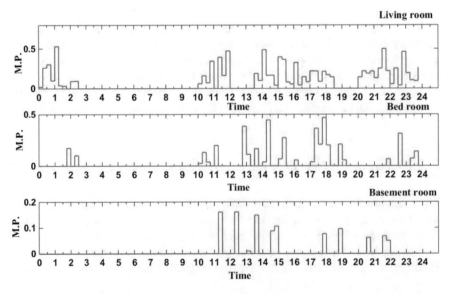

Fig. 3.3 Motion probability within 15 min interval in three different rooms (Living room, bed room, and basement)

Table 3.2 Input features for short-term load forecasting

Inputs	Descriptions
1	Date type: weekday is represented by 1 and weekend is represented by 0
2–25	Eg(d-7,t), Eg(d-6,t), Eg(d-5,t), Eg(d-4,t), Eg(d-3,t), Eg(d-2,t), Eg(d-1,t): Energy of the 7 days preceding to the forecasted day at the same hour
26–121	Mo(d-7,t), Mo(d-6,t), Mo(d-5,t), Mo(d-4,t), Mo(d-3,t), Mo(d-2,t), Mo(d-1,t): Occupants motion of the 7 days preceding to the forecasted day at the same hour
122–169	Te(d-7,t), Te(d-6,t), Te(d-5,t), Te(d-4,t), Te(d-3,t), Te(d-2,t), Te(d-1,t): Temperature and Humidity of the 7 days preceding to the forecasted day at the same hour
169-t	C(t), C(t-1), C(t-2),…,: Prior to *time t*, new collected data temperature, humanity, energy and occupants motion

3.4 Least-Squares Solver Based Training Algorithm

In this section, we firstly reformulate a regularized least-squares problem. Then square-root-free Cholesky decomposition is discussed to reduce the complexity. Final, an incremental least-squares method is introduced to further simplify the operation to basic linear algebra subprograms (BLAS).

3.4.1 Regularized ℓ_2-Norm

Considering (3.8), a better generalized training method is to minimize the training error and the norm of the output weights, which can be defined as a regularized ℓ_2-norm as follows:

$$\min ||\mathbf{H}\boldsymbol{\Gamma} - \mathbf{T}||_2 + \lambda ||\boldsymbol{\Gamma}||_2 \tag{3.11}$$

where $\mathbf{H}$ is the hidden-layer output matrix generated from the Sigmoid function for activation; and λ is a user defined parameter that biases the training error and output weights [11]. This problem can be reformulated as

$$\min ||\tilde{\mathbf{H}}\boldsymbol{\Gamma} - \tilde{\mathbf{T}}||_2$$
$$\text{where } \tilde{\mathbf{H}} = \begin{pmatrix} \mathbf{H} \\ \sqrt{\lambda}\mathbf{I} \end{pmatrix} \quad \tilde{\mathbf{T}} = \begin{pmatrix} \mathbf{T} \\ \mathbf{0} \end{pmatrix} \tag{3.12}$$

where $\mathbf{I} \in \mathbb{R}^{L \times L}$ and $\tilde{\mathbf{H}} \in \mathbb{R}^{(N+L) \times L}$. This is a standard least-squares problem with general solution:

$$\boldsymbol{\Gamma} = (\tilde{\mathbf{H}}^T \tilde{\mathbf{H}})^{-1} \tilde{\mathbf{H}}^T \tilde{\mathbf{T}}, \ \tilde{\mathbf{H}} \in \mathbb{R}^{N \times L} \tag{3.13}$$

where $\tilde{\mathbf{T}} \in \mathbb{R}^{(N+L) \times M}$ and M is the number of classes. The new training algorithm is summarized in Algorithm 1. The complexity of solving output weight will be reduced by the square-root-free Cholesky decomposition and incremental least-squares solutions.

Algorithm 1 The proposed training algorithm of neuron network

---

Input: Training Set $(x_i, t_i), x_i \in \mathbf{R}^n, t_i \in \mathbf{R}^M, i = 1, \ldots N$, activation function $G(a_i, b_i, x_j)$, maximum number of hidden neuron node L and accepted training error ϵ.
Output: Neuron Network output weight Γ
1: Randomly assign hidden-node parameters
    $(a_{ij}, b_{kj}), \ i = 1, 2, \ldots, n, \ j = 1, \ldots, l, \ k = 1, 2, \ldots, N;$
2: Calculate the hidden-layer output matrix $\mathbf{H}$
    $\mathbf{preH} = \mathbf{XA} + \mathbf{B}, \ \mathbf{H} = 1/(1 + e^{-\mathbf{preH}})$
3: Form regularized ℓ_2-norm

$$\tilde{\mathbf{H}} = \begin{pmatrix} \mathbf{H} \\ \sqrt{\lambda}\mathbf{I} \end{pmatrix} \tilde{\mathbf{T}} = \begin{pmatrix} \mathbf{T} \\ \mathbf{0} \end{pmatrix}$$

4: Calculate the output weight
    $\Gamma = (\tilde{\mathbf{H}}^T \tilde{\mathbf{H}})^{-1} \tilde{\mathbf{H}}^T \tilde{\mathbf{T}}$
5: IF $(l \leq L$ or $error > \epsilon)$
    Increase number of hidden node
    $l = l + 1$, repeat from Step 1
6: ENDIF

3.4.2 Square-Root-Free Cholesky Decomposition

The main step for a direct solution of the training problem is the standard least-squares problem of minimizing $||\tilde{\mathbf{T}} - \tilde{\mathbf{H}}\Gamma||_2$. This can be the solution using SVD, QR, and Cholesky decomposition. The computational cost of SVD, QR, or Cholesky decomposition for the problem is $O(4(N + L)L^2 - \frac{4}{3}L^3)$, $O(2(N + L)L^2 - \frac{2}{3}L^3)$, and $O(\frac{1}{3}L^3)$, respectively [17]. Therefore, we use Cholesky decomposition to solve the least-squares problem. Moreover, its incremental and symmetric property reduces the computational cost and hence saves half of memory required [17]. Here, we use H_L to represent the matrix with L number of hidden neuron nodes, which decomposes the symmetric positive definite matrix $\tilde{\mathbf{H}}^T \tilde{\mathbf{H}}$ into

$$\tilde{\mathbf{H}}^T \tilde{\mathbf{H}} = \mathbf{QDQ^T} \tag{3.14}$$

where $\mathbf{Q}$ is a lower triangular matrix with diagonal elements $q_{ii} = 1$ and $\mathbf{D}$ is a positive diagonal matrix. Such method can maintain the same space as Cholesky factorization but avoid the extracting the square roots as the square root of $\mathbf{Q}$ is resolved by diagonal matrix $\mathbf{D}$ [18].

$$
\begin{aligned}
\tilde{\mathbf{H}}_L^T \tilde{\mathbf{H}}_L &= \begin{bmatrix} \tilde{\mathbf{H}}_{L-1} & h_L \end{bmatrix}^T \begin{bmatrix} \tilde{\mathbf{H}}_{L-1} & h_L \end{bmatrix} \\
&= \begin{pmatrix} \tilde{\mathbf{H}}_{L-1}^T \tilde{\mathbf{H}}_{L-1} & \mathbf{v}_L \\ \mathbf{v}_L^T & g \end{pmatrix}
\end{aligned} \tag{3.15}
$$

where $(\mathbf{v}_L, g)$ is a new column generated from new data $h_L^T h_L$, compared to $\tilde{\mathbf{H}}_{L-1}^T \tilde{\mathbf{H}}_{L-1}$. Therefore, we can find

$$
\mathbf{Q}_L \mathbf{D}_L \mathbf{Q}_L^T = \begin{pmatrix} \mathbf{Q}_{L-1} & 0 \\ \mathbf{z}_L^T & 1 \end{pmatrix} \begin{pmatrix} \mathbf{D}_{L-1} & 0 \\ 0 & d \end{pmatrix} \begin{pmatrix} \mathbf{Q}_{L-1}^T & \mathbf{z}_L \\ 0 & 1 \end{pmatrix} \tag{3.16}
$$

Therefore, we can easily calculate the vector $\mathbf{z}_L$ and scalar d for Cholesky decomposition as

$$
\mathbf{Q}_{L-1} \mathbf{D}_{L-1} \mathbf{z}_L = \mathbf{v}_L, \ d = g - \mathbf{z}_L^T \mathbf{D}_{L-1} \mathbf{z}_L \tag{3.17}
$$

where $\mathbf{Q}_L$ and $\mathbf{v}_L$ are known from (3.15), which means that we can continue to use previous factorization result and only update according part. Algorithm 2 shows more details on each step. Note that Q_1 is 1 and D_1 is $\tilde{\mathbf{H}}_1^T \tilde{\mathbf{H}}_1$.

3.4.3 Incremental Least-Squares Solution

The optimal residual for least-squares problem $\tilde{\mathbf{H}} \boldsymbol{\Gamma} = \mathbf{T}$ is defined as r:

$$
r = \mathbf{T} - \tilde{\mathbf{H}} \boldsymbol{\Gamma}_{ls} = (\tilde{\mathbf{H}}(\tilde{\mathbf{H}}^T \tilde{\mathbf{H}})^{-1} \tilde{\mathbf{H}}^T - \mathbf{I})\mathbf{T} \tag{3.18}
$$

Therefore, r is orthogonal to $\tilde{\mathbf{H}}$, where the projection of r to $\tilde{\mathbf{H}}$ is

$$
< r, \tilde{\mathbf{H}} > = \mathbf{T}^T (\tilde{\mathbf{H}}(\tilde{\mathbf{H}}^T \tilde{\mathbf{H}})^{-1} \tilde{\mathbf{H}}^T - \mathbf{I})^T \tilde{\mathbf{H}} = 0 \tag{3.19}
$$

Similarly, for every iteration of Cholesky decomposition, x_{l-1} is the least-squares solution of $\mathbf{T} = \tilde{\mathbf{H}}_{\Lambda_{l-1}} * \boldsymbol{\Gamma}$ with the same orthogonality principle, where Λ_l is the selected column sets for matrix $\tilde{\mathbf{H}}$. Therefore, we have

$$
\begin{aligned}
\mathbf{T} &= r_{l-1} + \tilde{\mathbf{H}}_{\Lambda_{l-1}} * x_{l-1} \\
\tilde{\mathbf{H}}_{\Lambda_l}^T \tilde{\mathbf{H}}_{\Lambda_l} x_l &= \tilde{\mathbf{H}}_{\Lambda_l}^T (r_{l-1} + \tilde{\mathbf{H}}_{\Lambda_{l-1}} * x_{l-1})
\end{aligned} \tag{3.20}
$$

where x_{l-1} is the least-squares solution in the previous iteration. By utilizing superposition property of linear systems, we can have

$$\begin{bmatrix} \tilde{\mathbf{H}}_{\Lambda_l}^T \tilde{\mathbf{H}}_{\Lambda_l} x_{tp1} \\ \tilde{\mathbf{H}}_{\Lambda_l}^T \tilde{\mathbf{H}}_{\Lambda_l} x_{tp2} \end{bmatrix} = \begin{bmatrix} \tilde{\mathbf{H}}_{\Lambda_l}^T r_{l-1} \\ \tilde{\mathbf{H}}_{\Lambda_l}^T \tilde{\mathbf{H}}_{\Lambda_{l-1}} * x_{l-1} \end{bmatrix} \tag{3.21}$$

$$x_l = x_{tp1} + x_{tp2} = x_{tp1} + x_{l-1}$$

where the second row of equation has a trivial solution of $[x_{l-1}\ 0]^T$. Furthermore, this indicates that the solution of x_l is based on x_{l-1} and only x_{tp} is required to be computed out from the first row of (3.21), which can be expanded as

$$\tilde{\mathbf{H}}_{\Lambda_l}^T \tilde{\mathbf{H}}_{\Lambda_l} x_{tp1} = \begin{bmatrix} \tilde{\mathbf{H}}_{\Lambda_{l-1}}^T r_{l-1} \\ h_l^T r_{l-1} \end{bmatrix} = \begin{bmatrix} 0 \\ h_l^T r_{l-1} \end{bmatrix} \tag{3.22}$$

Due to the orthogonality between the optimal residual $\tilde{\mathbf{H}}_{\Lambda_{l-1}}$ and r_{l-1}, the dot product becomes 0. This clearly indicates that the solution x_{tp1} is a sparse vector with only one element. By substituting square-root-free Cholesky decomposition, we can find

$$\mathbf{Q}^T dx_{tp} = h_l^T r_{l-1} \tag{3.23}$$

where x_{tp} is the same as x_{tp1}. The other part of Cholesky factorization $\mathbf{Q}$ for multiplication of x_{tp1} is always 1 and hence is eliminated. The detailed algorithm including Cholesky decomposition and incremental least-squares is shown in Algorithm 2. By utilizing Cholesky decomposition and incremental least-squares techniques, the computational complexity is reduced with only 4 basic linear algebra operations per iterations.

3.5 Least-Squares Based Machine Learning Accelerator Architecture

3.5.1 Overview of Computing Flow and Communication

The top level of proposed VLSI architecture for training and testing is shown in Fig. 3.4. The description of this architecture will be introduced based on testing flow. The complex control and data flow of the neural network training and testing is enforced by a top level finite state machine (FSM) with synchronized and customized local module controllers.

For the neural network training and testing, an asynchronous first-in first-out (FIFO) is designed to collect data through AXI4 light from PCIe Gec3X8. Two buffers are used to store rows of the training data $\mathbf{X}$ to perform ping-pong operations.

Algorithm 2 Fast incremental least-squares solution

Input: Activation matrix $\tilde{\mathbf{H}}_L$, target matrix $\tilde{\mathbf{T}}$ and number of hidden nodes L
Output: Neuron Network output weight x

1: Initialize $r_0 = \tilde{\mathbf{T}}$, $\Lambda_0 = \emptyset$, $d = \mathbf{0}$, $x_0 = \mathbf{0}$, $l = 1$,
2: While $\|r_{l-1}\|_2^2 \le \epsilon^2$ or $l \le L$
3: $c(l) = h_l^T r_{l-1}$, $\Lambda_l = \Lambda_{l-1} \cup l$
4: $\mathbf{v}_l = \tilde{\mathbf{H}}_{\Lambda_l}^T h_l$
5: $\mathbf{Q}_{l-1} w = \mathbf{v}_l(1:l-1)$. $\mathbf{z}_l = w./diag(\mathbf{D}_{l-1})$
6: $d = g - \mathbf{z}_l^T w$
7: $\mathbf{Q}_l = \begin{bmatrix} \mathbf{Q}_{l-1} & \mathbf{0} \\ \mathbf{z}_l^T & 1 \end{bmatrix}$, $\mathbf{D}_l = \begin{bmatrix} \mathbf{D}_{l-1} & \mathbf{0} \\ \mathbf{0} & d \end{bmatrix}$
   $(\mathbf{Q}_1 = 1, \mathbf{D}_1 = h_1 * h_1^T)$
8: $\mathbf{Q}_l^T x_{tp} = \begin{bmatrix} \mathbf{0} \\ c(l)/d \end{bmatrix}$
9: $x_l = x_{l-1} + x_{tp}$, $r_l = r_{l-1} - \tilde{\mathbf{H}}_{\Lambda_l} x_{tp}$, $l = l + 1$
10: END While

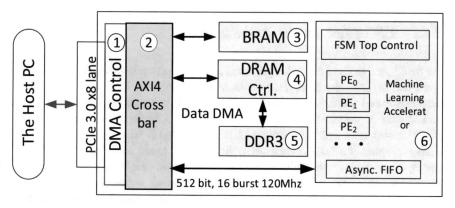

Fig. 3.4 Accelerator architecture for training and testing

These two buffers will be re-used when collecting the output weight data. To maintain high training accuracy, floating point data is used with parallel fixed point to floating point converter. As the number indicated on each block in Fig. 3.4, data will be firstly collected through PCIe to DRAM and Block RAM. Block RAM is used to control the core to indicate the read/write address of DRAM during the training/testing process. The core will continuously read data from block RAM for configurations and starting signal. Once data is ready in DRAM and the start signal is asserted, the core will process computation for neural network testing or training process. An implemented FPGA block design on Vivado is shown in Fig. 3.9.

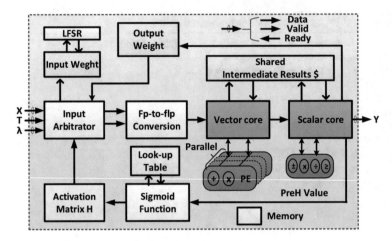

Fig. 3.5 Detailed architecture for online learning

3.5.2 FPGA Accelerator Architecture

As mentioned in the Sect. 3.3, operations in neural network are performed serially from one layer to the next. This dependency reduces the level of parallelism of accelerator and requires more acceleration in each layer. In this chapter, a folded architecture is proposed as shown in Fig. 3.5. Firstly, the *input arbitrator* will take input training data and input weight. A pipeline stage is added for activation after each multiplication result. Then depending on the mode of training and testing, the *input arbitrator* will decide to take label or output weight. For testing process, output weight is selected for calculation neural network output. For training, label will be taken for output weight calculation based on Algorithm 2. To achieve similar software-level accuracy, floating-point data is used during the computation process and 8-bit fixed point is used for data storage.

3.5.3 ℓ_2-Norm Solver

As mentioned in the reformulated ℓ_2-norm Algorithm 2, Step 5 requires forward substitutions. Figure 3.6 provides the detailed mapping for forward substitutions on our proposed architecture. For the convenient purposes, we use $\mathbf{QW} = \mathbf{V}$ to represent Step 5, where $\mathbf{Q}$ is a triangular matrix. Figure 3.7 provides the detailed equations in each PEs and stored intermediate values. To explore the maximum level of parallelism, we can perform multiplication at the same time on each row to compute $w_i, i \neq 1$ as shown in the right of Fig. 3.7. However, there is different number of multiplication and accumulations required for different w_i. In the first round, to have the maximum level of parallelism, intuitively we require $L-1$ parallel

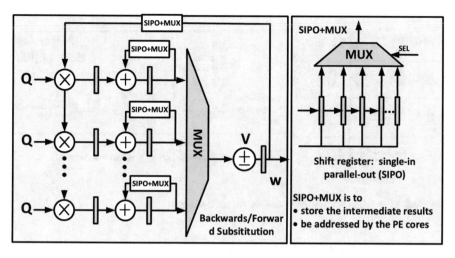

Fig. 3.6 Computing diagram of forward/backward substitution in L2-norm solver

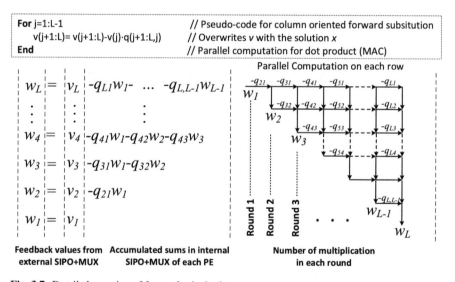

Fig. 3.7 Detailed mapping of forward substitution

PEs to perform the multiplication. After knowing w_2, we need $L - 2$ parallel PEs for the same computations in the second round. However, if we add a shift register, we can store the intermediate results in the shift register and take it with a decoder of MUX. For example, if we have parallelism of 4 for $L = 32$, we can perform 8 times parallel computation for the round 1 and store them inside registers. This helps improve the flexibility of the process elements (PEs) with better resource utilization.

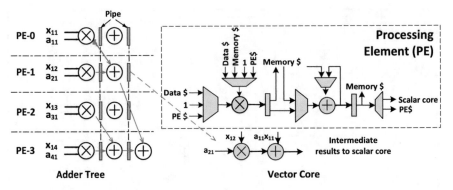

Fig. 3.8 Computing diagram of matrix–vector multiplication

3.5.4 Matrix–Vector Multiplication

All the computation relating to vector operation is performed on processing elements (PEs). Our designed PE is similar as [19] but features direct instruction to perform vector–vector multiplications for neural network. Figure 3.8 gives an example of vector–vector multiplication (dot product) for (3.7) with parallelism of 4. If the vector length is 8, the folding factor will be 2. The output from PE will be accumulated twice based on the folding factor before sending out the vector–vector multiplication result. The adder tree will be generated based on the parallelism inside vector core. The output will be passed to scalar core for accumulations. In the PE, there is a bus interface controller. It will control the multiplicand of PE and pass the correct data based on the top control to PE.

3.6 Experiment Results

In this section, we firstly discuss the machine learning accelerator architecture and resource usage. Then details of FPGA implementation with CAD flow are discussed. The performance of proposed scalable architecture is evaluated for regression problem and classification problem, respectively. Finally, the energy consumption and speed-up of proposed accelerator are evaluated in comparison with CPU, embedded CPU and GPU.

3.6.1 Experiment Setup and Benchmark

To verify our proposed architecture, we have implemented in on Xilinx Virtex 7 with PCI Express Gen3x8 [20]. The HDL code is synthesized using Synplify and

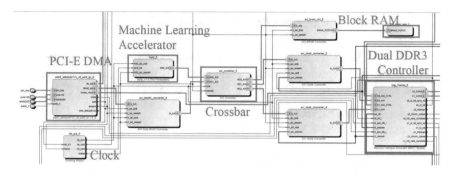

Fig. 3.9 Vivado block design for FPGA least-squares machine learning accelerator

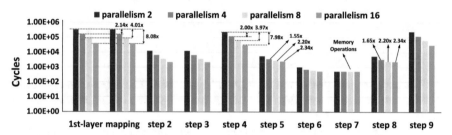

Fig. 3.10 Training cycles at each step of the proposed training algorithm with different parallelisms ($N = 74$; $L = 38$; $M = 3$ and $n = 16$)

the maximum operating frequency of the system is 53.1 MHz under 128 parallel PEs. The critical path is identified as the floating-point division, where 9 stages of pipeline are inserted for speedup. We develop three baselines (x86 CPU, ARM CPU, and GPU) for performance comparisons.

Baseline 1: General Processing Unit (x86 CPU). The general CPU implementation is based on C program on a computer server with Intel Core -i5 *3.20GHz* core and *8.0GB* RAM.

Baseline 2: Embedded processor (ARM CPU). The embedded CPU (Beagle-Board-xM) [21] is equipped with 1GHz ARM core and 512MB RAM. The implementation is performed using C program under Ubuntu 14.04 system.

Baseline 3: Graphics Processing Unit (GPU). The GPU implementation is performed by CUDA C program with cuBLAS library. A Nvidia GeForce GTX 970 is used for the acceleration of learning on neural network.

The dataset for residential load forecasting is collected by Singapore Energy Research Institute (ERIAN). The dataset consists of 24-henergy consumptions, occupants motion, and environmental records such as humidity and temperatures from 2011 to 2015. Features for short-term load forecasting is summarized in Table 3.2. Please note that we will perform hourly load forecasting using real-time environmental data, occupants motion data, and previous hours and days energy consumption data. Model will be retrained sequentially after each hour with new generated training data.

3.6.2   FPGA Design Platform and CAD Flow

The ADM-PCIE-7V3 is a high-performance reconfigurable computing card intended for high speed performance applications, featuring a Xilinx Virtex-7 FPGA. The key features of ADM-PCIE 7V3 are summarized as below [20]

– Compatible with Xilinx OpenCL compiler
– Supported by ADM-XRC Gen 3 SDK 1.7.0 or later and ADB3 Driver 1.4.15 or later.
– PCIe Gen1/2/3 x1/2/4/8 capable
– Half-length, low-profile x8 PCIe form factor
– Two banks of DDR3 SDRAM SODIMM memory with ECC, rated at 1333 MT/s
– Two right angle SATA connectors (SATA3 capable)
– Two SFP+ sites capable of data rates up to 10 Gbps
– FPGA configurable over JTAG and BPI Flash
– XC7VX690T-2FFG1157C FPGA

The development platform is mainly on Vivado 14.4. The direct memory access (DMA) bandwidth is 4.5GB/s. The DDR3 bandwidth is 1333 MT/s with 64 bits width.

The CAD flows for implementing the machine learning accelerator on the ADM-PCIE 7V3 are illustrated in Fig. 3.11. The Xilinx CORE Generator System is first used to generate the data memory macros that are mapped to the BRAM resources on the FPGA. The generated NGC files contain both the design netlist, constraints files and Verilog wrapper. Then, these files together with the RTL codes of the machine learning accelerator are loaded to Synplify Premier for logic synthesis.

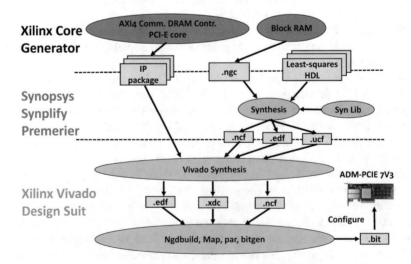

Fig. 3.11 CAD flows for implementing least-squares on ADM-PCIE 7V3

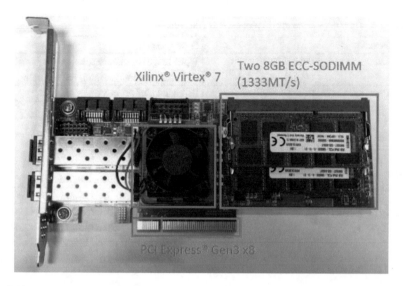

Fig. 3.12 Alpha-Data PCIe 7V3 FPGA board

Note that the floating-point arithmetic units used in our design are from the Synopsys DesignWare library. The block RAM is denoted as black box for Synplify synthesis. The EDF file stores the gate-level netlist in an electronic data interchange format (EDIF), and the UCF file contains user-defined design constraints. Next, the generated files are passed to Xilinx Vivado Design Suite to merge with other IP core such as DRAM controller and PCI-E core. In the Vivado design environment, each IP is packaged and connected. Then, we synthesize the whole design again under Vivado environment. Specifically, the "ngbbuild" command reads in the netlist in EDIF format and creates a native generic database (NGD) file that contains a logical description of the design reduced to Xilinx NGD primitives and a description of the original design hierarchy. The "map" command takes the NGD file, maps the logic design to a specific Xilinx FPGA, and outputs the results to a native circuit description (NCD) file. The "par" command takes the NCD file, places and routes the design, and produces a new NCD file, which is then used by the "bitgen" command for generating the bit file for FPGA programming. Figure 3.12 shows Alpha-Data PCIe FPGA board.

3.6.3 Scalable and Parameterized Accelerator Architecture

The proposed accelerator architecture features great scalability for different applications. Table 3.3 shows all the user-defined parameters supported in our architecture. At circuit level, users can adjust the stage of pipeline of each arithmetic to satisfy the speed, area, and resource requirements. At architecture level, the parallelism of PE

Table 3.3 Tunable parameters on proposed architecture

Parameters		Descriptions
Circuits	{MAN EXP}	Word-length of mantissa, exponent
	$\{P_A, P_M, P_D, P_C\}$	Pipe. stages of adder, mult, div and comp
Architectures	P	Parallelism of PE in VC
	n	Maximum signal dimensions
	N	Maximum training/test data size
	H	Maximum number of hidden nodes

Table 3.4 Resource utilization under different parallelism level ($N = 512$, $H = 1024$, $n = 512$ and 50 Mhz clock)

Paral.	LUT	Block RAM	DSP
8	52,614 (12%)	516 (35%)	51 (1.42%)
16	64,375 (14%)	516 (35%)	65 (1.81%)
32	89,320 (20%)	516 (35%)	96 (2.67%)
64	139,278 (32%)	516 (35%)	160 (4.44%)
128	236,092 (54%)	516 (35%)	288 (8.00%)

can be specified based on the hardware resource and speed requirement. The neural network parameters n, N, H can be also reconfigured for specific applications.

Figure 3.10 shows the training cycles on each step on proposed training algorithms for synthesized dataset. Different parallelism P is applied to show the speed-up of each steps. The speed-up of 1st-layer for matrix–vector multiplication is scaling up with the parallelism. The same speed-up improvement is also observed in the Step 3, 4, and 9 in Algorithm 2, where the matrix–vector multiplication is the dominant operation.

However, when the major operation is the division for the backward and forward substitution, the speed-up is not that significant and tends to saturate when the division becomes the bottleneck. We can also observe in Step 7, the memory operations do not scale with parallelism. It clearly shows that matrix–vector multiplication is the dominant operation in the training procedure (1st Layer, Step 3, Step 4, and Step 9) and our proposed accelerator architecture is scalable to dynamically increase the parallelism to adjust the speed-up.

The resource utilization under different parallelism is achieved from Xilinx ISE after place and routing. From Table 3.4, we can observe that LUT and DSP are almost linearly increasing with parallelism. However, Block RAM keeps constant with increasing parallelism. This is because Block RAM is used for data buffer, which is determined by other architecture parameters (N, H, n). Figure 3.13 shows the layout view of the FPGA least-squares solver.

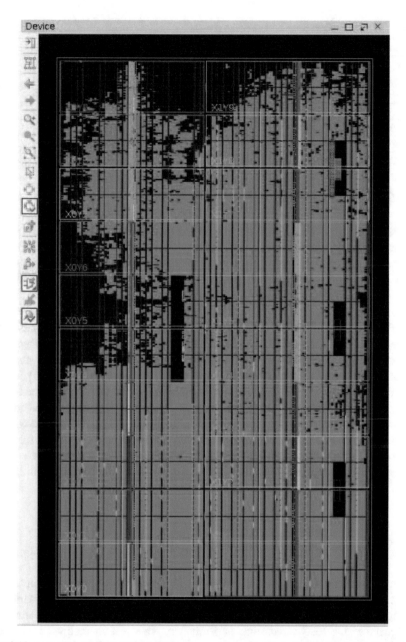

Fig. 3.13 Layout view of the FPGA with least-squares machine learning accelerator implemented

Table 3.5 UCI Dataset Specification and Accuracy

Benchmarks	Data size	Dim.	Class	Node No.	Acc. (%)
Car	1728	6	4	256	90.90
Wine	178	13	3	1024	93.20
Dermatology	366	34	6	256	85.80
Zoo	101	16	7	256	90.00
Musk1	476	166	2	256	69.70
Conn. Bench	208	60	2	256	70

3.6.4 Performance for Data Classification

In this experiment, six datasets are trained and tested from UCI dataset [22], which are wine, car, dermatology, zoo, musk and Connectionist Bench (Sonar, Mines vs. Rocks). The details of each dataset are summarized in Table 3.5. The architecture is set according to the training data set size and dimensions to demonstrate the parameterized architecture. For example, $N = 128(74)$ represents that the architecture parameter (training size) is 128 with the actual dataset wine size of 74. The accuracy of the machine learning is the same comparing to Matlab result since the single floating-point data format is applied for the proposed architecture.

For speed-up comparison, our architecture will not only compare to the time consumed by least-squares solver (DS) training method, but also SVM [23] and BP based method [24] on CPUs. For example, in dataset dermatology, the speed-up of training time is lower comparing to CPU based solution when the parallelism is 2. This is mainly due to the high clock speed of CPU. When the parallelism increases to 16, $4.70\times$ speed-up can be achieved. For connectionist bench dataset, the speed-up of proposed accelerator is as high as $24.86\times$, when compared to the least-squares solver software solution on CPUs (Table 3.6). Furthermore, $801.20\times$ and $25.55\times$ speed-up can be achieved comparing to BP and SVM on CPUs.

3.6.5 Performance for Load Forecasting

Figure 3.14 shows the residential load forecasting with FPGA and CPU implementation. Clearly, all the peaks period are captured. It also shows that approximation by number representation (fixed point) will not degrade the overall performance. To quantize the load forecasting performance, we use two metrics: root mean square error (RMSE) and mean absolute percentage error (MAPE). Table 3.7 is the summarized performance with comparison of SVM. We can observe that our proposed accelerator has almost the same performance as CPU implementation. It also shows an average of 31.85% and 15.4% improvement in average on MAPE and RMSE comparing to SVM based load forecasting (Table 3.7).

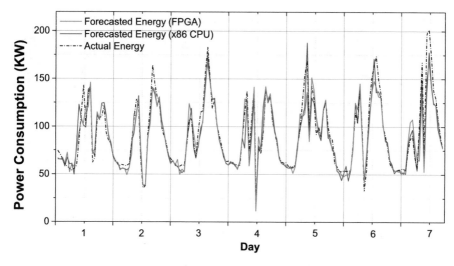

Fig. 3.14 7-Day residential load forecasting by proposed architecture with comparison of CPU implementation

Table 3.6 Parameterized and scalable architecture on different dataset with speed-up comparison to CPU

Benchmarks	FPGA (ms)	CPU (ms)	BP (ms)	SVM (ms)	Imp. (CPU)
Car	44.3	370	36,980	1182	8.35×
Wine	207.12	360	11,240	390	1.74×
Dermatology	19.45	160	17,450	400	8.23×
Zoo	22.21	360	5970	400	16.21×
Musk	24.09	180	340,690	3113	7.47×
Conn. Bench	14.48	360	11,630	371	24.86×

Table 3.7 Load forecasting accuracy comparison and accuracy improvement comparing to FPGA results to SVM result

	MAPE			RMSE		
Machine Learning	Max	Min	Avg	Max	Min	Avg
NN FPGA	0.12	0.072	0.92	18.87	6.57	12.80
NN CPU	0.11	0.061	0.084	17.77	8.05	12.85
SVM	0.198	0.092	0.135	20.91	5.79	15.13
Imp. (CPU)(%)	4.28	−18.03	−9.52	−6.19	18.39	0.39
Imp. (SVM)(%)	41.01	21.74	31.85	9.76	−13.47	15.40

3.6.6 Performance Comparisons with Other Platforms

In the experiment, the maximum throughput of proposed architecture is 12.68 Gflops with 128 parallelism for matrix multiplication. This is slower than GPU based implementation 59.78 Gflops but higher than x86 CPU based implementation 5.38 Gflops.

Table 3.8 Proposed architecture performance in comparison with other computation platform

Platform	Type	Format	Time (ms)	Power (W)	Energy	Speed-up	E. Imp.
x86 CPU	Train	Single	1646	84	138.26 J	2.59×	256.0×
	Test		1.54	84	0.129 J	4.56×	450.2×
ARM CPU	Train	Single	32,550	2.5	81.38 J	51.22×	150.7×
	Test		30.1	2.5	0.0753 J	89.05×	261.9×
GPU	Train	Single	10.99	145	1.594 J	0.017×	2.95×
	Test		0.196	145	0.0284 J	0.580×	98.92×
FPGA	Train	Single+ Fixed	635.4	0.85	0.540 J	–	–
	Test		0.338	0.85	0.287 mJ	–	–

To evaluate the energy consumptions, we calculate the energy for a given implementation by multiplying the peak power consumption of corresponding device. Although this is pessimistic analysis, it is still very likely to reach due to intensive memory and computation operations. Table 3.8 provides detailed comparisons between different platforms. Our proposed accelerator on FPGA has the lowest power consumption ($0.85W$) comparing to GPU implementation ($145W$), ARM CPU ($2.5W$) and x86 CPU implementation ($84W$). For training process, although GPU is the fastest implementation, our accelerator still has $2.59\times$ and $51.22\times$ speed-up for training comparing to x86 CPU and ARM CPU implementations. Furthermore, our proposed method shows $256.0\times$, $150.7\times$, and $2.95\times$ energy saving comparing to CPU, ARM CPU, and GPU based implementations for training model. For testing process, it is mainly on matrix–vector multiplications. Therefore, GPU based implementations provide better speed-up performance. However, our proposed method still has and $4.56\times$ and $89.05\times$ speed-up for testing comparing to x86 CPU and ARM CPU implementations. Moreover, our accelerator is the most low-power platform with $450.1\times$, $261.9\times$ and $98.92\times$ energy saving comparing to x86 CPU, ARM CPU and GPU based implementations. In summary, our proposed accelerator provides a low-power and fast machine learning platform for smart-grid data analytics.

3.7   Conclusion

This chapter presents a fast machine learning accelerator for real-time data analytics in smart micro-grid of buildings with consideration of occupants behavior. An incremental and square-root-free Cholesky factorization algorithm is introduced with FPGA realization for training acceleration when analyzing the real-time sensed data. Experimental results have shown that our proposed accelerator on Xilinx Virtex-7 has a comparable forecasting accuracy with an average speed-up of $4.56\times$ and $89.05\times$, when compared to x86 CPU and ARM CPU for testing. Moreover, $450.2\times$, $261.9\times$, and $98.92\times$ energy saving can be achieved comparing to x86 CPU, ARM CPU, and GPU.

Acknowledgements This work is sponsored by grants from Singapore MOE Tier-2 (MOE2015-T2-2-013), NRF-ENIC-SERTD-SMES-NTUJTCI3C-2016 (WP4) and NRF-ENIC-SERTD-SMES-NTUJTCI3C-2016 (WP5).

References

1. L.D. Harvey, *Energy and the New Reality 1: Energy Efficiency and the Demand for Energy Services* (Routledge, London, 2010)
2. H. Ziekow, C. Goebel, J. Strüker, H.-A. Jacobsen, The potential of smart home sensors in forecasting household electricity demand, in *2013 IEEE International Conference on Smart Grid Communications (SmartGridComm)* (IEEE, New York, 2013), pp. 229–234
3. H.S. Hippert, C.E. Pedreira, R.C. Souza, Neural networks for short-term load forecasting: a review and evaluation. IEEE Trans. Power Systems **16**(1), 44–55 (2001)
4. W. Mielczarski, G. Michalik, M. Widjaja, Bidding strategies in electricity markets, in *Power Industry Computer Applications, 1999. PICA'99. Proceedings of the 21st 1999 IEEE International Conference* (IEEE, New York, 1999), pp. 71–76
5. E.A. Feinberg, D. Genethliou, Load forecasting, in *Applied Mathematics for Restructured Electric Power Systems* (Springer, Berlin, 2005), pp. 269–285
6. C. Sandels, J. Widén, L. Nordström, Forecasting household consumer electricity load profiles with a combined physical and behavioral approach. Appl. Energy **131**, 267–278 (2014)
7. S.J. Russell, P. Norvig, J.F. Canny, J.M. Malik, D.D. Edwards, *Artificial Intelligence: A Modern Approach*, vol. 2 (Prentice Hall, Upper Saddle River, NJ, 2003)
8. S. Theodoridis, K. Koutroumbas, Pattern recognition and neural networks, in *Machine Learning and Its Applications* (Springer, Berlin, 2001), pp. 169–195
9. S. Li, P. Wang, L. Goel, Short-term load forecasting by wavelet transform and evolutionary extreme learning machine. Electr. Power Syst. Res. **122**, 96–103 (2015)
10. A. Ahmad, M. Hassan, M. Abdullah, H. Rahman, F. Hussin, H. Abdullah, R. Saidur, A review on applications of ann and svm for building electrical energy consumption forecasting. Renew. Sust. Energ. Rev. **33**, 102–109 (2014)
11. G.-B. Huang, Q.-Y. Zhu, C.-K. Siew, Extreme learning machine: theory and applications. Neurocomputing **70**(1), 489–501 (2006)
12. Y.-H. Pao, G.-H. Park, D.J. Sobajic, Backpropagation, part iv learning and generalization characteristics of the random vector functional-link net. Neurocomputing **6**(2), 163–180 (1994). [Online]. Available http://www.sciencedirect.com/science/article/pii/0925231294900531
13. M.D. Martino, S. Fanelli, M. Protasi, A new improved online algorithm for multi-decisional problems based on mlp-networks using a limited amount of information, in *Proceedings of 1993 International Joint Conference on Neural Networks, 1993. IJCNN '93-Nagoya*, vol. 1 (1993), pp. 617–620
14. I. Richardson, M. Thomson, D. Infield, A high-resolution domestic building occupancy model for energy demand simulations. Energy Buildings **40**(8), 1560–1566 (2008)
15. N.-Y. Liang, G.-B. Huang, P. Saratchandran, N. Sundararajan, A fast and accurate online sequential learning algorithm for feedforward networks. IEEE Trans. Neural Netw. **17**(6), 1411–1423 (2006)
16. Y. Pang, S. Wang, Y. Peng, X. Peng, N.J. Fraser, P.H. Leong, A microcoded kernel recursive least squares processor using fpga technology. ACM Trans. Reconfigurable Technol. Syst. **10**(1), 5 (2016)
17. L.N. Trefethen, D. Bau III, *Numerical Linear Algebra*, vol. 50 (SIAM, Philadelphia, 1997)
18. A. Krishnamoorthy, D. Menon, Matrix inversion using cholesky decomposition. Preprint (2011). arXiv:1111.4144

19. F. Ren, D. Marković, A configurable 12237 kS/s 12.8 mw sparse-approximation engine for mobile data aggregation of compressively sampled physiological signals. IEEE J. Solid State Circuits **51**(1), 68–78 (2016)
20. Adm-pcie-7v3 [Online]. Available http://www.alpha-data.com/dcp/products.php?product= adm-pcie-7v3 (2016)
21. Beagleboard-xm [Online]. Available http://beagleboard.org/beagleboard-xm (2015)
22. M. Lichman, UCI machine learning repository (2013). [Online]. Available http://archive.ics. uci.edu/ml
23. J.A. Suykens, T. Van Gestel, J. De Brabanter, B. De Moor, J. Vandewalle, J. Suykens, T. Van Gestel, *Least Squares Support Vector Machines*, vol. 4 (World Scientific, Singapore, 2002)
24. R. Hecht-Nielsen, Theory of the backpropagation neural network, in *International Joint Conference on Neural Networks, 1989. IJCNN* (IEEE, New York, 1989), pp. 593–605

Chapter 4
Compute-in-Memory Architecture for Data-Intensive Kernels

Robert Karam, Somnath Paul, and Swarup Bhunia

4.1 Introduction

In modern computing systems, energy efficiency has become a major design challenge. This is especially true in the deep nanoscale regime, where transistors no longer enjoy the same cubic reductions in energy at new process nodes that they once did [2]. In this era, alternative approaches to improve energy efficiency have become the subject of intense research, including new algorithms, highly optimized accelerators, smarter hardware–software partitioning, and sophisticated power management techniques [4]. The demand for improved energy efficiency is not limited to any one application domain; rather, it spans diverse areas, including scientific computations, web serving, multimedia storage, and analytics. Many of these applications call for the processing of large volumes of data, and are therefore referred to as *data-intensive applications*. Such applications exacerbate the problem of energy scaling by placing an additional constraint on the system: memory bandwidth, especially from secondary storage, and the associated transfer energy become bottlenecks to system performance. In the past decade, integrated graphics processing units (GPUs) and other on-chip accelerators have largely addressed the issue of compute energy for these applications, while cache hierarchies have partially alleviated the memory bottleneck.

However, technology scaling gives rise to two main barriers to energy scalability. The first is the fact that Dennard scaling, which called for the cubic reduction in energy at new process nodes, no longer holds, and the fact that for many workloads,

R. Karam • S. Bhunia (✉)
University of Florida, Gainesville, FL 32611, USA
e-mail: robkaram@ufl.edu; swarup@ece.ufl.edu

S. Paul
Intel Corporation, Hillsboro, OR 97124, USA
e-mail: somnath.paul@intel.com

© Springer International Publishing AG 2017
A. Chattopadhyay et al. (eds.), *Emerging Technology and Architecture for Big-data Analytics*, DOI 10.1007/978-3-319-54840-1_4

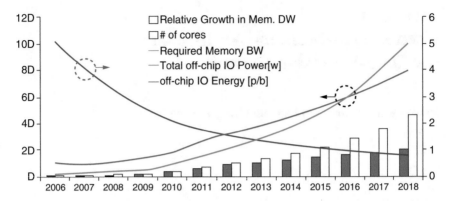

Fig. 4.1 Scaling trends for off-chip bandwidth and power suggest that a large gap exists between the technology projections and system requirements. Alternative architectures which can reduce this effect will be critical for efficient data-intensive computing in the coming years [11]

additional cores in a multicore system do not provide the same benefits they once did, due to physical (e.g., thermal) constraint [3]. These energy-constrained, over-provisioned systems are unable to meet the performance requirements for emerging data-intensive applications. The second barrier is posed by the ever-increasing gap between on-chip memory, processor frequency, and external data rates (Fig. 4.1). Even advanced caching techniques are unable to fully address (i.e., hide) the off-chip access latency incurred especially during data-intensive workloads.

As datasets grow from petabytes (10^{15} B) to exabytes (10^{18} B) and beyond, attention has turned to off-chip computing frameworks, which physically bring the processing and memory closer together. In many cases, this helps alleviate some of the issues facing data-intensive computing by reducing the required data transfer energy, while providing additional opportunities for optimization, such as increased area- and energy-efficiency. The *Malleable Hardware Accelerator*, or MAHA, has emerged as a suitable off-chip computing framework [9, 10]. MAHA can be described as *memory-centric*, because it leverages dense, 2D memory arrays for both *storage and computation*, in the form of multi-input, multi-output lookup tables (LUTs). Computation is *mixed-granular*, in that it supports bit-sliceable and fused logic operations. MAHA supports both spatial and temporal computing models to achieve optimal energy efficiency: *spatial* because operations are executed—often in parallel—across a number of interconnected processing elements (PEs); and *temporal* because, at their core, each PE is an ultra-light-weight and extensible instruction processor.

This chapter describes the basic MAHA architecture and two of its specific embodiments, namely *memory-centric computing*, in which computations occur in close proximity to, though still physically *outside* of the memory, and *in-memory computing*, where computation occurs *inside* the memory array. Both versions make use of the memory for computation, though the amount of data movement, and the associated transfer energy, will differ. Both versions also have a smaller logic datapath than a typical processor, but are nevertheless extensible, so that they are able to support new instructions using the available LUT memory. A custom-designed software tool is used to generate the configuration file for the MAHA fabric. It can therefore be used to accelerate a wide range of data-intensive kernels from various application domains. Furthermore, the architecture is amenable to *domain customization*, where certain features, including the datapath and interconnect fabric, can be modified to better suit the specific needs of a target domain, improving area- and energy-efficiency over more general purpose accelerators, such as Field Programmable Gate Array (FPGA) or General Purpose Graphics Processing Units (GPGPU), and certainly over a General Purpose Processor (GPP). Finally, as a memory-centric architecture, it can benefit from using emerging nanoscale memory technologies, such as Resistive RAM (RRAM) or Spin-Transfer Torque RAM (STTRAM), to improve area- and energy-efficiency [5, 6].

The rest of the chapter is organized as follows: Sect. 4.2 will describe the MAHA architecture for both memory-centric and in-memory computing models, the application mapping tool flow, and the general process for domain customization. Next, Sect. 4.3 will present a number of case studies which take the general framework and procedure for domain customization, and demonstrate how they can be used to achieve extreme energy-efficiency for data-intensive kernels in numerous domains, using both memory-centric and in-memory computing models.

4.2 Malleable Hardware Acceleration

This section describes the general MAHA architecture for data-intensive applications and the corresponding software framework.

4.2.1 Hardware Architecture

At a high level, the Malleable Hardware Accelerator consists of a set of interconnected processing elements called *Memory Logic Blocks*, or MLBs. A single MLB consists of components typically found in a microprocessor, such as instruction memory, data memory, a register file, and a datapath, as shown in Fig. 4.2. In MLB terminology, the instruction memory is referred to as the *schedule table*, while the data memory, partially used for LUT responses, is called the *function table*. Instruction execution occurs over multiple cycles, just as in a typical processor, and constitutes the *temporal computing* aspect. MLBs communicate using a hierarchical interconnect, which emphasizes lower latency, more energy-efficient local communication with other MLBs. This constitutes the *spatial computing* aspect. By finding the right balance of spatial and temporal computing, optimal energy efficiency can be achieved for a particular application under area, power, and delay constraints.

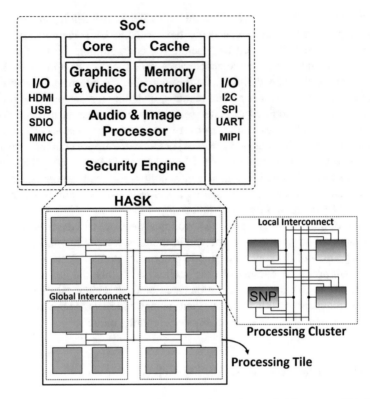

Fig. 4.2 Integration of a reconfigurable security accelerator in a typical System-on-Chip [1]

Within this general framework, two implementation classes, or embodiments of the MAHA architecture, are possible:

Memory-Centric   A paradigm in which processing occurs in close proximity to, but still distinctly separates from the memory.

In-Memory   A paradigm in which the memory array is augmented with additional circuitry, enabling the memory itself to either store data or process it in situ. The memory can perform on-demand context switching between these two modes.

For memory-centric accelerators, no modifications are required to the memory array, but additional peripheral logic is needed. Accelerators implementing memory-centric computing can typically be retrofitted into existing computing systems with additional software support, such as operating system (OS) integration, or device drivers. Conversely, in-memory accelerators require significant design-time modifications of the memory array, as well as the peripheral logic. They can also leverage the extremely high bandwidth available within the memory array, rather than relying on a read/write interface to the memory [10]. Accelerators implementing in-memory computing require more drastic changes to the system architecture; for example, an in-memory accelerator can be integrated into the CPU cache, or into the last level memory device, such as a solid state drive (SSD). Such changes would require replacement of an existing CPU, or swapping out of old SSDs for one with the in-memory accelerator. This is in addition to the necessary OS and driver support.

In both cases, internal accelerator parameters such as the MLB's schedule and function table sizes, the logic datapath and functional units, as well as the connectivity and number of levels in the MLB interconnect hierarchy can be customized to a specific domain. This can result in significantly improved area- and energy-efficiency over a general accelerator framework, as described in the Case Studies (Sect. 4.3).

4.2.2 Application Mapping

Given this general framework, it is important to have a parameterized software tool with which to map applications into a particular hardware configuration. The MAHA *mapper* tool achieves this goal, and is capable of mapping diverse applications of varying complexity given a certain input configuration. The configuration consists of several components, as well as the input parameters listed in Table 4.1.

4.2.2.1 Application Description Using an Instruction Set Architecture

The *mapper* requires an instruction set definition that includes common control and data flow operations.

bits
: Bit-sliceable operations which do not require a carry bit. These are generally logic functions such as AND, OR, or XOR.

bitswc
: Bit-sliceable operations which require a carry bit. For example, a 32-bit ADD, SUBTRACT, or COMPARE operation which can be bit-sliced into two 16-bit or four 8-bit sub-operations.

mult
: Represents a signed, 2-input multiplication operation, $A = B \times C$.

shift/rotate
: A 3-input operation, with a bit indicating SHIFT or ROTATE, the amount by which to shift or rotate, and the input operand.

sel
: An N-input to 1-output selection operation, which takes as input N operands and $\log_2(N)$ select bits.

complex
: A variable instruction which represents the general class of LUT operations. The number of inputs and the bit-widths of each input are constrained by the LUT size.

load/store
: Represents a read or write operation to the data memory.

In order to provide greater flexibility, the ISA description must define which of these operations are supported on the target architecture. Determining which operations are suitable for datapath execution and when a single memory read

Table 4.1 Generic design parameters for malleable hardware accelerator

Category	Parameter	Description
MLB parameters	Granularity	Minimum number of bits that can be processed by the MLB
	Datapath width	Width of adder/multiplier/other functional units
	Schedule table size	Maximum number of instructions that can be mapped to one MLB
	MLB in/out ports	Maximum number of ports to which other MLBs can connect
Scheduling parameters	Max datapath ops.	Maximum number of parallel datapath instructions per cycle
	Max LUT ops.	Maximum number of parallel lookup instructions per cycle
Interconnect parameters	Max levels	Maximum number of hierarchy levels in the interconnect
	Buswidth	Width of interconnect bus at each level of hierarchy

may be more advantageous can help to improve energy efficiency. For example, in a given technology, it may be advantageous to perform a single table lookup of several fused and bit-sliced logic operations than to execute the same operations sequentially with a coarser operand granularity. Such tradeoffs are evaluated by the tool using energy models described by input configuration files. These files define the power and latency for datapath and lookup operations, as well as inter-MLB communications through different levels of the interconnect.

4.2.2.2 Application Mapping to the General Framework

The mapping process consists of two stages. First, the decomposition of operations with varying granularity. A number of supported operations can be decomposed into smaller sub-operations; these include *bits, bitswc, mult, sel, shift/rotate*, and *load/store*. The decomposition is followed by the judicious fusion of multiple lookup and datapath operations. Three fusion routines are incorporated into the tool: first, fusion of random LUT-based operations; next, fusion of bit-sliceable operations; and finally, fusion of custom datapath operations. Fusion is performed using graph traversals on the input, a control and dataflow graph (CDFG), following the decomposition. The level of decomposition and fusion is in part determined by the input functions, as well as the available schedule table memory and the MLB interconnect. Scheduling of instruction execution is performed following a modulo scheduling policy in which vertices are mapped into a particular MLB.

Given the number of modules in each level of the memory hierarchy and their I/O bandwidth, the software tool places the MLBs in a hierarchical fashion such that the number of inputs and outputs crossing each module is minimized. To realize this, a bi-partitioning approach is followed, first allocating MLBs to the first level modules, then distributed among second level modules. This continues until each MLB has been mapped to the lowest memory module. Signal routing is also performed hierarchically, beginning first by routing primary inputs to each MLB, then routing signals which cross each level of the memory hierarchy, and finally routing primary outputs from each MLB.

Finally, the bitfile generation routine accepts the placed and routed netlist and generates the control or select bits for the programmable routing switches, the schedule table, and the LUT entries. Functionality is then validated by applying inputs to the datain bus and noting the outputs available at the dataout bus of the instrumented memory.

4.2.3 Domain Customization for Efficient Acceleration

One of the greatest benefits of accelerator platforms such as FPGA and GPU is their flexibility to map diverse applications—FPGA using a purely spatial reconfigurable fabric, and GPUs using hundreds or thousands of lightweight cores with a very particular execution model that is geared towards graphics processing. However, this flexibility comes at a price of significantly lower area- and energy-efficiency compared to a dedicated hardware accelerator for a given application. For reconfigurable platforms, *domain-specific* accelerators represent a happy medium between application specific accelerators such as application specific integrated circuits (ASICs) and general purpose FPGAs and GPUs. Rather than optimizing the circuitry for a specific application, these accelerators customize the reconfigurable architecture to a given *domain* of applications, such as security, signal processing, analytics, or machine learning. The general procedure for customizing an architecture to a given domain can follow a top-down approach:

1. Identify typical applications belonging to the domain.
2. Identify kernel functions or algorithms within these applications.
3. Identify atomic operations common to the majority of the kernel functions.

While this approach cannot guarantee that future applications in this domain will enjoy the same level of acceleration that modern kernels (off of which the architecture is based), balancing domain-specificity with general-purpose computing can in some cases provide a measure of future-proofing, especially if the operations are more general, i.e. preprocessing that is common to all algorithms, rather than specific functions found in a given subset. This is discussed in more detail in Sect. 4.3.2.

4.3 Case Studies for Memory-Centric Computing

The following case studies demonstrate the advantages of memory-centric computing for two specific domains: security/cryptography and text analytics.

4.3.1 *MAHA for Security Applications*

Security is an important design metric for many systems, including embedded, personal, and enterprise level computing. Pure software implementations of diverse security algorithms, including encryption and hashing, can be costly. Therefore, the inclusion of hardware cryptographic modules has become a *de facto* standard for many devices, including embedded System-on-Chip, as shown in Fig. 4.2. This case study explores how the memory-centric MAHA architecture can be customized to the security domain of applications, resulting in the *Hardware Accelerator for Security Kernels*, or HASK [1].

4.3.1.1 Domain Exploration

To develop the security accelerator, a number of algorithms were analyzed, including the Advanced Encryption Standard (AES), Blowfish, CAST-128, IDEA, MD5, and SHA-1. These applications make use of addition, mixed-granular bitwise logical operations (such as AND, OR, and XOR), circular and logical shifts, and finally non-linear functions such as substitution boxes (S-Boxes), which differ among each application and are amenable to implementation as lookup tables (LUTs). In addition, a number of complex logic operations such as $Z = (A \wedge B) \vee (A \wedge C)$ are common in cryptographic hashing algorithms.

In general, security algorithms are deterministic, and therefore amenable to parallelism and pipelining, depending on the application and energy/latency requirements. Most algorithms operate on the input data, maintain an internal state, and iterate over the internal state using different key inputs for a given number of rounds. Therefore, latency can also be improved by applying instruction-level and data-level parallelism within each round; this in turn affects the type and number of communications required for a given application mapping, and ultimately the choice of interconnect network topology selected for the accelerator.

4.3.1.2 Architecture Description

The Security Nano-Processor (SNP), shown in Fig. 4.3, represents the smallest unit of the Hardware Accelerator for Security Kernels (HASK) [1]. Each SNP operates independently, with its own local data and instruction memory. Based on the observed communication patterns, a two level hierarchical interconnect was chosen for HASK; the first level contains groups of four SNPs and is called the *cluster*; the second level, called the *tile*, contains four clusters, as well as a central controller responsible for writing to the instruction memory of each SNP, as well as facilitating data transfer between the main memory and local SNP memory.

Security Nano-Processors

Each SNP contains components typical to a RISC-style processor, including an SRAM memory array, a register file, and a program counter. A lightweight, customized datapath, and two-way execution engine help reduce latency and improve efficiency for security kernels. The register file is designed to have a large number of read ports (8) and write ports (4) to support the wide execution engine. Each SNP has a dedicated schedule table that holds the 128×80-bit (2×40) wide instructions which are preloaded when the SNP is configured. A number of standard operations are supported in the datapath, including add, shift, simple logical operations, and load/store. A number of domain-specific optimizations are made to improve area- and energy-efficiency:

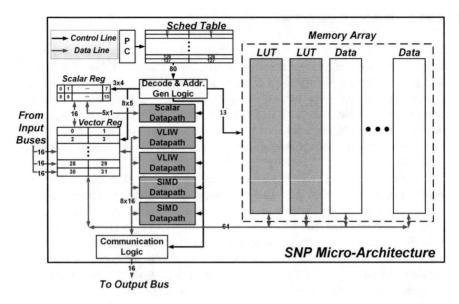

Fig. 4.3 Block diagram of a single Security Nano-processor (SNP) showing two-way execution engine, dedicated lookup and data memories, as well as typical RISC-style processor features, including the instruction memory (sched. table) and register file [1]

- hardware support for variable width vectorized lookup table (LUT) operations
- a fused logical unit for arbitrary 3-input functions
- a byte-addressable register file
- support for SIMD-style datapath operations.

Each SNP supports 8-bit input lookup operations with variable output sizes, including 8, 16, and 32 bits to map nonlinear functions (e.g., AES S-BOX) efficiently. The first 4 kB of memory in each SNP is reserved for LUTs and uses an asymmetric memory design to achieve a 40% reduction in read energy [8], as well as fine-grained wordline segmentation allowing efficient access to the variable width LUTs. The remaining memory is used as a byte addressable scratchpad-memory that stores inputs and resultant data. Complex logical operations such as $(A \wedge B) \vee (A \wedge C)$ are mapped to a novel reconfigurable logic datapath which is capable of implementing arbitrary logical functions of up to three inputs. This is realized and encoded using a Reed-Muller expansion, which results in substantially fewer required transistors than an equivalent canonical representation [15].

Interconnect Architecture

To support the typical communication requirements of security kernels, HASK employs a sparse hierarchical interconnect that leverages data locality and the spatio-temporal MAHA mapping to improve optimize energy efficiency. SNPs within the same Cluster have a fully connected shared bus. Similarly, a 16-bit shared bus is used for inter-cluster communication; however, unlike the fully connected intra-cluster bus, the intercluster bus can only be reached from one SNP per cluster, termed the Gateway SNP, or gSNP, through which all communications must be routed. A gSNP can broadcast to multiple inter-Tile buses in the same cycle, allowing the architecture to scale to an arbitrary size while maintaining limited connectivity between any two nodes.

These communication buses form a time-multiplexed programmable interconnect. Because the communication requirements for the security applications are both constant and known a priori, they can be scheduled at compile time. Routing information is stored in the instructions as an immediate value and decoded at runtime to control communication buses. If a buffer is enabled on a given cycle, output data from the SNP's operation is written to the appropriate bus. Subsequently, other SNPs can read the data into their local register files.

4.3.1.3 Results and Comparison to Other Platforms

The representative cryptographic applications mentioned in Sect. 4.3.1.1 were mapped to the accelerator, and latency, area, energy, and energy-efficiency were estimated from careful simulation at 32nm. Results were compared with FPGA implementations (at the same process node) and CPU implementations using a highly optimized cryptography library. Compared to the CPU, FPGA and HASK significantly improved latency, though the latency improvement from FPGA was greater. As a result, both platforms improve iso-area throughput substantially over CPU. Similarly, both accelerators see an order of magnitude improvement in energy efficiency, with HASK seeing greater improvements in energy than FPGA. These improvements for both FPGA and HASK relative to CPU are due to the following common factors:

- they perform their computations in lightweight processing elements
- they contain domain-specific hardware, with few to no extraneous functional units
- they leverage different types of parallelism within the algorithms to improve latency and throughput
- they utilize dedicated hardware structures and lookup tables to evaluate complex and fused logical functions

For the majority of the benchmark applications, HASK latency is slightly worse than that of FPGA. However, since HASK implementations generally use less die area than FPGA, the iso-area throughput is better than that of FPGA, and HASK uses less energy than FPGA on average for the same set of benchmarks. This energy improvement is sufficiently high such that an improvement in energy efficiency, measured as the energy delay product (EDP) is observed. The primary reasons behind the improvement in energy efficiency for HASK over FPGA are as follows:

- it supports LUT operations of different bitwidth
- it has dedicated hardware for fused logic operations, which reduces the total number of operations (energy)
- it uses a spatio-temporal mapping rather than purely spatial interconnect, which greatly reduces mapping complexity and routing energy
- the highly customizable memory structure of FPGAs results in energy inefficient memory accesses [13]

In addition, the HASK configuration bitstream is generally smaller than the equivalent FPGA implementation. This is an important consideration for embedded or remotely configured platforms, where a smaller bitstream size is advantageous due to aggressive area, power, and storage constraints.

4.3.2   MAHA for Text Mining Applications

Text mining applications use statistical methods to find relevant information within data sources. Analyzing textual data sources is crucial to many businesses, and it is widely accepted that the majority of business intelligence data is found in unstructured datasets. Meanwhile, the amount of data needing analysis grows exponentially, making energy efficiency for data-intensive text mining kernels a challenge. Due to physical limitations—power consumption, space, and cooling requirements, among others—as well as processor-to-memory bandwidth bottlenecks within the system itself, the practice of increasing computational power simply by adding more of the same processing elements (PEs) will inevitably reach a fundamental limit [3]. A dedicated, yet flexible, accelerator for text analytics is therefore attractive; however, due to the breadth of the text analytics field, the number and complexity of different algorithms, approaches, and techniques for performing analysis, and the existing infrastructure of data warehouses and processing centers, the options for such an accelerator are limited. We note three critical requirements for such a system:

- It must support the acceleration of multiple kernels found in a variety of common text analytics techniques.
- It must be amenable to hardware retrofitting and seamless operating system integration with systems at existing data warehouses while requiring minimal host-side software development and compiler overhead
- It must function independent of the input data representation (encoding) and potentially accept a variety of languages as input.

In this case study, we describe a reconfigurable hardware accelerator residing at the interface to the last level memory device, as shown in Fig. 4.4, which is designed specifically to accelerate text analytics data processing applications. The accelerator connects to the last level memory (LLM) and host system using the existing peripheral interface, such as SATA or SAS. By employing massively parallel kernel execution in close proximity to the data, the processor-to-memory bandwidth

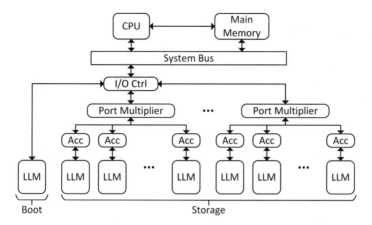

Fig. 4.4 System architecture showing the location of the accelerator and the last level memory device. Due to the close proximity of the two, the system bus is only used to transfer the results to the CPU, rather than an entire data set. This also allows the use of port multipliers without significant bandwidth requirements, as data processing occurs before the communication bottleneck is reached [7]

bottleneck, which plagues data-intensive processing systems today, is effectively mitigated by minimizing required data transfers, thus reducing transfer latency and energy requirements. Several architectural customizations enable efficient, hardware accelerated text processing for ubiquitous kernels. In particular, the hardware is capable of accelerating character downcasting and filtering for string tokenization, as well as the token frequency analysis, a basic operation found in many text analytics and natural language processing techniques. A data engine that interfaces the LLM to the accelerator also ensures the input matches the expected or supported languages and encodings while mapping characters to an internal compressed representation to reduce power consumption.

4.3.2.1   Domain Exploration

Examples of text mining applications can include text indexing and search, pattern mining, summarization, sentiment analysis, and classification, among others. These applications require similar preprocessing steps on the text, including the following:

Tokenization	Dividing a text stream into a series of tokens based on specific punctuation rules.
Change of Case	Changing all characters to lowercase.
Stemming	Removal of suffixes from tokens, leaving the stem.
Stop Word Removal	Removal of common, potentially
Frequency Analysis	Counting the number of times each (stemmed) token has appeared in an individual document and the corpus as a whole.

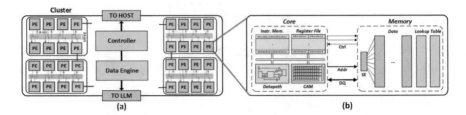

Fig. 4.5 (**a**) Overall accelerator architecture, showing eight interconnected PEs, the controller for SATA/SAS configuration, as well as an on-chip data distribution engine. (**b**) Microarchitecture of the PE, showing separate lookup and data memories with an output SE for bit-specific access. The core consists of an instruction table, register file, datapath, and a small CAM [7]

Tokenization, change of case, and frequency analysis represent three kernels, which are trivial relative to the complexity of the full application, but are nevertheless necessary and time-consuming tasks. Stemming and stop word removal can be considered extensions of frequency analysis, and are potentially more complex depending on the target language.

4.3.2.2   Architecture Description

A single processing element (PE) is shown in Fig. 4.5. Similar to the HASK architecture, each PE in the text mining accelerator operates independently and has its own local data and instruction memory. The interconnect fabric differs in that it is a two level hierarchy, with clusters of 8 PEs at the lowest level, which is advantageous for many text mining kernels, especially token frequency counting. Several customizations are also made to the datapath and PE architecture to satisfy the requirements of text mining applications.

The Text Mining Processing Element

The text mining PE is designed to accelerate a wide range of text analytics tasks while minimizing data movement, as shown in Fig. 4.5. This is accomplished by accelerating individual text mining primitives, as listed in Sect. 4.3.2.1. Here, we describe how one of the more complex operations, token frequency counting, can be accelerated (Fig. 4.6).

Analyzing how often a term appears in a document is one of the most common operations performed on textual data. Therefore, dedicated hardware is added to each PE which is capable of efficiently counting the occurrence of terms in the data set [5]. This hardware utilizes two kinds of memory: a CAM, which enables single-cycle lookup of terms already encountered, and an SRAM array, which stores the corresponding term counts. Rather than relying on a software hash table or a similar data structure, the CAM enables single-cycle, parallel lookup, without collisions or

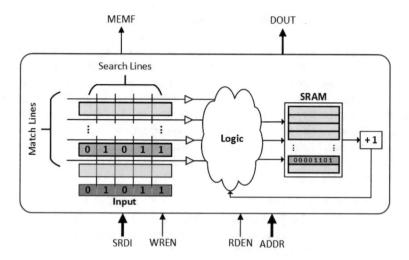

Fig. 4.6 Hardware implementation of TF counting using CAM and SRAM for storage. Data input, output, and some control signals are shown for reference [5]

requiring support for chaining or other collision handling techniques. However, the CAM is a limited hardware resource, and so it must be carefully sized. In general, when the dictionary memory is full, either the entries (terms/frequencies) can be consolidated by iteratively merging with terms/frequencies in other PEs, making use of the high-bandwidth local interconnect, or instead writing terms and frequencies to file and merging later (i.e., on the host CPU).

Interconnect Architecture

PEs are organized in a two level hierarchy. The lowest level of hierarchy consists of eight PEs connected by a shared bus, with 8 bit dedicated per PE. The choice of an eight PE cluster helps accommodate the frequency counting application, which benefits from high-bandwidth intra-cluster communication as previously mentioned. However, inter-cluster communication is far less frequent, and so requires significantly less bandwidth. Therefore, the architecture may be scaled by adding multiple clusters connected by a 16-bit wide mesh interconnect. Similar to the concept of gateway SNPs from HASK, the routerless design is made possible by providing a dedicated connection for one of every four PEs in the cluster to the shared bus, making it responsible for reading and writing data from and to other clusters.

4.3.2.3 Results and Comparison to Other Platforms

The text preprocessing kernels were mapped to the accelerator, and latency, area, energy, and energy-efficiency were estimated from careful simulation at 32nm. Results were compared with optimized CPU and GPU implementations, two platforms which are commonly employed to process textual data. The accelerator and GPU significantly outperformed the CPU, and the GPU slightly outperformed the accelerator for most benchmarks. However, from the perspective of energy-efficiency, the accelerator was two orders of magnitude more efficient than the GPU, and up to four orders of magnitude more efficient than the CPU when area efficiency is factored in. This is due to several reasons:

- The accelerator contains dedicated hardware for operations like term frequency counting, which are expensive on the CPU and GPU
- The accelerator does not contain extraneous hardware which is not used by the text mining kernels, improving area-efficiency
- The accelerator has close proximity to the data, which significantly reduces transfer energy and latency.

Many of these are shared with HASK's improvements over CPU, namely, the domain-specific hardware and lack of unnecessary functional units. However, the overall system integration of the text mining accelerator differs significantly from HASK, in that it is intended to be retrofitted into existing data warehouses, rather than used as an embedded security accelerator. This type of integration is more appropriate for data intensive workloads, such as text mining, when small kernels (e.g., downcasting) with little computational complexity must be applied to an entire massive dataset.

4.4 Case Studies for In-Memory Computing

The previous case studies detailed two *memory-centric* accelerator architectures, which saw an overall reduction in data movement, though it still had to be read out of the last level of memory storage device. This alone had a significant, positive impact on energy efficiency for data-intensive kernels in both security and text mining domains. However, it is possible to take this one step further by actually computing *within the memory array*, rather than just doing so in close proximity. This section presents two case studies which detail such in-memory computing architectures; the first case study details a compute-in-Flash memory architecture for general data-intensive applications from various domains, including analytics-style applications; the second describes a novel *multifunctional memory* design where resistive memory elements (RRAM) are used both for traditional storage, and for acceleration of complex neuromorphic computing tasks.

4.4.1 Flash-Based MAHA

Flash memory, as well as emerging non-volatile memory technologies such as resistive memory (RRAM) and spin-transfer torque (STTRAM), promises several improvements over traditional volatile SRAM storage, including access energy, read/write latency, and integration density. Their nonvolatility also makes them excellent candidates for off-chip, in-memory computing frameworks. This case study presents such a framework based on NAND Flash memory.

4.4.1.1 Domain Exploration

Unlike the domain explorations performed for the security and text mining memory-centric accelerators, this architecture instead targets general data-intensive applications. Therefore, the domain exploration begins by analyzing the extent to which applications with certain properties will benefit from the off-chip compute framework, when compared to a software-only solution. In order to compare between a software-only solution and a hybrid system with off-chip in-memory accelerator, application characteristics and the system configuration are expressed using a set of primitives:

g fraction of load/store (memory reference) instructions.
f fraction of instructions amenable to acceleration in the off-chip framework
c fraction of instructions in the host processor's ISA that are translated to that of the off-chip accelerator.
o fraction of the host processor's original instructions which result in an output; $f \times c \times o$ thus produces outputs which needs to be transferred back to the host processor.
e_{offchip} average energy per instruction in the off-chip compute engine
e_{txfer} energy expended in the transfer of an output from the off-chip framework to the host processor.
t_{offchip} ratio of cycle time for the off-chip compute framework to that for the host processor.
n fraction of the application which can be accelerated by exploiting parallelism in the off-chip framework.
t_{txfer} time taken in terms of processor clock cycles to transfer an output from the off-chip compute framework to the host processor.

With these primitives, the average time to execute an instruction in a system with a host processor and the off-chip compute framework can be formulated as:

$$T_{\text{sys}} = T_{\text{offchip}} + T_{\text{proc}} + T_{\text{txfer}} \tag{4.1}$$

$$T_{\text{offchip}} = t_{\text{offchip}} \times (I_{f>g}(f) \times (f - g + f \times c \times n) + I_{f \leq g}(f) \times f \times c \times n) \tag{4.2}$$

$$T_{\text{proc}} = (1 - f) \times t_{\text{proc}} \tag{4.3}$$

$$T_{\text{txfer}} = t_{\text{txfer}} \times (I_{f>g}(f) \times ((f - g) \times o + f \times c \times o) + I_{f\leq g}(f) \times f \times c \times o) \quad (4.4)$$

$$I_A(x) = \begin{cases} x, & x \in A \\ 0, & x \notin A \end{cases} \quad (4.5)$$

Here, T_{offchip}, T_{proc}, and T_{txfer} denote the fraction latencies in the off-chip compute framework due to processor execution and in the transfer of the resultant output from the off-chip platform to the processor. A similar expression for the energy of the resultant system is given below which shows the transfer energy increasing and the processor energy decreasing with increasing f:

$$E_{\text{sys}} = E_{\text{offchip}} + E_{\text{proc}} + E_{\text{txfer}} \quad (4.6)$$

$$E_{\text{offchip}} = e_{\text{offchip}} \times (I_{f>g}(f) \times (f - g + f \times c) + I_{f\leq g}(f) \times f \times c) \quad (4.7)$$

$$E_{\text{proc}} = (1 - f) \times e_{\text{proc}} \quad (4.8)$$

$$E_{\text{txfer}} = e_{\text{txfer}} \times (I_{f>g}(f) \times ((f - g) \times o + f \times c \times o) + I_{f\leq g}(f) \times f \times c \times o) \quad (4.9)$$

By substituting typical values (based on 45 nm technology) for e_{offchip}, e_{txfer}, t_{offchip}, n, and t_{txfer}, as 50 pJ, 10,000 pJ, 15, 0.01, and 10,000, respectively, the general system trends can be derived. Figure 4.7a shows the three components and the total system energy with $g = 0.7$ and $c = o = 0.05$, respectively. As shown in Fig. 4.7a, for the values of c and o selected, values of f close to g yield the lowest energy. When c and o differ by an order of magnitude, the total energy is dependent on f. A similar dependence on c, o, and f was observed for total execution latency. Combining these trends enables the derivation of system's energy efficiency for different inputs, as shown in Fig. 4.7b–d: in short, for applications which are not data-intensive (lower g), the overall energy efficiency of the hybrid system is reduced. On the other hand, applications that are data intensive (high value of g), with a small output size (low value of o), are most likely to see improvements in energy-efficiency through off-chip, in-memory computing.

Using the SimpleScalar architectural simulator, a set of 10 data-intensive benchmark applications were compiled and executed, after which runtime parameters based on the previously defined primitives were logged and used as input into the model. These benchmarks include the Advanced Encryption Standard (AES), Automatic Target Recognition (ATR), the Secure Hashing Algorithm (SHA-1), Motion Estimation (ME), Smith-Waterman (SW), 2D Discrete Cosine Transform (DCT), Discrete Wavelet Transform (DWT), Maximum Entropy Calculation (MEC), Color Interpolation (CI), and Census (histogram calculation). It was observed that over 75% of the energy expended in program execution was due to data transport. This suggests that optimizing the compute model for data-intensive tasks can improve energy efficiency. As previously observed in the security and text mining accelerators, physically relocating compute resources closer to the last level of memory can significantly reduce the data transfer energy and latency overhead. Or, in the more drastic case, the conventional software pipeline and caching mechanisms

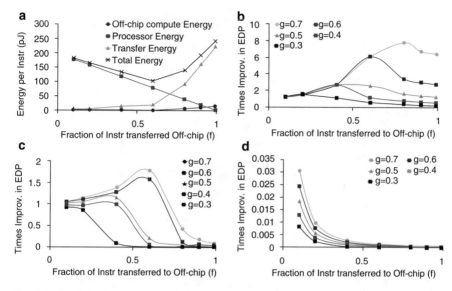

Fig. 4.7 Energy and performance for a hybrid system with a host processor and off-chip accelerator. (**a**) Energy per operation in the hybrid system with $c = o = 0.05$ and $g = 0.7$. Improvement in energy-efficiency (EDP) for the hybrid system with: (**b**) $c = o = 0.005$; (**c**) $c = o = 0.05$; (**d**) $c = o = 0.5$ [10]

can be replaced with a distributed compute and memory framework, which can better leverage local computation to improve energy efficiency. This second case is presented here.

4.4.1.2 Architecture Description

The in-memory MAHA architecture is similar to the general architecture presented in Sect. 4.2. The primary difference is that, instead of operating in close proximity to the memory with no modifications to the memory array, the Flash is augmented with additional circuitry that leverages the extremely high internal memory bandwidth and effectively processes data in situ.

Overview of Current Flash Organization

The typical organization of NAND Flash memory is shown in Fig. 4.8a, including the Flash memory array, as well as several peripheral structures which control the read and write operations from and to the memory array. A special component, called the Flash Translation Layer (FTL), converts the logical address of a location to its corresponding physical address—this is done to support operations like wear leveling, since overusing the same physical addresses will reduce the lifetime of

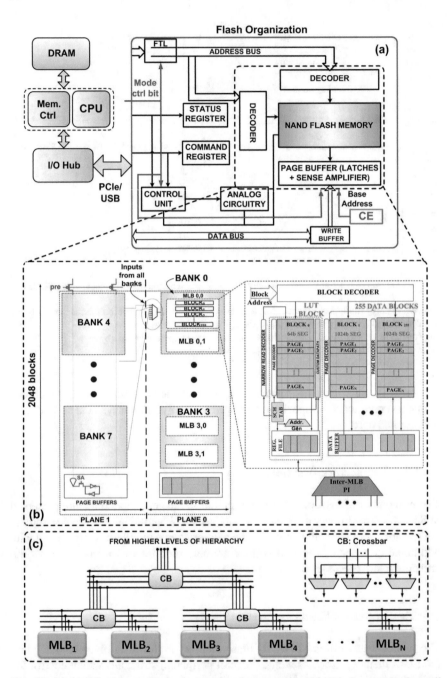

Fig. 4.8 (**a**) Modifications to Flash memory interface to realize MAHA framework. A small control engine (CE) outside the memory array is added to initiate and synchronize parallel operations inside the memory array; (**b**) Modified Flash memory array for on-demand reconfigurable computing. The memory blocks (called MLB) are augmented with local control and compute logic to act as a hardware reconfigurable unit. (**c**) MLBs are interconnected in a hierarchy, leveraging the existing Flash memory hierarchy for area- and energy-efficient inter-MLB communication [10]

the device. NAND Flash is organized in units of *pages*, and a number of pages are combined to form a *block*. During a read operation, the page content is first read into the page register, and subsequently transferred serially to the Flash external interface.

Modifications to Flash Array Organization

By introducing certain design modifications to the Flash memory, it is transformed into a resource which can be converted on-demand into a set of processing elements (MLBs), which communicate over a hierarchical interconnect.

Compute Logic Modifications

A group of *N* Flash blocks is logically clustered to form a single MLB. The custom datapath is implemented using static CMOS logic, and a custom dual ported asynchronous read register file is used to store temporary values. All operations within a given MLB are scheduled beforehand and stored in the schedule table, which is implemented using a 2-D flip-flop array.

Typical NAND Flash requires that an entire page be read at once during a read operation. However, this framework must operate on smaller data sizes; therefore, a narrow-read scheme is used for the memory blocks to allow reading a fraction of a page at a time. This is realized using *wordline segmentation*, whereby AND gates are inserted into the wordline and selectively enabled by the controller to read a portion of the page at once. This necessarily incurs hardware overhead, but can improve energy efficiency when small operands (relative to the size of the page) are stored in contiguous regions of memory within a page. For example, a 2 kB page can be segmented such that 512 bytes can be read, rather than all 2 kB of memory at once.

Routing Logic Modifications

The Flash-based MLB uses a set of hierarchical buses in order to minimize the inter-MLB interconnect overhead. This hierarchy is analogous to the existing Flash memory organization in 4 levels of hierarchy: banks, subbanks, mats, and subarrays. Because higher levels of hierarchy have more sparse connections, the applications running on the Flash-based MAHA framework must be mapped in a way that exploits local communication. This in turn improves execution latency and reduces power consumption.

4.4.1.3 Results and Comparison to Other Platform

The additional circuitry and peripheral logic required to implement the MLBs and interconnect result in relatively small (∼5%) area overhead compared to a baseline Flash design. When compared against a software-only system, the CPU with MAHA off-chip accelerator demonstrates considerable improvements for data-intensive applications:

- Data-intensive applications which are not compute intensive, and where the output size is similar to the input size, see the lowest improvement from the off-chip compute framework.
- Data- and compute-intensive applications where the output datasize is similar to the input (e.g., AES) can see slightly better, but still small, improvements.
- Data-intensive applications that are less compute-intensive, and where the output data size is similar to the input (e.g., time/frequency domain transforms) see moderate improvements.
- Map/Reduce-style applications (e.g., Census), where the output size is small compared to the input, see the greatest improvement. This is especially promising for many analytics applications, which fall into this category.

Therefore, the in-memory computing model is amenable to data-intensive applications. The use of NAND Flash in this case study can be extended to other emerging nonvolatile memory technologies as well.

4.4.2 MultiFunctional Memory

Emerging memory devices such as Resistive RAM (RRAM) or Spin-Transfer Torque RAM (STTRAM) have interesting properties which can be leveraged for increased efficiency and integration density for several applications [5]. RRAM demonstrates promising properties, including low programming voltage, fast switching speed, high on/off ratio, excellent scalability, reasonable programming endurance, high data retention, and compatibility with silicon CMOS fabrication processes [14]. Most promising is the fact that RRAM can not only serve as a traditional data storage medium, it has also found utility as a means to *process* data in a brain-inspired computing system. This *multifunctional memory* can be used within the MAHA framework to accelerate the weighted-sum operation which is common to machine learning algorithms in artificial intelligence applications. This weight is represented as the conductance of each cell in an RRAM array [12], and thus requires analog components to read out and digitize the output.

4.4.2.1 Architecture Description

The standard 1-transistor, 1-resistor (1T1R) memory array is shown in Fig. 4.9a. By rotating the bitlines by 90°, as shown in Fig. 4.9b, a "pseudo-crossbar array" can be implemented, and the memory array can switch on-demand between storage mode and computation mode. Figure 4.9c demonstrates how this multifunctional memory array can be integrated into the MAHA framework, replacing the LUT and

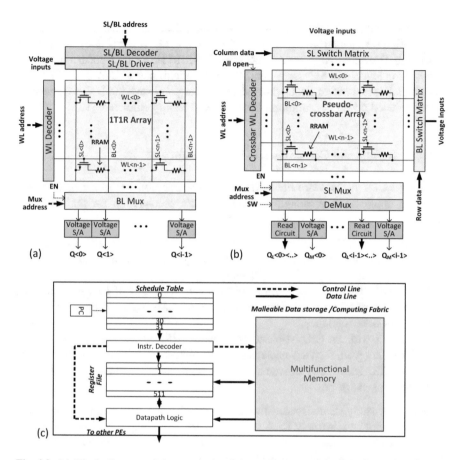

Fig. 4.9 (**a**) Block diagram of the conventional 1-transistor, 1-resistor (1T1R) memory array; (**b**) the multifunctional memory (MFM) with a rotated, pseudo-crossbar array; (**c**) system level block diagram showing a single processing element (PE) containing the MFM—note the replacement of LUT and Data memory found in other instances of the MAHA framework with the single, multifunctional memory block [12]

Data memories found in other instances of MAHA with the MFM to realize one processing element.

In this context, each PE consists of a 256×256 MFM array responsible for data storage as well as computation. Meanwhile, the schedule table is considerably smaller than other instances of MAHA, holding just 32 entries; however, each instruction is 545 bits wide—this comes from a 2-bit opcode selecting between standard computing mode, memory mode, and no-op, a 1 bit read/write flag, an 8 bit value controlling the wordline selection, 6-bits for mux and datapath logic, 16 bits for data source and destination registers, and the remaining 512 bits to indicate the status of the 256 bit lines and 256 select lines in the MFM array.

This system was tested using a number of applications which are amenable to acceleration with a neuromorphic computing framework. These include the Iterative Shrinking Threshold Algorithm, Stochastic Gradient Descent, K-means clustering (centroid calculation), Gaussian Blur, Sobel Gradient, and Radial Basis Function in artificial neural networks. Compared with other platforms, the MFM framework vastly reduced power consumption and improved energy efficiency [12].

4.4.2.2 Results and Comparison to Other Platforms

Compared with the standard Flash MAHA implementation, MFM shows improvements in applications amenable to neuromorphic acceleration for the following reasons:

- MFM increases the hardware resource reusage by using the same memory for storage and computation, rather than having separate data and lookup memories. This improves energy and area efficiency.
- The use of RRAM allows storing 4 bits per cell, for a total of 16 states, reducing the amount of required memory. In addition, RRAMs higher integration density improves area efficiency.
- Compared with an SRAM-based framework, RRAM has significantly better leakage energy; in fact, unselected cells have almost no leakage whatsoever. Conversely, SRAM, which must remain constantly powered, contributes significantly to system leakage power.

This case study demonstrated the feasibility of using an emerging nonvolatile memory technology, namely RRAM, as an alternative computing fabric. This can be used as an accelerator for data-intensive machine learning applications, which can be applied to a number of domains, including analytics.

4.5 Conclusion

In this chapter, novel memory-based or in-memory computing architectures have been presented, which can vastly improve energy efficiency compared to existing platforms for data-intensive applications. The basic hardware architecture—a set of interconnected, lightweight processing elements—was customized to meet the needs of various domains, including security and text mining and analytics, showing significant improvements in energy efficiency. It was shown how this framework, which relies heavily on memory elements for both storage and computation, can exploit the properties of existing (e.g., Flash) and emerging (e.g., RRAM) memory technologies for further improved performance and efficiency, for both general data-intensive kernels, as well as for learning algorithms with a novel, multifunctional memory framework. By focusing on energy-efficiency and scalability in the design, this framework and its embodiments can arise to the challenge of efficient computing for data-intensive kernels in the future.

Acknowledgements Robert Karam, Somnath Paul, and Swarup Bhunia would like to acknowledge the contributions of Christopher Babecki (Intel), Pai-Yu Chen (Arizona State University), Dr. Ligang Gao (Arizona State University), Dr. Ruchir Puri (IBM T. J. Watson Research Lab), Dr. Wenchao Qian (Xilinx), and Dr. Shimeng Yu (Arizona State University).

References

1. C. Babecki, W. Qian, S. Paul, R. Karam, S. Bhunia, An embedded memory-centric reconfigurable hardware accelerator for security applications. IEEE Trans. Comput. (99):1–1 (2016). doi:10.1109/TC.2015.2512858
2. M. Bohr, A 30 year retrospective on dennard's mosfet scaling paper. IEEE Solid-State Circuits Soc. Newsl. **12**(1), 11–13 (2007)
3. H. Esmaeilzadeh, E. Blem, R.S. Amant, K. Sankaralingam, D. Burger, Dark silicon and the end of multicore scaling, in *2011 38th Annual International Symposium on Computer Architecture (ISCA)* (IEEE, New York, 2011), pp. 365–376
4. Intl Solid State Circuits Conference (2013) ISSCC 2013 Tech Trends. http://isscc.org/trends
5. R. Karam, R. Puri, S. Ghosh, S. Bhunia, Emerging trends in design and applications of memory-based computing and content-addressable memories. Proc. IEEE **103**(8), 1311–1330 (2015). doi:10.1109/JPROC.2015.2434888
6. R. Karam, K. Yang, S. Bhunia, Energy-efficient reconfigurable computing using spintronic memory, in *2015 IEEE 58th International Midwest Symposium on Circuits and Systems (MWSCAS)* (2015), pp. 1–4. doi:10.1109/MWSCAS.2015.7282213
7. R. Karam, R. Puri, S. Bhunia, Energy-efficient adaptive hardware accelerator for text mining application kernels. IEEE Trans. Very Large Scale Integr. **24**, 3526–3537 (2016)
8. S. Paul, S. Chatterjee, S. Mukhopadhyay, S. Bhunia, Energy-efficient reconfigurable computing using a circuit-architecture-software co-design approach. IEEE J. Emerg. Sel. Topics Circuits Syst. **1**(3), 369–380 (2011)
9. S. Paul, R. Karam, S. Bhunia, R. Puri, Energy-efficient hardware acceleration through computing in the memory, in *2014 Design, Automation Test in Europe Conference Exhibition (DATE)* (2014). doi:10.7873/DATE.2014.279
10. S. Paul, A. Krishna, W. Qian, R. Karam, S. Bhunia, Maha: an energy-efficient malleable hardware accelerator for data-intensive applications. IEEE Trans. Very Large Scale Integr. VLSI Syst. **23**(6), 1005–1016 (2015) doi:10.1109/TVLSI.2014.2332538
11. S. Pawlowski, Architectural considerations for todays technology trends (2013). http://eecs.oregonstate.edu/
12. W. Qian, P.Y. Chen, R. Karam, L. Gao, S. Bhunia, S. Yu, Energy-efficient adaptive computing with multifunctional memory. Trans. Circuits Systems (2016)
13. H. Wong, V. Betz, J. Rose, Comparing fpga vs. custom cmos and the impact on processor microarchitecture, in *Proceedings of the 19th ACM/SIGDA International Symposium on Field Programmable Gate Arrays* (ACM, New York, 2011), pp. 5–14
14. H.S.P. Wong, H.Y. Lee, S. Yu, Y.S. Chen, Y. Wu, P.S. Chen, B. Lee, F.T. Chen, M.J. Tsai, Metal–oxide rram. Proc. IEEE 100(6):1951–1970 (2012)
15. X. Wu, X. Chen, S. Hurst, Mapping of reed-muller coefficients and the minimisation of exclusive or-switching functions. IEE Proc. E-Comput. Digital Tech. **129**(1), 15–20 (1982)

Chapter 5
New Solutions for Cross-Layer System-Level and High-Level Synthesis

Wei Zuo, Swathi Gurumani, Kyle Rupnow, and Deming Chen

5.1 Introduction

The rise of the Internet of Things—billions of internet connected sensors constantly monitoring the physical environment has coincided with the rise of big data and advanced data analytics that can effectively gather, analyze, generate insights about the data, and perform decision making. Data analytics allows analysis and optimization of massive datasets: deep analysis has led to advancements in business operations optimization, natural language processing, computer vision applications such as object classification, etc. Furthermore, data-processing platforms such as Apache Hadoop [29] have become primary datacenter applications, but the rise of massive data processing also has a major impact on the increasing demand for both datacenter computation and data processing in edge devices to improve scalability of massive sensing applications.

This increase in computation demand is not simply an increase in aggregate computational throughput, but also a demand for energy efficiency. Datacenters now commonly consume multiple MW: each of the top ten supercomputers in the Top500 list consumes more than 2 MW (and up to 17.8 MW)—enough that some datacenters also require their own dedicated power plant. Edge devices are no less critical: devices with wired power but high power consumption impose restrictions on both

W. Zuo
Electrical and Computer Engineering, University of Illinois at Urbana-Champaign,
Champaign, IL, USA
e-mail: dchen@illinois.edu

S. Gurumani • K. Rupnow
Advanced Digital Sciences Center, Singapore

D. Chen (✉)
University of Illinois at Urbana Champaign, Champaign, IL, USA

© Springer International Publishing AG 2017
A. Chattopadhyay et al. (eds.), *Emerging Technology and Architecture for Big-data Analytics*, DOI 10.1007/978-3-319-54840-1_5

heat dissipation and aggregate power consumption, and devices on battery power have direct lifetime, cost, and physical size implications due to power consumption.

The rise in computation demand has corresponded to a rise in computation acceleration hardware. Graphics processing units (GPUs) have long been used for acceleration of data parallel algorithms, but the energy efficiency of GPU-based acceleration can still be an order magnitude less efficient compared to FPGA-based hardware acceleration [7, 8, 20]. FPGAs have seen recent adoption in datacenters; the Catapult project [24] brings FPGAs into in Microsoft's datacenters, accelerating Microsoft's Bing search. FPGAs are also tested for acceleration in Google, Intel, and IBM, and there are recent startups targeting Hadoop acceleration for FPGAs in the datacenter [10].

The adoption of FPGAs in the datacenter as well as for acceleration in edge devices has led to significant performance and energy efficiency advantages, particularly for low latency and high throughput processing of massive data analytics applications. However, development for FPGAs remains complex and challenging. The continuing growth in size and complexity of FPGA devices has driven effort in the Electronic Design Automation (EDA) industry to improve design flow productivity by improving high level design space exploration, reducing complexity and required expertise for design entry, improving debugging and verification, and automating tedious and error-prone low-level implementation details.

In the past, the pressure to improve EDA tools for hardware design led to the transition from transistor-level to gate-level design, and then from gate-level to first structural RTL and eventually behavioral RTL in Verilog and VHDL. These advances include development of now-standard technology for synthesis including placement and routing for low-level circuit implementations, and RTL and logic synthesis to translate RTL to circuits. The continued growth in design complexity is now driving the development and adoption of high-level synthesis (HLS), which translates C/C++, SystemC, CUDA, OpenCL, Java, or Haskell (among others) to RTL implementations automatically; manual RTL design, even in behavioral RTL is extremely challenging to simultaneously meet performance, area, power, and time-to-market constraints for typical large and complex applications. This is particularly emphasized for data analytics applications, the time-to-market pressure is extreme, where developers may desire iterative performance optimizations hundreds of times per year, and the same developers desire the performance and energy efficiency advantages of FPGA-based acceleration yet often lack low-level hardware design expertise to manually design accelerators.

For these reasons, HLS has proven a viable and rapidly adopted method for performing hardware development for FPGA-based acceleration. Although HLS-based design helps automate design and optimization of hardware accelerators, design in high-level languages also introduces several new challenges for design automation: the system level design must consider both implementations of each function, and efficiency of communications between functions. In addition, we must consider hardware/software codesign options to select the best method for offloading complex computation to the FPGA including both partial and fully hardware accelerated algorithm implementations. Next, issues such as design closure become more challenging—whereas manual RTL designers often have experience analyzing and optimizing the low-level circuit in order to meet latency and achievable frequency constraints, HLS-generated RTL is significantly more challenging (or entirely infeasible) to optimize manually. However, HLS users often have little insight into how modifications of high level code affect the critical path latency and achievable frequency in output RTL. Thus, although HLS is necessary to improve design productivity, there are several remaining bottlenecks that we will address in this article to ensure that HLS-based design flows can meet quality and productivity demands.

Cross-layer design methodology (CLDM) flows span multiple abstraction levels, with challenges and optimization opportunities at each of these levels. These unaddressed challenges represent limitations in quality and design productivity, and the challenges are yet more complex when considering the interactions between different design layers. In this chapter, we identify research directions and opportunities related to interactions among design layers in a HLS-based flow (Sect. 5.2), and present four representative techniques that use cross-layer (or stage) interaction information to optimize designs (Sect. 5.3).

5.2   ESL Design Flow Challenges

Our CLDM flow is an EDA flow that starts with a system-level specification in a high level language (HLL), and proceeds through a series of design stages to eventually produce a physical circuit implementation. Each design stage performs analysis, optimization, and refinement before generating hardware specifications at a lower level of design abstraction. For example, system synthesis optimizes and performs several design tasks before generating a system architecture and HLL code partitioned into software and hardware portions. Then, high level synthesis takes the hardware portion of HLL code and after its design tasks produces RTL. Finally, RTL synthesis and physical design take the RTL as input and produce circuit implementations and/or FPGA configuration files.

We decompose a CLDM flow into three main interacting stages as shown in Fig. 5.1: system synthesis, high level synthesis, and RTL synthesis and physical design. Typically, each stage of the design flow optimizes at one abstraction level while ignoring details from other stages. For example, high-level synthesis optimizes resource allocation, scheduling and binding of computation operations, but does not explore hardware/software codesign decisions made by system-level synthesis or optimization of blocks of combinational logic during logic synthesis. However, a CLDM flow provides superior quality of results and design productivity by integrating decision making and optimization across all design stages.

Each stage of the CLDM flow may be further decomposed into a set of tasks performed at the same level of abstraction. These tasks optimize different aspects of the hardware architecture and application implementation, and interact with each other to explore the design options at each level of abstraction. Before describing the interactions within each stage and between stages of the CLDM flow, we will first introduce each stage of the CLDM flow and the component tasks.

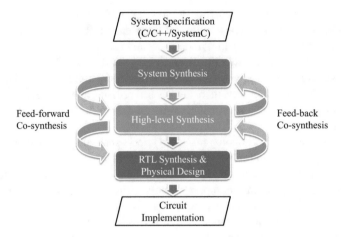

Fig. 5.1 Research opportunities on mixed-level co-synthesis in a CLDM flow: forward and backward paths

1. **System Synthesis** System synthesis starts from system-level specifications in C, C++, or other suitable high-level languages produced by software engineers. These system-level specifications define required functional behavior of the algorithms, as well as high-level constraints in area, latency, throughput, power consumption, and cost. System synthesis explores system level architectures to specify the major computational and communication components including CPUs, HW accelerators and the communications between them. Based on the algorithm features, system synthesis explores the task-level design space to perform HW/SW partitioning (CoDesign). HW/SW CoDesign examines options for mapping the algorithm to either a CPU (software) or custom accelerators (HW). CoDesign determines which HW/SW mappings best improve latency and throughput while minimizing area, power consumption, and required communications between software and hardware components.

2. **High-level Synthesis** For each hardware task, HLS automatically translates the C/C++ algorithm description into register transfer level (RTL) specifications. HLS performs standard optimization of the high level language input including dead code elimination, function inlining, and constant propagation, then performs a sequence of hardware optimizations. First, we allocate functional units that will perform computations, then schedule the applications' operations into control steps while respecting data dependence, latency, throughput, and area constraints. Finally, HLS uses the scheduling result and allocation to assign (bind) operations to functional units while minimizing required storage and multiplexing resources to implement the schedule.

3. **RTL Synthesis, Logic Synthesis, and Physical Design** Starting from the RTL specifications, the next stage of a CLDM flow optimizes the RTL, then performs elaboration and technology mapping to translate behavioral RTL to hardware structures into FPGA or ASIC-specific hardware components.

From this circuit structure, we then perform placement and routing to realize the physical implementation of the circuit. In addition, this stage may perform repeated optimization through floorplanning, retiming, and iterative enhancement of place and route results.

CLDM flows include both feed-forward and feedback paths both between synthesis stages and within individual stages. Feed-forward and feedback paths influence hardware quality of results significantly. RTL/logic synthesis and physical design quality is directly influenced by the quality of the input RTL, which is in turn influenced by the quality of system level synthesis. At a high level, this feed-forward dependence is intuitive, and directly influenced by two general factors: (1) the ability of synthesis algorithms to effectively consider design constraints and select the best solution given design constraints and quality estimations and (2) the ability to effectively estimate output hardware metrics (e.g., area, latency, throughput, power).

These two factors require significant independent implementation effort, yet are clearly highly interrelated: algorithms that select solutions clearly depend critically on the fidelity of hardware estimations. Although high-level design estimations can have good prediction fidelity, the interrelation of estimation quality and optimization efficacy leads to the need for common feedback paths.

Within a single synthesis stage, there are many common feedback paths: if HW/SW CoDesign cannot find a design partitioning that meets design constraints, it will provide feedback to architectural specification in order to select different system architecture components (e.g., higher performance CPU or larger FPGA if we cannot meet performance goals). In HLS, there may be feedback paths between allocation, scheduling, and binding stages—for example, requesting increased allocation of functional resources or different memory partitioning to improve schedule latency, or decreased allocation of resources to reduce design area. There are also strong interactions between logic synthesis and the place and route process. Retiming in logic synthesis requires accurate path delay information, which depends on the physical layout and connections of the gates, which are produced by place and route.

In addition, there are critical feedback interactions between synthesis stages: scheduling in HLS requires accurate delay estimation to determine the critical path and correctly divide a sequence of operations into control steps while also meeting constraints in achievable frequency. However, the delay of combinational paths is directly affected by RTL/logic synthesis and physical design: chaining and optimization of combinational paths can significantly affect combinational delay, and placement and routing of the circuit can also significantly impact delay—up to 70% of delay in modern FPGAs can be due to signal routing. Thus, a feedback path between RTL/logic synthesis and physical design and HLS is necessary to help achieve timing closure.

Similarly, there are feedback interactions between system synthesis and HLS: system synthesis optimizes communication between blocks using information about communication frequency and data access patterns to optimize buffers, communication channel bandwidth, and optimization priority. Furthermore, system synthesis makes high level estimates of the achieveable latency, frequency, and throughput of a hardware accelerator and decides between HW/SW mapping options based on performance increase relative to estimated area consumption of the hardware architecture. However, the area and performance metrics are only known after HLS or even RTL synthesis, thus requiring feedback to system synthesis in order to refine estimates and validate high level decision making.

These strong interactions within and between synthesis stages motivate the need for feedback paths that help to refine performance and area estimations and thus overall hardware quality. However, despite the importance of feedback paths to output quality, it is not feasible to simply perform the entire hardware design process iteratively—while it is inexpensive to explore alternative algorithm implementations during the system synthesis stage, it would be prohibitively expensive to perform the entire HLS, RTL/logic synthesis and physical design process for every algorithm option (and every optimization option within each algorithm). The design space for even a single algorithm and constrained set of optimization options may require multiple months of HLS and RTL/logic synthesis time to explore [8]. In fact, each stage of a CLDM flow can individually explore many thousands of design options for an algorithm—naively combining all stages is infeasible and a waste of computational effort.

In total, the interactions between synthesis stages are both challenges and opportunities. Effectively integrating feedback path interactions to refine design quality increases tool complexity and worst-case runtime complexity, yet there is significant optimization opportunity when tools leverage later analysis to feedback and refine earlier decisions. Furthermore, although worst-case runtime increases in iterative design, effective feedback, update heuristics, and guided optimization can substantially improve quality with minimal runtime overhead; despite runtime overhead, the overall design cycle may still be accelerated due to a reduction in designer effort. Thus, both feed-forward and feedback interactions can significantly improve design quality by integrating system-level, behavioral-level, and RTL information. In this chapter, we limit ourselves to cross-layer interactions between adjacent abstraction levels; iterative techniques that further integrate CLDM flows across all design stages will further increase design flow complexity, but may prove to be effective to further improve design quality.

5.3   System-/High-Level Synthesis Techniques

In this section, we propose and demonstrate four representative cross-layer techniques for improving productivity and both feed-forward and feedback interactions in CLDM flows. First, we begin with two system-level techniques for analyzing and optimizing high level language source code before high-level synthesis in order to improve the hardware suitability of high level language descriptions and applicability of typical HLS optimizations, which improves feed-forward interaction of system-level exploration and HLS. Then, we introduce two HLS techniques that integrate with RTL/logic synthesis and physical design: one that provides feed-forward interaction by analyzing generated RTL for multi-cycle paths and generating appropriate RTL synthesis constraints, and another that integrates feedback interaction by iteratively refining HLS timing estimates based on post-RTL synthesis timing analysis of combinational path latency.

5.3.1 Polyhedral Transformation to Improve HLS Optimization Opportunity

High level synthesis optimization techniques concentrate on parallelization of input C/C++ code, together with a variety of source code optimizations such as loop unrolling or memory partitioning that improve parallelization opportunity. HLS optimizations include specifications of communication between code blocks, for example, a user can denote that two loops can communicate through a first-in first-out (FIFO) buffer. However, despite the parallelization optimizations, complex applications with multiple communicating functions/modules commonly contain code organization that is intuitive for C/C++ coding style, but not compatible with HLS optimizations. Thus, for these applications, the lack of feed-forward interaction to transform HLL input for HLS optimization compatibility leads to significantly reduced quality of results compared to manual RTL design.

HLS tools provide a variety of powerful optimizations including loop unrolling, parallelization, data pipelining, memory optimizations (partitioning and/or merging), and fine-grained data communication. However, applicability of these powerful optimizations is limited by code structure and data dependence; in many cases, the optimizations cannot be applied because conservative analysis in HLS cannot guarantee that the transformed code would remain functionally correct. With more complex analysis and source transformation, however, we can produce functionally equivalent transformed HLL (C/C++) code that is also suitable for further HLS optimization.

The desired source transformations rely on data-dependence analysis, which has been well studied in the compiler community. In particular, the *Polyhedral model* has been an attractive method for advanced dependence analysis. Whereas traditional data-dependence analysis examines only the dependences within a single loop body, the polyhedral model analyzes the entire loop and series of data loads and stores. With this more complete analytical model for data-dependence relations, the polyhedral model can explore high-level loop transformations that produce functionally equivalent loop traversals with improved data dependence relations. The polyhedral model includes source-to-source transformations to generate improved loop nests. For example, the polyhedral model may transform a loop to improve buffer sizes (through loop tiling), or reduce dependence distance by transforming row-major to column-major or diagonal access patterns.

Therefore, to achieve high quality results for HLS of complex applications, we use a feed-forward interaction between compiler analysis and code optimization to HLS can effectively transform HLL input source in order to improve the applicability and effectiveness of HLS optimizations. In our prior work [35], we adopted polydral model transformations for HLS. We build an integrated poly-hedral framework which transforms C loops to expose optimization opportunities across multiple communicating modules. In particular, given an input application composed of several blocks, we target two types of important optimizations: (1) intra-block parallelization: improving the latency at block level by duplication and parallelization within a single block, and (2) inter-block pipelining: improving the latency at the application level by enabling data streaming between different blocks to overlap execution. Both of these optimizations are supported by existing HLS tools, but are challenging to apply in practice due to conservative data-dependence analysis.

In Fig. 5.2, we show a flowchart of our framework. We start with a data-dependent multi-block program as input, then perform analysis and source-to-source transformations with the objective of minimizing total application latency. Then, our framework automatically inserts the corresponding HLS directives into the transformed C code and performs HLS to produce an optimized RTL design. Our optimization framework consists of three major steps, described as follows:

5.3.1.1 Step 1

First, we systematically define and model a set of *data access patterns* using the Polyhedral model, classify them, and derive the associated loop transformations. The program inputs are composed of multiple data-dependent blocks where each block contains a single multi-dimensional loop nest.

Fig. 5.2 Overview of the framework [35]

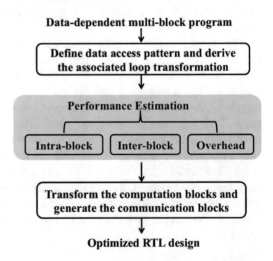

Given a loop nest of dimension D that accesses an N-dimensional array, the array access pattern is defined by matrix M whose size is $N \times D$, where the rows i represent the data access pattern in dimension of i of the data array, and columns j represent the access pattern in the loop level j. Given the array access pattern M, loop iteration vector i, and constant offset vector o, the array access vector S is defined as

$$\mathbf{s} = M\mathbf{i} + \mathbf{o} \tag{5.1}$$

We now describe how to define the array access patterns using M. For simplicity of demonstration, we illustrate the data access pattern for a two-dimensional array, although the same approach is applicable to higher dimension arrays.

Let $M = \begin{pmatrix} a_1 & b_1 \\ a_2 & b_2 \end{pmatrix}$, based on the value of a_1, a_2, b_1, and b_2, we classify three array access patterns as follows, based on how they traverse a multi-dimensional array.

- Column and reverse column: $M = \begin{pmatrix} \pm 1 & 0 \\ 0 & \pm 1 \end{pmatrix}$

- Row and reverse row: $M = \begin{pmatrix} 0 & \pm 1 \\ \pm 1 & 0 \end{pmatrix}$

- Diagonal access:
    In this category, the loops traverse in a diagonal line. Based on the slopes of the diagonal lines, we further divide this into two cases:
    $$\text{slope} \geq 1 \; M = \begin{pmatrix} \pm 1 & N \geq b_1 \geq 1 \\ 0 & \pm 1 \end{pmatrix} \text{ and slope} < 1 \; M = \begin{pmatrix} N \geq a_1 \geq 1 & \pm 1 \\ \pm 1 & 0 \end{pmatrix}$$

With a fixed array access pattern M, we can compose the array access vector using formula (5.1), and this is later used as input to the polyhedral model compilation framework to guide the code transformation and code generation.

5.3.1.2 Step 2

Next, the input array is tested with many loop transformations to determine whether the transformed loop is still functionally identical (i.e., it preserves all data-dependencies). For each loop transformation that preserves data-dependencies, we build a performance model to estimate optimization opportunity, and choose the loop tranformation(s) with the best estimated performance.

Our performance metric combines modeling of both intra- and inter-block speedup together with associated implementation overheads. A program is defined as a sequence of K blocks $b_1, b_2, \ldots, b_k$, where each block's latency is denoted lat_i, and due to linear dependence of the input code, the worst-case total program latency is the sum of blocks' latencies: $\text{lat}_{\text{base}} = \sum_i \text{lat}_i$.

The overall latency of a parallelized application lat_p after applying our optimization is then represented by:

$$\text{lat}_p = \frac{\text{lat}_{\text{base}}}{S_p^{\text{intra}} \times S_p^{\text{inter}}} + \text{cost}_p \tag{5.2}$$

S_p^{intra} and S_p^{inter} are the intra- and inter-block speedup and cost_p represents the implementation overhead. We now describe how to compute each of these items in detail.

- S_p^{intra} Given the fixed resource budget of the FPGA, we can derive the maximally allowed parallization degree, Max_{par}, and using this, we define S_p^{intra} as follows:

$$S_p^{\text{intra}} = \begin{cases} \text{Max}_{\text{par}} & \text{pattern } p \text{ enables parallelization} \\ & \text{for all the blocks} \\ 1 & \text{otherwise} \end{cases}$$

- S_p^{inter} Similarly, if we can transform all the blocks to follow the same data pattern for inter-block communication, we can fully enable inter-block pipelining. Therefore, we define S_p^{inter} as follows:

$$S_p^{\text{inter}} = \begin{cases} K & \text{pattern } p \text{ enables pipelining} \\ 1 & \text{otherwise} \end{cases}$$

  where K is the number of blocks in the program.

- cost_p The overhead is incurred when the access pattern is diagonal, where the loop is skewed. In this case, we first use the maximum loop bound for the loop boundary, and then use internal if condition to filter out false loop iterations, the loop body will be executed only when the if-statement is true. However, these preparations take extra cycles, and the number of empty cycles is proportional to the slope of the diagonal pattern (larger slopes produce more skewed loops, and therefore incur more empty loop iterations). Hence, we define the performance overhead as follows:

$$\text{cost}_p = \begin{cases} 0 & \text{slope} = 0 \text{ or slope} = \infty \\ C \times \text{slope} & 1 \leq \text{slope} < N \\ C \times \frac{1}{\text{slope}} & 0 < \text{slope} < 1 \end{cases}$$

where C is a constant.

Once these items are calculated, we can calculate the overall latency using formula (5.2) for each given access pattern. Therefore, we use this modeling method to evaluate each access pattern generated in Step 1, and select the access pattern with the minimum latency as the access pattern matrix to guide the polyhedral transformation.

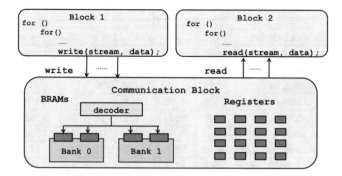

Fig. 5.3 Communication block [35]

5.3.1.3 Step 3

Finally, for the chosen transformation, we automatically perform the loop transformations at the source code level, insert high-level synthesis directives, and
generate the communication blocks that interface the data-dependent blocks. If
the communication block is a simple FIFO, we automatically insert FIFO high-
level synthesis directives; if the communication interface requires multiple reads
or a more complicated pattern, we customize the communication blocks. The
customized communication block is an automatically inserted reuse buffer that
can handle more complex stencil reuse patterns compared to a simple FIFO. The
implementation of the reuse buffer uses multi-ported BRAMs or registers depending
on the resource and buffer sizes, as shown in Fig. 5.3.

5.3.1.4 Evaluation

We evaluate the framework with a set of real-world applications with multiple data-
dependent computation blocks. The output of our framework is transformed C code
with HLS optimization pragmas, which is passed to the AutoPilot.[1] HLS tool for
synthesis to Verilog RTL. Table 5.1 shows the latency in clock cycles, resources,
and achieved frequencies of each different implementation of all the benchmarks.

We compare the performance of three different implementations. The first
implementation is the baseline—synthesis of the original source code without
polyhedral optimization or AutoPilot optimization. The second implementation is
an improved version—it applies the HLS directives of intra-block parallelization
and inter-block pipelining optimization without code transformation. We observed
that although the second implementation tries to use optimization, it fails whenever
the default data access pattern does not support it. The third implementation is our

---

[1] AutoPilot was acquired by Xilinx, and is now Vivado HLS.

Table 5.1 Performance and resource comparison of different implementations

Benchmark	Implementation	Cycles	LUT	FF	DSP	BRAM	Frequency (MHz)
Deconv	w/o trans, w/o opt	5408	3234	948	24	48	151
	w/o trans, w/ opt	1809	6433	2650	24	5	182
	w/ trans, w/ opt	257	13,819	13,826	108	17	182
Denoise	w/o trans, w/o opt	5408	3266	948	24	5	160
	w/o trans, w/ opt	1809	6503	2672	24	5	182
	w/ trans, w/ opt	250	13,817	13,824	108	17	230
Seg	w/o trans, w/o opt	9864	3735	1202	30	24	117
	w/o trans, w/ opt	9864	3735	1202	30	24	117
	w/ trans, w/ opt	500	48,796	9560	216	34	156
Seidel	w/o trans, w/o opt	64,803	1400	891	2	2	103
	w/o trans, w/ opt	1818	13,375	6626	32	6	134
	w/ trans, w/ opt	1130	47,402	20,040	96	14	134
Jacobi	w/o trans, w/o opt	5373	5563	1890	3	16	101
	w/o trans, w/ opt	1439	39,430	18,832	64	10	134
	w/ trans, w/ opt	482	38,877	18,664	64	10	133

proposed implementation: it transforms the source code and then enables the intra- and inter-block optimization. From Table 5.1, we observe that on average, HLS optimization without polyhedral transformation improves performance by 4.89×, but that increases to 29.59× with polyhedral transformation.

5.3.2 Polyhedral Code Generation for High-Level Synthesis

In addition to the benefits of code analysis and optimization, there is a strong interaction between code generation and HLS output quality. Although polyhedral models can dramatically enhance HLS optimization opportunity at the source code level, the polyhedral model transformed loops are often more complex and use CPU-oriented code features that are unsuitable for HLS. Table 5.2 lists some critical differences between C for CPUs and C for hardware. In order to fully take advantage of polyhedral optimizations, it is important that code generation for HLS avoids unsuitable features.

In [34], we tailor polyhedral code generation for HLS. Using the state-of-the-art polyhedral code generator CLooG [5], we explore code generation schemes tailored for efficient HLS. Particularly, we explore techniques to significantly improve resource utilization on FPGAs, and a technique designed for effective code generation for rectangular loop-tiling boundaries leading to further improvements in resource utilization, while retaining parallelization optimization opportunity exposed by our prior polyhedral optimizations.

Table 5.2 Polyhedral code generation for CPU vs. circuit

Property	CPU	Circuit
Performance	Cycle counts	Cycle and clock period
Code size	Memory/cache (cheap)	Logic gates (expensive)
Branches	High performance penalty for mis-prediction	No performance penalty for mis-prediction
Floating point OPs	FPU (cheap)	Dedicated FP module (expensive and slow)

In order to evaluate the effect of different code generation techniques, we start with the state-of-the-art code generator CLooG [5], and studied the area, performance, power, and energy of various code generation techniques. Our framework is based on PoCC, the Polyhedral Compiler Collection [22], which includes both CLooG and Pluto [6, 27]. We examine four representative stencil computations from PolyBench [23]—a suite of benchmarks with loop nests suitable for polyhedral optimizations. Jacobi-1D (J1D) is a 1D 3-point iterative Jacobi process, and Jacobi-2D (J2D) is a 2D 5-point iterative Jacobi process. Seidel-2D (Seid) is a 2D 9-point iterative Seidel process, and FDTD-2D (FDTD) is a finite-domain time difference discrete solver. In addition, we use a matrix-multiplication kernel GEMM. These five codes each benefit from loop tiling and have large data reuse potential. The four stencils pose numerous code complexity challenges when required tiling transformations are applied.

We primarily studied four code generation techniques for improved hardware quality, including: (1) Turning off polyhedra separation to restructure loops with less nesting but more (inner) conditionals to reduce code size and enable resource sharing. (2) Optimizing division operations to replace floating-point divisions by appropriate integer divisions and offsets. (3) Using hierarchical min/max operations to reconstruct a balanced min/max tree at the iteration boundaries to simplify boundary computation. (4) Simplifying loop bounds using sub-bounding box tiling to find a parallelogram hull to approximate the tile origin domain to further simplify boundary computation.

We will now discuss each of these techniques in detail, explain the key concept of each technique, and evaluate the effectiveness by applying to the five benchmarks.

5.3.2.1 Turning Off Polyhedra Separation

When generating the code, ClooG offers options to control polyhedra separation. That is, when on, CLooG generates code structures with more loop nests and fewer conditionals, where each loop nest represents a convex sub-polyhedra. In contrast, when no separation is used, the generated code contains fewer loop nests but more (inner) conditionals. The first option works well when running on CPUs because it effectively simplifies the control flow and makes branching behavior

Table 5.3 Impact of separation

	LUT	FF	DSP	CP (ns)	Cycle	Pwr (mW)
FDTD-sep	39,803	24,532	56	11.313	12,622,094	1378.49
FDTD-nosep	25,595	16,692	40	8.965	11,500,269	1452.56
Gemm-sep	1822	1532	14	7.609	14,567,698	1302.43
Gemm-nosep	1822	1532	14	7.609	14,567,698	1302.43
J1D-sep	11,926	7586	14	8.435	7,638,461	1384.48
J1D-nosep	11,327	7350	14	8.100	5,724,101	1411.92
J2D-sep	24,818	15,351	35	8.990	13,356,949	1435.77
J2D-nosep	14,216	10,103	27	8.582	21,800,977	1427.02
Seid-sep	32,561	19,599	9	8.763	159,281,806	1459.67
Seid-nosep	32,561	19,599	9	8.763	159,281,806	1459.67

more predictable. Thus, it reduces branch misprediction performance penalties. However, in hardware branch mispredictions cause no penalties, yet the "polyhedral separation" options result in a code with more complicated loop structures and thus more complex finite state machine control and less resource efficient hardware. Hence, we evaluate the impact of using this option; Table 5.3 shows the impact of separation on our five benchmarks.

We first observe that Seidel-2D and GEMM both have identical code output with and without polyhedra separation, but the other three benchmarks have a strong benefit in resource savings. In terms of latency, the effect varies: FDTD improves both achievable clock period and latency in cycles whereas Jacobi-2D increases latency in cycles. This latency tradeoff is intuitive; on one hand, turning off separation will reduce the number of loop nests, which reduces startup and completion costs for outer loop nests. However, without separation there will also be more branching conditions in the innermost loop nest. When these conditions are sufficiently simple to be implemented via predication, they will impose little performance degradation, but when they are complex they can increase the inner loop latency, increase the number of pipeline stages, or make the inner loop infeasible for pipelining.In the case of Jacobi-2D, complex conditions with divisions in the innermost loop increase the pipeline from 33 stages to 77 stages. We will later introduce other code generation techniques which can address this gap.

5.3.2.2 Division Optimization

Next, we optimize division operations. The original code generated by CLooG frequently uses floating-point divisions together with *floor* and *ceiling* operations for loop boundary control. With custom, dedicated, high-performance floating-point functional units, these operations perform well in CPUs, but impose significant resource use and latency overheads on FPGA hardware implementations. Therefore, we evaluate a series of alternative implementations for these operations that replace floating point division with integer division with appropriate scaling and bounds-checking in order to reduce both area and latency.

First, **IntDiv** replaces the expensive floating-point division by appropriate integer division and offsets. Precisely, we replace floor and ceiling operations with the following two operations: $\lceil \frac{x}{y} \rceil$ (`(x > 0)? (1 + (x - 1)/y): (x / y))`) and $\lfloor \frac{x}{y} \rfloor$ (`((x > 0)? (x / y): (1 + (x +1)/ y))`) Next, **MulUB** is a technique to scale the upper bound constraints to eliminate division whenever possible. That is, $cX \leq \lfloor \frac{x}{y} \rfloor$ is replaced by $y \cdot cX \leq x$. Finally, **LcmLB** is a technique to scale the lower bound constraints to eliminate division whenever possible. That is, $\lceil \frac{x}{y} \rceil \leq cX \wedge \lceil \frac{u}{v} \rceil \leq cX$ is replaced by $\left\lceil \frac{\max\left(\frac{\text{lcm}(y,v)}{y}x, \frac{\text{lcm}(y,v)}{v}u\right)}{\text{lcm}(y,v)} \right\rceil \leq cX \cdot \frac{\text{lcm}(y,v)}{y}$, $\frac{\text{lcm}(y,v)}{v}$ and $\text{lcm}(y, v)$ can be computed at compile time if y and v are constants.

Table 5.4 shows the results of applying these three techniques to our benchmarks, starting from the version already optimized by the first technique. We observe that the first optimization that replaces floating-point division with integer produces dramatic improvements in LUT and FF resource use for all of the stencil applications, with as much as a 4× reduction. This does lead to an increase in DSP use, but that increase is more than compensated for by the reduction in other resources. Optimizations that improve the upper and lower-bounds checking further improve resource use. In total, division optimization significantly reduces resource use which also produces minor improvements in cycle time, latency in cycles and power consumption.

5.3.2.3 Hierarchical Min/Max Operations

The third technique we have evaluated is a simple re-organization of conjunctions in a tree, versus a sequence. Currently, for a bound of the form $cX \leq A \wedge cX \leq B \wedge cX \leq C \wedge cX \leq D$, CLooG generates $cX \leq \min(\min(\min(A, B), C), D)$. We instead generate a balanced tree of operations $cX \leq \min(\min(A, B), \min(C, D))$ to leverage operation parallelism

Table 5.5 summarizes the results of applying this technique to applications already optimized by eliminating polyhedra separation and division transformations. These improvements, albeit marginal, further reduce resource usage in all cases.

5.3.2.4 Loop-Tiling Bound Simplification

Finally, we explore general optimization of loop tiling bounds. Loop-tiling is an essential loop transformation to improve data locality, data reuse, and hence the overall performance. The polyhedral model provides powerful techniques for effective loop tiling, but the default CLooG-generated code might have irregular tiling shapes that precisely capture the loop dimension, but create complex loop boundaries. Such code executes effectively on CPUs by precisely iterating over the loop space without any empty iterations. However, for HLS, complex loop bound-

Table 5.4 Impact of division optimization

	LUT	FF	DSP	CP (ns)	Cycle	Pwr (mW)
FDTD-nosep	25,595	16,692	40	8.965	11,500,269	1452.56
+ IntDiv	11,319	8864	100	8.948	11,487,327	1365.44
+ MulUB	9210	7456	62	8.722	11,486,905	1366.24
+ LcmLB	9988	7572	50	8.668	11,488,085	1364.21
Gemm-nosep	1822	1532	14	7.609	14,567,698	1302.43
+ IntDiv	1822	1532	14	7.609	14,567,698	1302.43
+ MulUB	1822	1532	14	7.609	14,567,698	1302.43
+ LcmLB	1822	1532	14	7.609	14,567,698	1302.43
J1D-nosep	11,327	7350	14	8.100	5,724,101	1411.92
+ IntDiv	2759	1897	17	8.148	3,194,601	1312.37
+ MulUB	2495	1787	16	8.853	3,094,421	1312.42
+ LcmLB	2496	1835	15	7.946	3,094,951	1312.11
J2D-nosep	14,216	10,103	27	8.582	21,800,977	1427.02
+ IntDiv	6312	4749	39	8.757	12,408,718	1318.99
+ MulUB	5221	4164	32	8.596	12,150,238	1319.23
+ LcmLB	4968	4148	31	8.428	12,148,886	1320.37
Seid-nosep	32,561	19,599	9	8.763	159,281,806	1459.67
+ IntDiv	8844	5909	33	8.460	159,265,028	1369.15
+ MulUB	7934	5227	18	7.908	159,265,296	1366.94
+ LcmLB	7573	5125	16	8.039	159,265,212	1368.42

Table 5.5 Impact of hierarchical min/max

	LUT	FF	DSP	CP (ns)	Cycle	Pwr (mW)
FDTD-ns-dopt	9988	7572	50	8.668	11,488,085	1364.21
+ hier	9746	7542	50	8.273	11,486,733	1365.45
Gemm-ns-dopt	1822	1532	14	7.609	14,567,698	1302.43
+ hier	1822	1532	14	7.609	14,567,698	1302.43
J1D-ns-dopt	2496	1835	15	7.946	3,094,951	1312.11
+ hier	2496	1835	15	7.946	3,094,951	1312.11
J2D-ns-dopt	4968	4148	31	8.428	12,148,886	1320.37
+ hier	4968	4148	31	8.428	12,148,886	1320.37
Seid-ns-dopt	7573	5125	16	8.039	159,265,212	1368.42
+ hier	7495	5097	16	8.348	159,265,212	1370.78

aries can cause inefficient scheduling and resource allocation, thus diminishing
the benefit of precise tiling boundaries. Thus, the last technique we evaluate is to
simplify loop bounds using sub-bounding box tiling to find a parallelogram hull to
approximate the tile origin domain to further simplify boundary computation. This
technique trades between preciseness in loop iteration (but less efficient hardware
implementations) and less precise loop iterations that have empty iterations, but
more efficient hardware implementations.

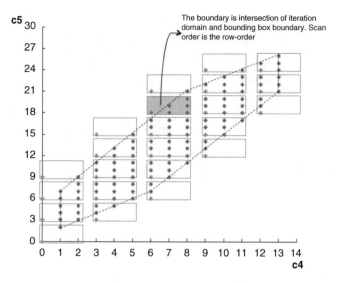

Fig. 5.4 Optimization for point loop [34]

We start with an input polyhedral representation using standard polyhedral tiling including a polyhedral schedule for all loops/dimensions. First, we generate the tile loop boundaries, and find a parallelogram that approximates the tile origin space. We combine the parallelogram domain with the input scheduling functions. Next, we generate the point loop boundaries to scan the iteration points within the tile in the order specified by the user-specified schedule. At this step, point loop boundaries are made of the intersection of the tile bounds and the iteration domain bounds.

Figure 5.4 illustrates our method with the loop iteration domain of the Seidel benchmark. The dashed lines are the boundaries of the loop iteration domain, and each rectangular box is a tile. Without our method, all tiles that intersect with the loop bounds require a complicated code structure to describe the loop shape. Using our method, we approximately generate only rectangular tiles, which are easy to describe in code and thus lead to more efficient hardware.

Table 5.6 shows the results comparing our technique with the default code generated by ClooG. In all cases, our sub-bounding box method uses fewer resources than the default code generation technique.

5.3.2.5 Experimental Results

In addition to the five benchmarks used during exploration, we tested all of our techniques on a set of eleven computation kernels. Results demonstrate that these optimizations can reduce area by 2× on average (up to 10×) without significant effect on latency or power consumption. Figure 5.5 represents the detailed comparison data, where HLSOpt includes all of the optimizations discussed above.

Table 5.6 Impact of bounding-box

	LUT	FF	DSP	CP (ns)	Cycle	Pwr (mW)
Seid-s-itr	7133	4366	9	7.704	159,135,759	1361.17
+ BB	5882	3637	5	7.874	159,143,274	1352.77
J2D-s-itr	4037	3237	24	8.244	11,761,063	1320.35
+ BB	3236	2514	19	7.943	11,752,537	1300.22
J1D-s-itr	2004	1558	15	8.068	3,094,952	1294.78
+ BB	1808	1475	15	8.12	3,074,938	1295.32
FDTD-s-itr	7259	5780	51	8.986	11,357,291	1369.41
+BB	6956	5549	50	8.768	11,354,663	1368.06

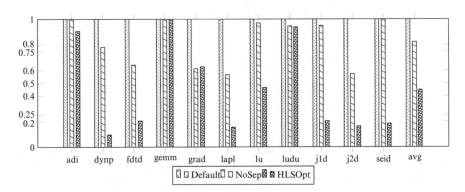

Fig. 5.5 Area ratio (normalized to default)

5.3.3 Multi-Cycle Path Analysis for High-Level Synthesis

HLS tools take high level language input specifications and design constraints
in latency, throughput, area, and power and generate RTL that attempts to meet
these constraints. The RTL is then passed to RTL synthesis and physical design—a
feed-forward interaction where the area and achievable clock period determined by
critical path delay is directly influenced by the quality of HLS-produced RTL. Due
to resource and/or dependency constraints, HLS tools cannot always deeply pipeline
datapaths to achieve a single cycle initiation interval in all designs [17]. In this
case, the tools generate **multi-cycle paths**—paths with more than one clock cycle
available for signal propagation. However, timing analysis during RTL synthesis
assumes that every register-to-register path has only one clock cycle for signal
propagation. RTL synthesis considers multi-cycle paths to be *timing exceptions*,
which critical path analysis may report as difficult to detect *false positives* [13].
Ideally, to prevent clock period degradation, HLS tools can insert registers to break
long paths, but these registers waste resources without improving execution latency.
In the worst case, inserting registers can significantly increase the number of control
steps and hurt execution latency due to pipeline imbalance.

To overcome this challenge, we introduce a feed-forward interaction between HLS and RTL synthesis to improve circuit quality and guide optimization effort towards critical paths. In our prior work, we perform behavioral level multi-cycle path analysis (BL-MCPA) [31, 33] and identify multi-cycle paths in a hardware design by calculating the interval (in cycles) between the state-transitions of the source and sink register of combinational paths. Thus, instead of breaking long combinational paths into single-cycle segments, we construct multi-cycle paths to reduce latency and register usage and then pass this information to RTL synthesis. With the multi-cycle path information, RTL synthesis improves the accuracy of timing analysis by filtering false positive critical paths.

Earlier works on multi-cycle path analysis detect paths with cumulative latency greater than one cycle, but rather than finding these **required** multi-cycle paths, we identify **available** multi-cycle paths—paths where the amount of time available for computation is greater than the necessary time, thereby providing optimization opportunity. Traditionally, precise multi-cycle path analysis must analyze the state-transition graph (STG) in which each state is the state combination of all flip-flops (FFs) in the circuit. Thus, the number of states in an STG scales exponentially with the number of FFs. Multi-cycle path analyses which attempt to analyze the whole STG are not practical [4, 26], while heuristic approaches [16, 19] deliver pessimistic results that do not improve circuit timing [13]. With the **behavioral-level** information from HLS, we avoid exponential time complexity of the prior exhaustive enumeration techniques by dividing circuit states into equivalent groups and analyzing the relationships between equivalent groups instead of individual circuit states.

BL-MCPA detects multi-cycle paths using the hardware architecture generated after binding and the **cycle-accurate** behavioral model generated by scheduling. We extract behavioral information from the CDFG and build an STG. Although we operate at both the behavioral and register transfer level, we consider our analysis **behavioral-level** because it relies on behavioral-level information for both accuracy and time complexity. We prove that our behavioral-level STG represents all reachable circuit states without exponential growth in the number of nodes or edges between nodes, and present our multi-cycle analysis, which considers all reachable circuit states by considering both control flow and guarding conditions (register assignment enables).

Our multi-cycle analysis consists of three high-level steps: (1) Construct an RTL datapath to identify combinational paths. (2) Construct a behavioral-level state transition graph (STG) and annotate with control flow and define-use chains [12]. (3) Analyze the STG to identify multi-cycle paths and determine available cycles for each path.

5.3.3.1 Circuit States and Control-States

We formally define circuit states, control-states, and reachable states as in [13]. For a circuit with N-bits of flip-flops(FFs) $(r_0, \ldots, r_{N-1})$, we represent a **circuit state** as a bit-vector $\mathbf{b} = (b_o, \ldots, b_{N-1})$, where b_i is the value of FF r_i in state $\mathbb{S}$ and the set of possible states $\mathbb{P}$ has 2^N states. The N bits of state FFs contain both N_C control FFs (**control-states**) and N_D data FFs (**data-states**) where $N = N_C + N_D$ and $N_C \ll N_D$. Any combination of control and data state can be represented by a bit-vector **b**.

HLS can use behavioral information to precisely track the set of control FFs and data FFs, but RTL synthesis tools cannot distinguish between control and data FFs, particularly because there are data-dependent conditional state transitions. We define a circuit state $\mathbb{S}$ as a **reachable state** if and only if the circuit can reach state $\mathbb{S}$ from an initial circuit state $\mathbb{S}_0$ through a sequence of state transitions. The set of reachable states $\mathbb{R}$ is a subset of all **possible** states $\mathbb{P}$. We can thus also consider *reachable* control (and data) states $\mathbb{R}_C$ and $\mathbb{R}_D$, where $\mathbb{R}$ is a subset of $\mathbb{R}_C \times \mathbb{R}_D$.

We use the STG defined in [11], which is generated after scheduling, to model circuit behavior in the control-state space, and denote it as *Control-STG*. We now formally connect the Control-STG and circuit STG, then extend the Control-STG with data-dependent behavior and control-flow information. Leveraging the precise control-state information from the Control-STG, we divide the circuit states into control-equivalent state sets (CES), and capture state transitions on data FFs by register assignments.

5.3.3.2 Capturing Conditional Behavior in the STG

The Control-STG captures control-flow behavior, but does not explicitly represent conditional behavior of register assignments. Thus, we first explain register assignment guarding conditions and present a *guarding condition aware* STG (GA-STG).

In a HLS flow, guarding conditions of register assignments are generated from the following 2 cases: (1) Predicated operations produced by If-conversion [1, 21, 28], and (2) Control-flow dependent register assignments of ϕ-nodes.

In the first case, the guarding condition is uniquely derived from the execution condition of the parent basic block (BB)[2] of the register assignment. In the second case, the guarding condition is uniquely derived from the incoming BB of the ϕ-node. Hence, the guarding condition of a register assignment can be represented by its parent BB. In both cases, different BBs in the control flow derive different guarding conditions and thus we use BBs to distinguish different guarding conditions that appear in the same control-state.

We build the *guarding condition aware* STG (GA-STG) for our multi-cycle analysis by extending the STG construction approach in [11]: Starting from a scheduled CDFG, we build control-states for each BB and the linear state transitions between states in the same BB. We then generate (virtual) register assignments from operations, and attach them to the corresponding control-states.

---

[2]A basic block is a portion of the code within a program with only one entry point, only one exit point and without conditional branches.

We then connect linear control-flow of different BBs by adding merge-edges between control-states from different BBs. Specifically, for each control-flow edge, we add a merge-edge from the last control-state in the source BB to the first control-state in the sink BB, and associate the merge-edge with the branching condition of the control-flow edge. Such a merge-edge indicates its source and sink control-states refer to the same control-state in the final control-STG, but the execution of the register assignment and state-transition in the sink control-state is **predicated** by the associated condition. If a BB has multiple predecessors, we need to duplicate its first state as well as the underlying register assignments, then connect the duplicated states with each of its predecessor's last state. These duplicated states belong to different control-states in the final control-STG.

The predication produces an intermediate control-STG which can be complex: states from different BBs may refer to the same control-state, but are guarded by different conditions. Thus we eliminate such ambiguities by replacing the nodes in the intermediate control-STG with guarding condition equivalent states (GESs). During the replacement, we keep the edges unchanged, thus we obtain a GA-STG after the replacement.

Finally, we perform register allocation and binding to determine the target register of register assignments. It is important to note that the GA-STG represents the relation between state-transitions guarded by different conditions without the need to evaluate the conditions. We assign a weight of 1 to edges that represent control-state transitions, and 0 to merge-edges that represent predicated execution. This edge weighting scheme, together with the SDC scheduler, ensures that the distance between two GESs in the GA-STG represents the number of cycles for a sequence of transitions between the two GESs. The GA-STG also represents the datapath between register assignments, similar to FSMD [14], to support multi-cycle path analysis.

5.3.3.3 Data Dependency Analysis

We first identify data dependencies between register assignments in the GA-STG. For each data dependency edge, the HLS tool will generate combinational paths between the registers to implement the computation. Thus, we implicitly identify underlying combinational paths between registers by identifying data dependencies between register assignments.

Our SSA-form [12] CDFG explicitly represents data dependencies with use-define chains. For a given register assignment A, we perform a depth-first search on the Directed-Acyclic Graph (DAG), which consists of operations and data dependence edges until we reach operations that are annotated as register assignments. Thus, we collect all data dependencies of A.

5.3.3.4 Available Cycles Calculation

For each edge $(d, u) \in E_A$, which represents a data dependency from register assignment d to u, we calculate the minimal number of cycles available for the state-transition made by d to propagate across combinational paths until it reach the sink assignment u. We denote the number of cycles as $k(d, u)$, which is the number of cycles of the transition sequence from $S_C(d)$ to $S_C(u)$ with the minimal number of transitions, i.e. the shortest path distance from $S_C(d)$ to $S_C(u)$ in the GA-STG [denoted by $\hat{d}(S_C(\text{def}), S_C(\text{use}))$]. Formally: $k(d, u) = \hat{d}(S_C(d), S_C(u))$.

Now, we have $k(d, u)$ computed for every edge $(d, u) \in E_A$ by applying the all-nodes shortest path algorithm to the GA-STG. However, there may be multiple register assignments that assign register $R(u)$ and $R(d)$ in different state transitions (from different GESs), which may imply different shortest-paths, one for each GES. Thus, we must compute the final number of cycles for combinational paths as the minimum number of cycles for all data dependency edges between register $R(u)$ and $R(d)$.

5.3.3.5 Multi-cycle Constraints Generation

During the annotation process, there may be multiple constraints on a combinational node. To ensure correct timing constraints generation, we only generate the constraint with smallest cycle count [9]. Then we write these constraints to a TCL file and pass it together with corresponding RTL to the RTL synthesis tool. These constraints filter out falsely identified critical paths (i.e. paths that have long delay, but do not affect the critical path) during the RTL synthesis, so as to guide synthesis efforts to attain timing closure, and eventually improve the quality of the final RTL synthesis output.

5.3.3.6 Evaluation

We evaluate BL-MCPA on CHStone [15], which contains both control and data-intensive benchmarks. We compare the number of circuit states and the equivalent-state sets (CESs and GESs) in Table 5.7. They represent the problem size of the multi-cycle path analysis at RT-level and behavioral-level, respectively. The table suggests that our BL-MCPA needs to analyze a much smaller STG than prior multi-cycle path analysis making this technique feasible for practical designs.

Next, we compare hardware quality of RTL synthesis with and without BL-MCPA, which primarily affects timing quality and clock period. Using an Altera Stratix IV FPGA, we evaluate each design in both execution time and area. We incrementally apply multi-cycle constraints and multi-cycle chaining to demonstrate the effects of the analysis and chaining independently. The **baseline** flow disables both BL-MCPA and multi-cycle chaining. Then, the **+constraints** flow enables BL-MCPA and generates multi-cycle path constraints to guide synthesis effort,

Table 5.7 Comparison on the problem size of MCPAs

Benchmark	Behavioral-level		RT-/gate-level	
	CESs	GESs	FFs	$|\mathbb{P}|$[a]
dfmul	46	129	1848	2^{1848}
dfdiv	158	273	3484	2^{3484}
gsm	289	630	9088	2^{9088}
mpeg	78	129	2120	2^{2120}
aes	251	301	9575	2^{9575}
mips	251	332	4562	2^{4562}
dfadd	49	201	3905	2^{3905}
dfsin	421	906	11,348	$2^{11,348}$
adpcm	160	191	13,412	$2^{13,412}$
sha	170	220	7318	2^{7318}
blowfish	489	522	21,749	$2^{21,749}$
jpeg	830	1234	23,194	$2^{23,194}$

[a] $\mathbb{P}$ is the set of possible circuit states defined in [31]

but multi-cycle chaining is still disabled. Thus, this flow leverages differences in *available* cycles even though all combinational paths are still limited to single cycle cumulative latency. Finally, the +**constraints**+**mcc** flow enables BL-MCPA and allows unlimited multi-cycle path chaining (given data dependence limitations).

For all three flows, we run all CHStone benchmarks targeting 200 MHz, 300 MHz, and 400 MHz. Results indicate that this analysis allows a reduction in the critical path analysis without actually modifying the circuit; thus BL-MCPA reduces achieved design latency by 12% by improved analysis without actual circuit changes. Combining BL-MCPA with multi-cycle chaining, we achieved 25% latency reduction as well as 29% register usage reduction at 300 MHz. We observed that the effects of multi-cycle constraints and multi-cycle chaining are more significant at higher target f_{max}. For example, the +constraints and +constraints+mcc flows are able to achieve 24% and 26% CP reduction at 400 MHz, respectively. These reductions are larger than the ones that are made at 200 MHz, which are 10% and 16%, respectively. Similarly, the latency and register usage reductions are also larger at 400 MHz than 200 MHz. This is because the RTL timing analysis is more sensitive at a higher f_{max}, hence providing extra multi-cycle information improves the quality of RTL timing analysis more significantly at a higher f_{max} (Table 5.8).

5.3.4 Layout-Driven High-Level Synthesis for FPGAs

HLS-based flows attempt to optimize the area, latency (in clock cycles) and achieveable frequency (f_{max}) of designs in order to meet designers' performance and cost objectives. In particular, f_{max} is limited by critical path delay, which is significantly impacted by low-level circuit implementation details including technology

Table 5.8 Normalized
performance and register
usage at different f_{max}

		200 MHz	300 MHz	400 MHz
CP	+c	0.90	0.88	0.76
	+c+m	0.84	0.83	0.74
Cycles	+c	1.00	1.00	1.00
	+c+m	0.94	0.90	0.86
Latency	+c	0.90	0.88	0.76
	+c+m	0.79	0.75	0.64
Regs	+c	1.09	1.09	1.08
	+c+m	0.84	0.71	0.66

mapping, placement, and routing. However, these details are difficult to estimate at the behavioral level; HLS tools accept constraints including target f_{max} and micro-benchmarking based pre-characterized component latencies [30] to estimate datapath latency and allocate operations to control cycles. However, actual datapath delay is significantly affected by logic synthesis optimization, technology mapping, register packing, placement, and routing. Furthermore, variations in optimizations and interconnect delay, which can be up to 70% of circuit delay [18] can affect different implementations of the same operation. Inaccuracies in behavioral level delay estimates can affect both f_{max} and latency in clock cycles, and thus we must integrate post-synthesis timing information in order to minimize latency in cycles while maximizing f_{max}.

Currently, HLS tools depend on logic synthesis tools to optimize to identify and optimize the design's critical path instead of modifying timing estimates and regenerating RTL to improve implementation quality. Logic synthesis tools attempt to fix timing closure using retiming and/or resynthesis [25]. However, although these tools can optimize some paths, they are limited to transformations that maintain the exact RTL behavior. Thus, these limitations force users to undergo design iterations to improve performance when logic synthesis optimizations are insufficient—challenging in RTL, but even more challenging with HLS because the mapping between the HLL and RTL implementation is obfuscated by HLS transformations.

Thus, we introduce a feed-back path from logic synthesis to HLS[32] that updates timing estimates based on Altera Quartus' [2] post placement and routing (PAR) delay estimates. We iteratively enhance the RTL and back-annotate post-PAR delays to improve both f_{max} and latency in cycles. This work reduces user design iteration and improves ability to meet user design objectives. In order to keep the iteration time feasible, we use Quartus' fast-PAR, and we demonstrate that our benchmarks converge in few iterations with fast overall runtime.

Our LLVM-based HLS framework begins with a control-dataflow graph (CDFG), and performs analysis and optimization both in LLVM-IR and our FPGA-specific IR. In our layout-driven flow, we start with pre-characterized component delays as in other HLS flows, but then iteratively refine the delays with back-annotated post-synthesis and PAR information. In each iteration, HLS optimizes design

Fig. 5.6 Layout-driven HLS flow [32]

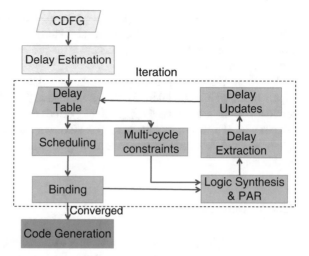

performance considering the f_{max} constraint and annotated delay estimates. This flow reduces design iteration while improving timing closure and still maintaining reasonable synthesis time. For synthesis time specifically, we examine Quartus' synthesis options including Quartus' fast-PAR tool which can reduce iteration runtime by up to $10\times$ [3, pp. 2–35]; all tested benchmarks converge in just three iterations.

We divide our flow into three stages: Initialization, Iteration, and Finalization: Initialization performs an initial synthesis based on pre-characterized delay estimates, then Iteration repeats a cycle of back annotation and regeneration of optimized RTL; Finalization performs a final RTL code generation and full Quartus synthesis. An overview of our flow is shown in Fig. 5.6. During Iteration, there are two different options for how to update the delay information as well as optional iterative constraints generation.

5.3.4.1 Component Pre-characterization

As a one-time task for each target platform, we perform component pre-characterization [30] for 188 micro-benchmarks of elementary operations, MUXs and storage elements of varying bit-widths. Each micro-benchmark is synthesized in Quartus for the target platform, and we measure resource use, critical delay, and power consumption.

5.3.4.2 Initialization Stage

Initialization begins with a simplified baseline synthesis similar to a non-iterative HLS flow. Firstly, we refine the CDFG to build a modified CDFG which contains only non-computation operations, and chains of computational operations are replaced by data dependence edges (data edges). Then we estimate the delay of the data edges in the modified CDFG by accumulating the delay of individual computational operations in the original chains using the pre-characterized, platform-specific component delays as in [30].

Based on these delay estimates, we build a delay table and the corresponding inputs to the SDC scheduler. Using the estimates, the scheduler assigns operations to clock cycles, preserving the f_{max} (clock period) constraint and CDFG dependency constraints. We then generate an initial RTL, we perform an initial Quartus logic synthesis, and fast-PAR to generate timing information and then continue to the Iterate stage.

5.3.4.3   Iteration Stage

After Initialization, Iteration will repeatedly extract the Quartus timing information, back-annotate to update the CDFG, repeat HLS scheduling and binding, generate constraints, and finally repeat synthesis in Quartus. We will now present details for the steps in our iteration stage: RTL synthesis, timing extraction, back annotation, and constraints generation.

RTL Synthesis

We integrate with Altera Quartus tool to perform logic synthesis and PAR on the generated RTL. Though this is a generic step in HLS flows, we explore three different synthesis options to study the impact of synthesis quality on delay estimation.

We use Fast PAR option and run logic synthesis (also with hierarchy flattening and logic optimization) followed by fast placement and routing of the design using the "early timing estimation" option to perform a quick PAR, and timing analysis.

Timing Extraction

Every data edge in our CDFG corresponds to a chain of computational operations; there may be many gate-level paths in the post-synthesis implementation that correspond to a single data edge in the CDFG. However, among the set of gate-level paths, only the critical paths, i.e. the gate-level paths with the largest delay, have an impact on the data edge delay and HLS scheduling.

To find the set of gate-level paths that affect the delay of data edges in our CDFG, we take advantage of a property of HLS: each node in our CDFG is allocated a register to store the input and/or output operands. Thus, the set of paths that affect the critical path are the set of paths from the source register to the sink register, a set of paths easily extracted from post-synthesis timing reports. We generate "path filter" for each data edge in the CDFG, so that Quartus correctly reports the corresponding critical gate-level path delay for the data edge according to the filter. A path filter is described by the source and sink register of the path, and the MUX input for the path. However, we have to apply the *keep* synthesis attribute to the MUX inputs, otherwise we cannot exactly match the MUX input with the filter in the post-synthesis implementation.

Back Annotation

After we have extracted the relevant timing information from the post-synthesis timing reports, we now back-annotate this information to the CDFG to refine timing estimates for the following HLS RTL generation. For the back annotation *update function*, we explore two options that will be compared in detail in evaluation.

Moving Average (MAVG)—This update function is a simple filter that averages the current extracted timing value with the timing value of the previous iteration. Specifically, $\text{est}_{n+1} = (\text{est}_n + \text{delay}_n)/2$, where delay_n is the extracted timing value from Quartus; est_n is the timing estimate of iteration n calculated by the same equation recursively, starting from pre-characterization based delay estimation (est_0).

Average (AVG)—This update function simply performs the average of the delay estimates of all iterations up to the current point. Whereas the moving average has exponentially decreasing importance of older datapoints, this update function linearly decreases importance.

Constraint Generation

In order to improve synthesis quality and quality of back-annotated delays, we also generate multi-cycle constraints for multi-cycle operations. Over-relaxed timing requirements may lead to sub-optimal implementations of gate-level paths; over-estimated delays from sub-optimal gate-level paths may generate low-performance hardware. Instead, we implement a more appropriate technique: for each data edge, we round the estimated delay up to the next integer number of cycles and generate an appropriate constraint. Although our approach is a coarser-grained approach to multi-cycle constraint generation, it gives Quartus sufficient optimization opportunity and allows the iterative process to properly track variation in delay estimates of long paths. The simple multi-cycle constraints are generated based on the estimated critical delay from the CDFG; given an estimated delay est_n, we generate the multi-cycle constraint $\lceil \text{est}_n - 0.5 \times T_{\text{clk}} \rceil$ cycles. The value 0.5 tightens constraints by a half cycle on average in order to encourage optimization.

Table 5.9 Geometric means of performance metrics (normalized to non-iterative, the smaller the better)

		AVG	AVG+C	MAVG	MAVG+C
400	CP	0.84	0.83	0.84	0.84
	Cycles	1.00	0.97	0.99	0.95
	Latency	0.84	0.81	0.83	0.80
450	CP	0.81	0.81	0.80	0.81
	Cycles	0.98	0.96	0.99	0.97
	Latency	0.80	0.78	0.79	0.78
500	CP	0.88	0.87	0.86	0.89
	Cycles	0.99	0.95	0.98	0.95
	Latency	0.87	0.83	0.84	0.84

5.3.4.4   Evaluation

For our experimental evaluation, we use our HLS platform and QuartusII 13.0 synthesis and fast PAR tool targeting a Stratix IV device (EP4SGX70), speedgrade 2 and we use the CHStone [15] set of benchmarks. We report f_{max} and also calculate the latency of the hardware by multiplying the number of cycles and the minimal achievable clock period, i.e. the reciprocal of f_{max}. We set the number of iterations to 10; as we will see, in practice, all benchmarks converge in three iterations.

Using FAST-PAR synthesis, we now investigate the impact of the two update functions, each with or without constraints. We denote the average update function AVG and average with multi-cycle constraints as AVG+C. Similarly, we denote the moving average function options as MAVG and MAVG+C. We synthesize each benchmark from the CHStone set for iterations with each of 4 00MHz, 450 MHz, and 500 MHz f_{max} constraints.

Table 5.9 shows the geometric mean cycles, clock period, and execution latency for each of AVG, AVG+C, MAVG, and MAVG+C for each of the user constraints. In all cases, all four update function combinations provide overall improvement in both f_{max} and total execution latency. Both the AVG and MAVG functions produce similar quality results on average, at the end of third iteration. When we let our flow continue to run for ten iterations, we observed that AVG flow improves execution latency by 2% more than the MAVG flow. However, this small improvement is at the cost of 3× additional runtime. Figure 5.7 compares the convergence property of AVG and MAVG for the AES benchmark as a histogram of the magnitude of estimation updates of ten iterations. The tighter distribution of estimation updates in the AVG function lets the flow further refine the design to get additional improvement.

In Fig. 5.8, we show the final achieved frequency using MAVG of each of the application for each user constraint; we see that on average we achieve over 400 MHz and we commonly reach physical limitations of the device such as the f_{max} of the 36 × 36 MULT. In ADPCM, GSM, and SHA, execution latency improvement is mainly due to reductions in the number of clock cycles while the f_{max} remains unchanged. In other benchmarks such as JPEG, MIPS, and MPEG2, execution

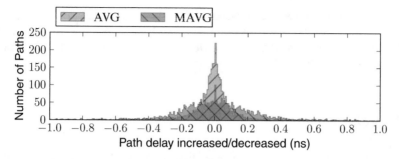

Fig. 5.7 Convergence comparison on AES (450 MHz) [32]

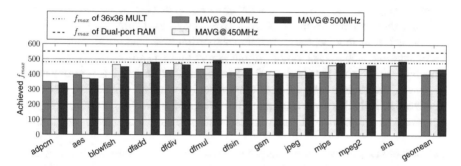

Fig. 5.8 Achieved f_{max} [32]

latency improvement is primarily due to increased f_{max} and unchanged number of clock cycles. In addition, our flow is able to simultaneously optimize both cycles and f_{max} in benchmarks such as AES and BLOWFISH.

Improve f_{max} by Increasing Cycles

Finally, in certain cases, our flow increases the number of cycles in order to meet a target f_{max} similar to the other frequency improvement techniques. Particularly in the floating point benchmarks, DFMUL, DFADD, DFDIV, and DFSIN, we need to increase the latency in clock cycles in order to meet f_{max} through a finer-grained division of operations into clock cycles. In most cases, our improvement to f_{max} offsets the cycles increment to achieve an overall execution latency improvement.

For resource consumption, there is little difference in resource usage between iterative and non-iterative flows or between different options of the iterative flows. We also compare runtime of different flows including the run-time of the initial non-iterative (pre-characterization based) flow. The flows with fast-PAR, but without multi-cycle constraints generation have average runtime of ∼12 min: 3–4× slower than the no PAR integrated flow. With the addition of filtered multi-cycle constraints, there is a minimal overhead over the no-constraints FAST-PAR flows. There was also no significant difference between AVG and MAVG. Thus, we used the flow to reduce the latency of the hardware by up to 22% and increase the f_{max} by upto 24%, compared to a non-iterative flow. Our flow achieved from 65% to 91% of the theoretical f_{max} on the Stratix IV device.

5.4 Conclusion

As design complexity continues to grow, EDA tool complexity grows correspondingly. With this increasing complexity, the quality impact of interactions between different stages in tool flow can no longer be ignored. This chapter identifies a set of challenges and opportunities in cross-layer design methodology (CLDM) flows. We present four representative techniques that improve Quality of Results (QoR) by integrating analyses and optimizations across multiple abstraction levels. Experimental results show that each of these representative techniques is promising to improve QoR through integration of cross-layer information. We believe that such a new design methodology is important for fast and high-quality hardware design for both the edge devices and datacenter accelerators in the era of Internet of Things.

Acknowledgements This work was supported in part by the CFAR Center, one of six centers of STARnet, a Semiconductor Research Corporation program sponsored by MARCO and DARPA, and by A*STAR under the Human-Centered Cyber-Physical Systems (HCCS) grant.

References

1. J.R. Allen, K. Kennedy, C. Porterfield, J. Warren, Conversion of control dependence to data dependence, in *Proceedings of the 10th ACM SIGACT-SIGPLAN Symposium on Principles of Programming Languages* (ACM, Austin, 1983), pp. 177–189
2. Altera Inc., *Quartus II Software*, http://www.altera.com/products/software/
3. Altera Inc., *Quartus II Handbook* (2013). https://www.altera.com/content/dam/altera-www/global/en_US/pdfs/literature/hb/qts/archives/quartusii_handbook_archive_131.pdf
4. P. Ashar, S. Dey, S. Malik, Exploiting multicycle false paths in the performance optimization of sequential logic circuits. IEEE Trans. Comput. Aided Des. **14**(9), 1067–1075 (1995)
5. C. Bastoul, Code generation in the polyhedral model is easier than you think, in *Proceedings of the 13th International Conference on Parallel Architectures and Compilation Techniques, PACT '04* (IEEE Computer Society, Washington, DC, 2004), pp. 7–16
6. U. Bondhugula, A. Hartono, J. Ramanujam, P. Sadayappan, A practical automatic polyhedral parallelizer and locality optimizer, in *PLDI* (2008), pp. 101–113
7. Y. Chen, S.T. Gurumani, Y. Liang, G. Li, D. Guo, K. Rupnow, D. Chen, Fcuda-noc: a scalable and efficient network-on-chip implementation for the cuda-to-fpga flow. IEEE Trans. Very Large Scale Integr. VLSI Syst. **24**(6), 2220–2233 (2016)
8. Y. Chen, T. Nguyen, Y. Chen, S. Gurumani, Y. Liang, K. Rupnow, J. Cong, W.M. Hwu, D. Chen, FCUDA-bus: hierarchical and scalable bus architecture generation on FPGAs with high-level synthesis. IEEE Trans. Comput. Aided Des. Integr. Circuits Syst. **PP**(99), 1–1 (2016)

9. L. Cheng, D. Chen, M.D. Wong, M. Hutton, J. Govig, Timing constraint-driven technology mapping for fpgas considering false paths and multi-clock domains, in *ICCAD* (2007), pp. 370–375
10. F. Computing, Falcon computing solutions, http://www.falcon-computing.com
11. J. Cong, Z. Zhang, An efficient and versatile scheduling algorithm based on sdc formulation, in *DAC* (2006), pp. 433–438
12. R. Cytron, J. Ferrante, B.K. Rosen, M.N. Wegman, F.K. Zadeck, Efficiently computing static single assignment form and the control dependence graph. ACM Trans. Program. Lang. Syst. **13**(4), 451–490 (1991)
13. V. D'silva, D. Kroening, Fixed points for multi-cycle path detection, in *DATE* (2009), pp. 1710–1715
14. D.D. Gajski, N.D. Dutt, A.C.-H. Wu, S.Y.-L. Lin, *High-Level Synthesis: Introduction to Chip and System Design* (Kluwer Academic, Norwell, MA, 1992)
15. Y. Hara, H. Tomiyama, S. Honda, H. Takada, K. Ishii, Chstone: a benchmark program suite for practical c-based high-level synthesis, in *ISCAS* (2008), pp. 1192–1195
16. H. Higuchi, Y. Matsunaga, Enhancing the performance of multi-cycle path analysis in an industrial setting, in *ASP-DAC* (2004), pp. 192–197
17. M. Lam, Software pipelining: an effective scheduling technique for vliw machines, in *Proceedings of the ACM SIGPLAN 1988 Conference on Programming Language Design and Implementation, PLDI '88* (ACM, New York, NY, 1988), pp. 318–328
18. V. Manohararajah, G.R. Chiu, D.P. Singh, S.D. Brown, Predicting interconnect delay for physical synthesis in a fpga cad flow. IEEE Trans. Very Large Scale Integr. **15**(8), 895–903 (2007)
19. K. Nakamura, K. Takagi, S. Kimura, K. Watanabe, Waiting false path analysis of sequential logic circuits for performance optimization, in *ICCAD* (1998), pp. 392–395
20. T. Nguyen, S. Gurumani, K. Rupnow, D. Chen, Fcuda-soc: platform integration for field-programmable soc with the cuda-to-fpga compiler, in *Proceedings of the 2016 ACM/SIGDA International Symposium on Field-Programmable Gate Arrays, FPGA '16* (ACM, New York, NY, 2016), pp. 5–14
21. J.C. Park, M. Schlansker, On predicated execution. Technical report, Technical Report HPL-91-58, HP Labs (1991)
22. Pocc, The polyhedral compiler collection, http://www.cse.ohio-state.edu/~pouchet/software/pocc/
23. Polyhedral benchmark suite v3.1, http://www.cse.ohio-state.edu/~pouchet/software/poly_bench/
24. A. Putnam, A. Caulfield, E. Chung, D. Chiou, K. Constantinides, J. Demme, H. Esmaeilzadeh, J. Fowers, J. Gray, M. Haselman, S. Hauck, S. Heil, A. Hormati, J.-Y. Kim, S. Lanka, E. Peterson, A. Smith, J. Thong, P.Y. Xiao, D. Burger, J. Larus, G.P. Gopal, S. Pope, A reconfigurable fabric for accelerating large-scale datacenter services, in *41st Annual International Symposium on Computer Architecture (ISCA)* (2014)
25. R. Ranjan, V. Singhal, F. Somenzi, R. Brayton, On the optimization power of retiming and resynthesis transformations, in *ICCAD* (1998), pp. 402–407
26. A. Saldanha, H. Harkness, P. McGeer, R. Brayton, A. Sangiovanni-Vincentelli, Performance optimization using exact sensitization, in *DAC* (1994), pp. 425–429
27. The Cloog code generator, http://www.cloog.org
28. R.A. Towle, Control and data dependence for program transformations. PhD thesis, University of Illinois at Urbana-Champaign (1976)
29. T. White, *Hadoop: The Definitive Guide*, 1st edn. (O'Reilly Media, Sebastopol, 2009)
30. Z. Zhang, Y. Fan, W. Jiang, G. Han, C. Yang, J. Cong, Autopilot: a platform-based esl synthesis system, in *High-Level Synthesis: From Algorithm to Digital Circuit* (Springer, Dordrecht, 2008), pp. 99–112
31. H. Zheng, S. Gurumani, L. Yang, D. Chen, K. Rupnow, High-level synthesis with behavioral level multi-cycle path analysis, in *2013 23rd International Conference on Field Programmable Logic and Applications (FPL)* (2013), pp. 1–8

32. H. Zheng, S.T. Gurumani, K. Rupnow, D. Chen, Fast and effective placement and routing directed high-level synthesis for FPGAs, in *Proceedings of the 2014 ACM/SIGDA International Symposium on Field-Programmable Gate Arrays* (ACM, New York, 2014), pp. 1–10
33. H. Zheng, S.T. Gurumani, L. Yang, D. Chen, K. Rupnow, High-level synthesis with behavioral-level multicycle path analysis. IEEE Trans. Comput. Aided Des. Integr. Circuits Syst. **33**(12), 1832–1845 (2014)
34. W. Zuo, P. Li, D. Chen, L.-N. Pouchet, S. Zhong, J. Cong, Improving polyhedral code generation for high-level synthesis, in *2013 International Conference on Hardware/Software Codesign and System Synthesis (CODES+ISSS)* (2013), pp. 1–10
35. W. Zuo, Y. Liang, P. Li, K. Rupnow, D. Chen, J. Cong, Improving high level synthesis optimization opportunity through polyhedral transformations, in *Proceedings of the ACM/SIGDA International Symposium on Field Programmable Gate Arrays, FPGA '13* (ACM, New York, NY, 2013), pp. 9–18

Part II
Approaches and Applications for Data Analytics

Chapter 6
Side Channel Attacks and Their Low Overhead Countermeasures on Residue Number System Multipliers

Gavin Xiaoxu Yao, Marc Stöttinger, Ray C.C. Cheung, and Sorin A. Huss

6.1 Introduction

Modular multiplications are the fundamental arithmetic operations for several popular public-key cryptography algorithms including Rivest-Shamir-Adleman (RSA) [23] and Elliptic Curve Cryptography (ECC) [14, 18]. The efficiency of the modular multiplier receives great attention from researchers and Residue Number System has therefore been introduced to perform this operation [21].

In RNS, a large integer is represented by several residues, whose size is much smaller than the original number. The operation is performed on these small residues independently and, therefore, a parallel architecture is available for RNS modular multiplication [13]. The parallel RNS architecture can thus provide a high speed for different public-key cryptography algorithms. It was first deployed to perform RSA [13, 20, 22], and then introduced for ECC [8, 24]. Recently, with the proposal of the bilinear pairing as a cryptographic primitive [11, 12], the high speed architectures using RNS have also been utilized for pairing-based computation [2, 27].

Besides speed performance, a cryptosystem should also resist various attacks. The Side Channel Attack or Analysis (SCA) [15, 16] uses the information other than the primary channel to reveal the secret, which bypasses the cumbersome cryptanalysis (SCA may also be combined with the cryptanalysis and thus makes the

G.X. Yao
NXP (China) Management Ltd., BM InterContinental Business Center,
100 Yu Tong Road, Shanghai, P.R. China

M. Stöttinger • S.A. Huss
Technische Universität Darmstadt, Darmstadt, Hessen, Germany

R.C.C. Cheung (✉)
City University of Hong Kong, Hong Kong S.A.R., China
e-mail: r.cheung@cityu.edu.hk

© Springer International Publishing AG 2017
A. Chattopadhyay et al. (eds.), *Emerging Technology and Architecture for Big-data Analytics*, DOI 10.1007/978-3-319-54840-1_6

cryptanalysis much more efficient), and has already demonstrated its powerfulness [17]. The side channel leakages include, but are not limited to, time [15], power [16], electro-magnetic emission [5], photon emission [25], and even sound [7]. In order to secure a cryptosystem, numerous countermeasures have been proposed to minimize and/or randomize the side channel leakage, e.g., Wave Dynamic Differential Logic (WDDL) [26], randomized projective coordinate [3], etc.

RNS has also been proposed as the underlying arithmetic structure, namely Leakage Resistant Arithmetic (LRA) [1], to secure a cryptosystem against SCA. This method employs the Mixed-Radix System (MRS) to perform one expensive computation called Base Extension (BE). However, MRS-based BE is not friendly to fully parallel architectures, and most RNS implementations perform BE by means of the parallelizable Chinese Reminder Theorem (CRT). Note that the RNS is attractable for its parallelism and, therefore, the LRA did not gain popularity although it promises better side channel resistance, at least theoretically. Recently, the work in [9] proposed to accommodate the LRA into the CRT-based BE context by precomputation. We will show in the sequel that LRA and CRT-LRA still suffer from several vulnerabilities.

In this work, we first examine the SCA resistance of current cryptosystems using LRA. By exposing the vulnerabilities, we propose appropriate countermeasures. The main contributions of this work are:

- The vulnerabilities are examined for the RNS-based modular multiplier, and the second order attacks are designed for the cryptographic architectures using LRA.
- Several low overhead countermeasures are proposed against various attacks, which can also be embodied independently to satisfy different security requirements.
- Experiments on implemented multipliers are performed in order to demonstrate the efficiency of the proposed methods.

The remaining of this paper is organized as follows: Sect. 6.2 recaps the backgrounds on SCA and the RNS modular multiplier with coverage of recent advances. Section 6.3 reviews the vulnerabilities of the current RNS modular multiplier, and demonstrates attacks accordingly. In Sect. 6.4 we propose countermeasures against these attacks. The implementation results are provided in Sect. 6.5. Further discussions about the RNS-based SCA countermeasures are included in Sect. 6.6 and Sect. 6.7 finally concludes the paper.

6.2 Preliminaries

Due to the nature of this work, we employ many symbols to formalize the presentation. For the ease of reference, we summarize them in Table 6.1, and the their definitions are provided in the appropriate context. The notations which only appear locally are excluded from the table.

Table 6.1 List of symbols

Symbol	Definition
avg	Operator to compute average
HD	Operator for Hamming distance
HW	Operator for Hamming weight
$\oplus$	Operator for bitwise xor
$\lvert x \rvert_y$	Operator for x mod y
a_i	One coprime candidate to form the RNS base
b_i	One modulus in base $\mathfrak{B}$
$b_{i,j}^{-1}$	$\lvert b_i^{-1} \rvert_{b_j}$
c_i	One modulus in base $\mathfrak{C}$
d_i	$d_i = 2^w - b_i, -2^{\lfloor w/2 \rfloor} < d_i < -2^{\lfloor w/2 \rfloor}$
$\mathfrak{B}$	1st RNS base
$\mathfrak{C}$	2nd RNS base
$M_{\mathfrak{B}}$	$\prod \mathfrak{B}$
$M_{\mathfrak{C}}$	$\prod \mathfrak{C}$
B_i	$M_{\mathfrak{B}}/b_i = \prod_{j=0,j\neq i}^{n-1} b_j$
$\tilde{B}_{i,j}$	$\prod_{j=0,j\neq i}^{n-1}(b_j - c_j)$
N	Application modulus
n	Number of moduli in one RNS base
w	Bitlength of one RNS modulus
α	$\mathrm{avg}(\mathrm{HW}(\lvert a_i \rvert_{b_i})), a_i \neq b_i$
β	Noise normalized to the HD/HW model
ξ_i	$\xi_i = \lvert x_i \cdot B_i^{-1} \rvert_{b_i}$
λ	$\lambda = \lfloor \sum_{i=0}^{n-1} \xi_i/b_i \rfloor$

6.2.1 Power Analysis and Related Countermeasures

6.2.1.1 Power Analysis

For simplicity and clarity, we mainly focus on the power analysis in this work. The attack by directly interpreting the power traces of a cryptosystem is known as Simple Power Analysis (SPA). A more sophisticated attack is the Differential Power Analysis (DPA) [17]. The basic principles of DPA are as follows: the power consumption for an unchanged bit value is different from that for a flipped one. More specifically, $0{\rightarrow}0$ and $1{\rightarrow}1$ cost less dynamic power than $0{\rightarrow}1$ and $1{\rightarrow}0$ [17]. For most processors, the registers are written after being pulled-up or -down. Suppose that the power consumption is uniform for the same kind of action (stay or flip) for every bit, then the dynamic consumption when writing the register is proportional to the Hamming weight of the written value. The Hamming weight, denoted as $\mathrm{HW}(\cdot)$, counts how many 1's are presented in the binary value representation.

Subsequently, an attacker can make hypotheses on the intermediate values using the Hamming weight model, whereas the hypothesis space is much smaller than the key space. The hypotheses are compared to the measured power traces using statistics analysis, e.g., maximum likelihood [16] or correlation coefficient [4]. A match indicates (a part of) the secret.

For ASICs or FPGAs, the registers are directly refreshed by the fan-in source without previous pull-up or -down operations. So, the model to be exploited should not be the Hamming weight, but the Hamming distance, which indicates how many bits have been flipped. The Hamming distance is denoted as $\mathrm{HD}(\cdot, \cdot)$, and equals to the Hamming weight of the bitwise XOR between the previous and the current value, i.e.:

$$\mathrm{HD}(x, y) = \mathrm{HW}(x \oplus y) \tag{6.1}$$

Subsequently, similar attack steps can be applied to the Hamming distance model.

6.2.1.2 Power Analysis Countermeasures

According to the mode of action, the SCA countermeasures can be classified into *hiding* and *masking* methods [17]. Techniques which to minimize the leakage or make the leakage undistinguishable belong to hiding class, e.g., one can increase the noise by adding a noise generator, or decease the leakage by using a balanced circuit, such as WDDL [26]. Techniques to cut off the dependency between the intermediate values and the secrecy are denoted as masking. For public-key cryptography, one can use the mathematical equivalent or congruence to randomize the intermediate values. Take ECC for example: the standard projective $(X : Y : Z)$ can be utilized to represent the affine projective (x, y) by defining $x = X/Z, y = Y/Z$. Then one can multiply the same random number to X, Y, Z, i.e., $(rX : rY : rZ)$, to represent the same point and all intermediate values are randomized [3]. The strength of the masking countermeasures is metered by the degree of randomness, i.e., the number of possible variances, and it is usually represented by its bitlength. For instance, the ECC coordinate has $\lceil \log_2 r \rceil$-bit randomness.

Another taxonomy of the countermeasures is related to the hierarchical level the techniques are applied to. A cryptosystem can be structured into several layers as depicted in Fig. 6.1. Thus, the WDDL countermeasure belongs to the logic level. Most of the masking methods for symmetric cryptography and hardware shuffling belong to the arithmetic level. Masking approaches for public-key cryptography usually fall into the algorithm level, etc. Note that the countermeasures on different layers are independent to each other and can be applied concurrently in order to enhance the SCA resistance.

6.2.2 RNS Modular Multiplier

6.2.2.1 Residue Number System

The RNS *base* $\mathfrak{B}$ is defined by n pairwise coprime constants $\{b_0, b_1, \ldots, b_{n-1}\}$. Each b_i is called an RNS *modulus*. Then any integer less than $M_{\mathfrak{B}} = \prod \mathfrak{B} =$

Fig. 6.1 Hierarchical levels of a cryptosystem

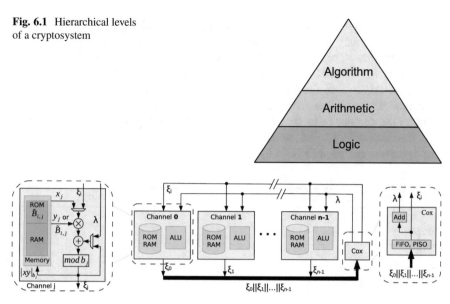

Fig. 6.2 The cox-rower architecture for the CRT-based RNS modular multiplier [13]

$\prod_{i=0}^{n-1} b_i$ can be represented uniquely by:

$$\{X\}_{\mathfrak{B}} = \{x_0, x_1, \ldots, x_{n-1}\} \tag{6.2}$$

where $x_i = |X|_{b_i}$ and $|X|_{b_i}$ represents $X \bmod b_i$. The original value X can be recovered by the Chinese Remainder Theorem (CRT) from RNS:

$$X = \left| \sum |x_i \cdot B_i^{-1}|_{b_i} \cdot B_i \right|_{M_{\mathfrak{B}}} \tag{6.3}$$

where $B_i = M_{\mathfrak{B}}/b_i = \prod_{j=0, j\neq i}^{n-1} b_j$ holds.

RNS enjoys a natural parallelism. Each RNS modulus b_i forms an RNS *channel*. For computations in the ring $\mathbb{Z}/M_{\mathfrak{B}}\mathbb{Z}$, the basic arithmetics are performed independently by the corresponding residues in the channels:

$$
\begin{aligned}
\{|X * Y|_{M_{\mathfrak{B}}}\}_{\mathfrak{B}} &= \{X\}_{\mathfrak{B}} * \{Y\}_{\mathfrak{B}} \\
&= \{|x_0 * y_0|_{b_0}, |x_1 * y_1|_{b_1}, \ldots, |x_{n-1} * y_{n-1}|_{b_{n-1}}\}
\end{aligned} \tag{6.4}
$$

where $* \in \{+, -, \times, \div\}$. Due to the fact that $|x_i * y_i|_{b_i}$ is independent on any other channels, parallel architectures similar to those depicted in Fig. 6.2 are available to accelerate the computation, and both the computation and the storage are assigned to each channel [2, 8, 13, 24].

Note that in RNS arithmetics Eq. (6.4), modulo-$M_\mathfrak{B}$ takes place implicitly. Also, the division is only available when Y has the multiplicative inverse, i.e., Y and $M_\mathfrak{B}$ are coprime. Then the RNS division can be performed by either the channel division or the channel multiplication by the inverse. In the channel reduction, modulo-b_i takes place for almost all arithmetics and, therefore, the efficiency of such modular reductions is significant to the resulting performance. In order to simplify the channel reduction, b_i is usually chosen as a pseudo-Mersenne number with the form $b_i = 2^w - d_i$, where $-2^{\lfloor w/2 \rfloor} < d_i < 2^{\lfloor w/2 \rfloor}$ holds. Then the channel multiplication can be performed by Algorithm 1.

Algorithm 1 Channel multiplication with pseudo-Mersenne modulus

---

Require: $x_i, y_i < 2^w - d_i, 0 < d_i < 2^{\lfloor w/2 \rfloor}$
Ensure: $R := |x_i y_i|_{2^w - d_i}$
 1: $R \leftarrow x_i \cdot y_i$
 2: $R \leftarrow |R|_{2^w} + d_i \cdot \lfloor R/2^w \rfloor$
 3: $R \leftarrow |R|_{2^w} + d_i \cdot \lfloor R/2^w \rfloor$
 4: **if** $R \geq 2^w - d_i$ **then**
 5:     $R \leftarrow |R + d_i|_{2^w}$
 6: **end if**
 7: **return** R

6.2.2.2 RNS Modular Multiplication

RNS cannot be utilized directly when the factorization of the modulus is either impossible or infeasible. Yet, in order to take the advantage of RNS, researchers have proposed to deploy RNS in the Montgomery modular algorithm [22]. This algorithm transforms the representation of X' to $X = |X'R|_N$, where N is the application modulus, and $R > N$. Then the modular multiplication of $|X'Y'|_N$ changes to $|XYR^{-1}|_N = |X'Y'R|_N$ in the transformed domain and it can be performed without the trial division with R a power of 2 [19]. For RNS Montgomery algorithm, R is set to $M_\mathfrak{B}$, so that modulo-R is automatically performed in the RNS representations. However, the operation $(T + Q \cdot N)/M_\mathfrak{B}$ cannot be taken in base $\mathfrak{B}$, and another base $\mathfrak{C} = \{c_0, \ldots, c_{n-1}\}$, where $M_\mathfrak{C} = \prod_{i=0}^{n-1} c_i$ is coprime to $M_\mathfrak{B}$, is introduced to execute the division as well as to enlarge the dynamic range. The procedure is as given in Algorithm 2, and the overhead results in two base extensions.

The Base Extension (BE) is aimed to compute the representation of X in base $\mathfrak{C}$ given its representation in base $\mathfrak{B}$ (and vice versa). This operation is performed by partially recovering X and then computing $|X|_{c_j}$. Using the CRT-based BE, Eq. (6.3) is written as

$$X = \left| \sum_{i=0}^{n-1} \xi_i \cdot B_i \right|_{M_\mathfrak{B}} = \sum_{i=0}^{n-1} \xi_i \cdot B_i - \lambda \cdot M_\mathfrak{B} \tag{6.5}$$

Algorithm 2 RNS Montgomery modular multiplication [22]

Require: RNS bases $\mathfrak{B}$ and $\mathfrak{C}$, $M_\mathfrak{B}, M_\mathfrak{C} > 2N$,
Require: $\gcd(N, M_\mathfrak{B}) = 1, \gcd(M_\mathfrak{B}, M_\mathfrak{C}) = 1$
Require: $\{X\}_{\mathfrak{B}\cup\mathfrak{C}}, \{Y\}_{\mathfrak{B}\cup\mathfrak{C}}, X, Y < 2N$
Require: $\{N'\}_\mathfrak{B} \leftarrow \{|-N^{-1}|_{M_\mathfrak{B}}\}_\mathfrak{B}$,
Require: $\{M'\}_\mathfrak{C} \leftarrow \{|M_\mathfrak{B}^{-1}|_{M_\mathfrak{C}}\}_\mathfrak{C}, \{N\}_\mathfrak{C}$
Ensure: $\{U\}_{\mathfrak{B}\cup\mathfrak{C}} : U \equiv T \times M_\mathfrak{B}^{-1} \bmod N, U < 2N$

1: $\{T\}_{\mathfrak{B}\cup\mathfrak{C}} \leftarrow \{X\}_{\mathfrak{B}\cup\mathfrak{C}} \times \{Y\}_{\mathfrak{B}\cup\mathfrak{C}}$ \\ in bases $\mathfrak{B}\&\mathfrak{C}$
2: $\{Q\}_\mathfrak{B} \leftarrow \{T\}_\mathfrak{B} \times \{N'\}_\mathfrak{B}$ \\ in base $\mathfrak{B}$
3: $\{Q\}_\mathfrak{C} \leftarrow \{Q\}_\mathfrak{B}$ \\ 1st BE
4: $\{S\}_\mathfrak{C} \leftarrow \{T\}_\mathfrak{C} + \{Q\}_\mathfrak{C} \times \{N\}_\mathfrak{C}$ \\ in base $\mathfrak{C}$
5: $\{U\}_\mathfrak{C} \leftarrow \{S\}_\mathfrak{C} \times \{M'\}_\mathfrak{C}$ \\ in base $\mathfrak{C}$
6: $\{U\}_\mathfrak{B} \leftarrow \{U\}_\mathfrak{C}$ \\ 2nd BE
7: **return** $\{U\}_{\mathfrak{B}\cup\mathfrak{C}}$

---

where $\xi_i = |x_i \cdot B_i^{-1}|_{b_i}$ and λ can be computed from

$$\lambda = \left\lfloor \frac{\sum_{i=0}^{n-1} \xi_i \cdot B_i}{M_\mathfrak{B}} \right\rfloor = \left\lfloor \sum_{i=0}^{n-1} \frac{\xi_i \cdot B_i}{M_\mathfrak{B}} \right\rfloor = \left\lfloor \sum_{i=0}^{n-1} \frac{\xi_i}{b_i} \right\rfloor \tag{6.6}$$

In [13], the computation of $\lfloor \sum_{i=0}^{n-1} \xi_i/b_i \rfloor$ is further simplified to the sum of the several significant bits of $\xi_i/2^w$ with enlarged dynamic range and error correcting. Then $|X|_{c_j}$ yields from:

$$|X|_{c_j} = \left| \sum_{i=0}^{n-1} \left| \xi_i \cdot |B_i|_{c_j} \right|_{c_j} - \lambda \cdot |M_\mathfrak{B}|_{c_j} \right|_{c_j} \tag{6.7}$$

The CRT-based BE is now friendly to a fully paralleled design and can reuse the datapath of the channel operations. In [13], the authors proposed an architecture named cox-rower as depicted in Fig. 6.2. The cox is an accumulator for λ computation, and each rower performs one channel in base $\mathfrak{B}$ and one channel in base $\mathfrak{C}$. On such a cox-rower architecture, the ξ computation is distributed to the n rowers and the computation of Eq. (6.7) is also performed by the parallel rowers in a multiply-accumulate fashion.

Recently, the authors of [6] proposed to merge several multiplication steps and to use $\{\xi_U\}_\mathfrak{C}$ instead of $\{U\}_\mathfrak{C}$ as the output. Consequently, the number of channel multiplications in one RNS modular multiplication is reduced from $2n^2 + 7n$ to $2n^2 + 4n$. In [2], the authors substitute $|B_i|_{c_j}$ with $\tilde{B}_{i,j} = \prod_{i=0,i\neq j}^{n-1}(b_i - c_j)$, and the bit-length of $\tilde{B}_{i,j}$ (denoted as v) is now much shorter than that of $|B_i|_{c_j}$ (denoted as w). Thus, n^2 $w \times w$-bit multiplications in one BE are replaced by n^2 $v \times w$-bit operations.

6.2.2.3 Leakage Resistant Arithmetic

Besides the CRT-based BE, another BE method stems from the Mix-Radix number System (MRS). Based on MRS-BE, Bajard et al. propose the Leakage Resistant Arithmetic (LRA) as an SCA countermeasure [1]. Instead of having fixed bases $\mathfrak{B}$ and $\mathfrak{C}$, the base $\mathfrak{B}$ in LRA is randomly formed by n constants out of $2n$ coprimes, whereas the remaining n constants form $\mathfrak{C}$. Therefore, the value of $M_{\mathfrak{B}}$ is randomized resulting in $|X'M_{\mathfrak{B}}|_N$, i.e., the Montgomery representation of X'. This method can be viewed as a masking technique, whereas the mask is $M_{\mathfrak{B}}$. The security introduced by LRA is determined by the number of possible combinations, i.e., $\binom{2n}{n}$. However, the MRS-based BE contains serial steps, which is in contradiction to the parallelism of RNS. Thus, the real effectiveness and efficiency of MRS-LRA is yet unclear.

On the other hand the CRT-based BE is more popular, thanks to the cox-rower architecture. The area overhead is only an accumulator for cox, and all the computations utilize the same parallel datapath. Therefore, the work in [9] proposed to move LRA to the CRT context. The basic idea of the CRT-LRA is the same: select randomly n coprimes out of $2n$ to form the base $\mathfrak{B}$. Next, one computes the parameters $|B_i|_{c_j}$ and $|M_{\mathfrak{B}}|_{c_j}$ with respect to the selected b_i during runtime. We call this step *mask initialization* in this paper. We will show in the next section that both MRS-LRA and CRT-LRA are vulnerable to attacks that are specifically targeting RNS designs.

6.3 Attacks on the RNS Modular Multiplier

6.3.1 Attack Assumptions

Before detailing the attacks, we first establish the hardware architecture and the assumptions on the attacks. We employ the cox-rower architecture as Fig. 6.2, whereas each rower performs one channel in base $\mathfrak{B}$ and one channel in base $\mathfrak{C}$.

The number of possible candidates for RNS moduli is limited to $2^{w/2}$ because one needs pseudo-Mersenne numbers for efficient channel reductions as Algorithm 1, and this number is usually less than 2^32, i.e., within the capability of exhaustive search. In fact, the searching space for suitable moduli are even constrained by the following fact:

- The moduli are pairwisely coprime.
- The value and/or the Hamming weight of d_i are small for the ease of multiplications in Algorithm 1.
- The interval of any two moduli is possibly small to reduce the computation complexity [2, 27].

- The moduli is probably directly provided because the expensive encoding and decoding of RNS may be performed on software instead of packed hardware.

Therefore, without loss of generality, we assume that the attacker has knowledge of the $2n$ coprime moduli candidates: $\{a_0, a_1, \ldots, a_{2n-1}\}$. If he or she can reveal the n moduli out of $2n$ which form base $\mathfrak{B}$, or the other n coprimes forming base $\mathfrak{C}$, then the mask is transparent, and the RNS-based modular multiplier is disarmed. Such attacks can be regarded as second order SCA.

We also assume that the attacker knows the scheduling strategy, that is, the attacker can align the side channel leakage measurements, and knows which operation is being performed at any given point in time. This implies that the countermeasures messing up timing, e.g., dummy clock, are disabled. We turn off other hiding protection methods such as noise generator or WDDL as well, so that the attacks can take place directly on the LRA-protected modular multiplier.

Another assumption is that the RNS modular multiplier uses the same unchanged mask in one attack set. Note that the RNS mask is not likely to change every modular multiplication, otherwise the time overhead increases dramatically. For MRS-LRA, changing mask takes twice the time of one modular multiplication, and CRT-LRA requires even more time because of the mask initialization. Note that the proposed method is still effective if the mask is changing all the time, since the statistic analyses, such as averaging, can filter the changed masks although more power traces are required.

6.3.2 Limited Randomness

Theoretically, the randomness provided by LRA is $\binom{2n}{n}$, which requires that n must be sufficiently large. In [1], the authors suggest $n = 18$ as the minimum, which is equivalent to 33-bit randomness. Note that 2^{33} is still within the ability of an exhaustive search and this implies that a system must change the mask quite often. Again, the overhead of changing masks is not negligible. Furthermore, with the shrunk size of the operand for algorithms such as ECC, the maximum value n is limited. For instance, n is usually not greater than 8 for 256-bit operands [8].

Things get even worse, as [27] suggested to use a relatively small n to reduce the computational complexity. For 256-bit operands, the optimal n is just 4 with $w = 64$, and $\binom{8}{4}$ results to only 70. Thus, one can easily find the mask by exhaustive search. Therefore, increasing the mask space is strongly recommended.

6.3.3 Zero Collision Attack

Zero values exhibit distinguishable behaviors in the side channel leakage, e.g., the power consumption would suddenly drop to a low value for a Hamming weight model. Therefore, if the intermediate value of one channel is zero, we assume that the attacker can observe this zero collision, and then derive the mask or the key according to the input, the operation when zero collision happens, and other information.

For an input X, the first operation is to encode X into the RNS representation, i.e., to compute $|X|_{b_i}$. We regard the time performing this operation as the point of interest. Then one straightforward attack is to feed all the $2n$ coprimes and, if the fed coprime is in a base, the point of interest should exhibit the zero collision behavior. For a_i, if the collision happens during the channels are generating the representation in $\mathfrak{B}$, then $a_i \in \mathfrak{B}$, otherwise $a_i \in \mathfrak{C}$ holds.

One may argue that a power consumption decrease caused by one channel to zero collision is not obvious comparing to noise and other effects. This argument is true if we simply use a_i as the input. Let α be the average Hamming weight of $|a_i - b_j|_{b_j}$ when $a_i \neq b_j$. Then α is usually a small value ($\alpha \ll w$), especially when the technique detailed in [2] is utilized. The overall Hamming weight is expected to be αn when $a_i \notin \mathfrak{B}$ and $\alpha(n-1)$ when $a_i \in \mathfrak{B}$. So, when one out of n values turns to zero causes little effect on the overall Hamming weight change. Let the general noise be β in the Hamming weight model. Then the Signal-to-Noise Ratio (SNR) is given by:

$$\frac{\alpha n - \alpha(n-1)}{\beta} = \frac{\alpha}{\beta} \tag{6.8}$$

Since α features a small value, the effectiveness of this method is not very good. Also, a small value checker can prevent this attack. All the inputs smaller than a certain bit-length can be considered as illegal and this constraint is able to counteract quite a few attacks besides the zero collision attack which also rely on small input or small intermediate results.

Therefore, we propose to use $r_z \times a_i$ as the input, where r_z is a random number satisfying $2^{(n-2)w} < r_z < 2^{(n-1)w}$. If $a_i = b_j$ (such a_i is denoted as $a_{(j)}$), the collision would happen in channel b_j. Otherwise, if the $a_i \neq b_j$, the average Hamming weight of the channel should satisfy:

$$\mathrm{avg}\left(\mathrm{HW}\left(|r_z \times a_i|_{b_j}\right)\right) = w/2, a_i \neq b_j \tag{6.9}$$

So, if $a_i \in \mathfrak{B}$, the overall Hamming weight should be $w(n-1)/2$. Otherwise, it should be $wn/2$. Thus, the SNR expression changes to:

$$\frac{nw/2 - (n-1)w/2}{\beta} = \frac{w}{2\beta} \tag{6.10}$$

Since $w \gg \alpha$, the performance of the zero collision is improved a lot. Figure 6.3 shows the power trace under the zero collision attack with input $r_z \times a_i$, and the single trace can already reveal whether $a_i \in \mathfrak{B}$ or $a_i \in \mathfrak{C}$. The experiment setup is elaborated in Sect. 6.5.

In addition, this zero value is propagating as shown in Fig. 6.3. For a standard Montgomery operation, the next operation is to multiply by $|M_{\mathfrak{B}}^2|_N$. So, after reduction the result is the representation in the Montgomery domain given by:

$$|XM_{\mathfrak{B}}|_N = |X \cdot |M_{\mathfrak{B}}^2|_N \cdot M_{\mathfrak{B}}^{-1}|_N \tag{6.11}$$

The following multiplication exploits this zero as input and after multiplication and channel reduction, this value will still be zero. In the CRT-based method, ξ_i and the

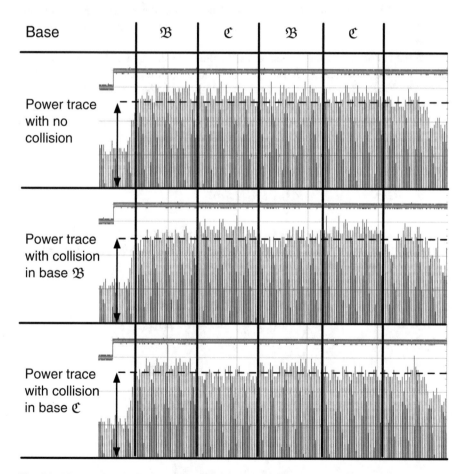

Fig. 6.3 SPA combined with the zero collision attack. When zero collision happens, the power consumption will decrease, and according to the position of such power consumption drop, the attacker can determine in which base the collision happened

$\xi_i \cdot |B_i|_{c_j}$ are also zero. Then, we can overlay the power points at the aforementioned operation and the SNR can be further improved by filtering the noise.

Even when doing so, the SNR value may be still unsatisfying when β is large. One can try to collide k coprimes instead of only one at a time. Then r_z should be $2^{(n-k-1)w} < r_z < 2^{(n-k)w}$, and the input is given by $r_z \cdot \prod_{i=0}^{k-1} a_{(i)}$. The downside of this approach is that the order of attempts increases to $\binom{2n}{k}$ and, therefore, k may not be too large.

The above discussion is referring to the Hamming weight model. For the Hamming distance model, the situation is slightly different: a single zero value does not have an obvious behavior in the power traces since the xor result with the previous or the next value is not zero any more. One naive method is to apply one normal input followed by an all-zero input. Then, the system degrades to the Hamming weight model. Anyhow, this attack can be prevented by the small value checker.

For the Hamming distance model, if the value is unchanged, the distance value is zero. Therefore, we can take $r_z \cdot a_i$ and $r_z' \cdot a_i$ as inputs consecutively, where $2^{(n-2)w} < r_z' < 2^{(n-1)w}$ is also a random number. Then, if $a_i = b_j$, channel b_j will stay on zero for two consecutive cycles, but other channels should feature a Hamming distance of $w/2$ on average. Note that we cannot input $r_z \cdot a_i$ twice, otherwise all the channels will have the zero Hamming distance, no matter whether $a_i \in \mathfrak{B}$ or not.

6.3.4 Attacks on Mask Initialization

In [9], Guillermin proposed the CRT-LRA by computing the required parameters during runtime when the bases $\mathfrak{B}$ and $\mathfrak{C}$ are determined. The following two matrices are precomputed and stored, whereas $a_{i,j}^{-1}$ stands for $|a_i^{-1}|_{a_j}$:

$$
A = \begin{pmatrix}
1 & |a_0|_{a_1} & \cdots & |a_0|_{a_{2n-1}} \\
|a_1|_{a_0} & 1 & \cdots & |a_1|_{a_{2n-1}} \\
\vdots & \vdots & \ddots & \vdots \\
|a_{2n-1}|_{a_0} & |a_{2n-1}|_{a_1} & \cdots & 1
\end{pmatrix}
\tag{6.12}
$$

$$
A^{-1} = \begin{pmatrix}
1 & a_{0,1}^{-1} & \cdots & a_{0,2n-1}^{-1} \\
a_{1,0}^{-1} & 1 & \cdots & a_{1,2n-1}^{-1} \\
\vdots & \vdots & \ddots & \vdots \\
a_{2n-1,0}^{-1} & a_{2n-1,1}^{-1} & \cdots & 1
\end{pmatrix}
\tag{6.13}
$$

Then the parameters needed by the CRT-based RNS modular multiplication can be computed as follows after the selection of $\mathfrak{B}$ and $\mathfrak{C}$, where $a_{(k)}$ stands for the selected a_i serving as the kth coprime in the bases:

$$|M_{\mathfrak{B}}|_{c_j} = \left| \prod_{k=0}^{n-1} b_k \right|_{c_j} = \left| \prod_{k=0}^{n-1} |a_{(k)}|_{a_{(n+j)}} \right|_{a_{(n+j)}} \tag{6.14}$$

$$|M_{\mathfrak{C}}|_{b_j} = \left| \prod_{k=0}^{n-1} c_k \right|_{b_j} = \left| \prod_{k=0}^{n-1} |a_{(n+k)}|_{a_{(j)}} \right|_{a_{(j)}} \tag{6.15}$$

$$|M_{\mathfrak{B}}^{-1}|_{c_j} = \left| \prod_{k=0}^{n-1} b_k^{-1} \right|_{c_j} = \left| \prod_{k=0}^{n-1} a_{(k),(n+j)}^{-1} \right|_{a_{(n+j)}} \tag{6.16}$$

$$|B_i^{-1}|_{b_i} = \left| \prod_{k=0,k\neq i}^{n-1} b_{k,i}^{-1} \right|_{b_i} = \left| \prod_{k=0,k\neq i}^{n-1} a_{(k),(i)}^{-1} \right|_{a_{(i)}} \tag{6.17}$$

$$|C_i^{-1}|_{c_i} = \left| \prod_{k=0,k\neq i}^{n-1} c_{k,i}^{-1} \right|_{c_i} = \left| \prod_{k=0,k\neq i}^{n-1} a_{(n+k),(n+i)}^{-1} \right|_{a_{(n+i)}} \tag{6.18}$$

$$|B_i|_{c_j} = \left| \prod_{k=0,k\neq i}^{n-1} b_k \right|_{c_j} = \left| |M_{\mathfrak{B}}|_{c_j} \cdot a_{(i),(n+j)}^{-1} \right|_{a_{(n+j)}} \tag{6.19}$$

$$|C_i|_{b_j} = \left| \prod_{k=0,k\neq i}^{n-1} c_k \right|_{b_j} = \left| |M_{\mathfrak{C}}|_{b_j} \cdot a_{(n+i),(j)}^{-1} \right|_{a_{(j)}} \tag{6.20}$$

There are in total $5n$ parameters for $|M_{\mathfrak{B}}|_{c_j}$, $|M_{\mathfrak{C}}|_{b_j}$, $|M_{\mathfrak{B}}^{-1}|_{c_j}$, $|B_i^{-1}|_{b_i}$, and $|C_i^{-1}|_{c_i}$, respectively, and for each parameter n multiplications have to be executed. To compute $|B_i|_{c_j}$ or $|C_i|_{b_j}$ only needs one multiplication each, but there are $2n^2$ of them. In total, $7n^2$ multiplications are required for mask initialization. (If techniques in [6] are deployed, $3n$ more multiplications are required.)

This initialization is performed with a fixed scheduling implied by [9], but this step leaks information about the selected a_i. Take $|M_{\mathfrak{B}}|_{c_j}$ for example. One only needs $4n^2$ hypotheses about the $|a_{(0)}|_{a_{(n+j)}}$ to fill in the multiplier. Once $|a_{(0)}|_{a_{(n+j)}}$ is spotted, c_j is derived, and one can recover $|a_{(1)}|_{a_{(n+j)}}$, $|a_{(2)}|_{a_{(n+j)}}$, etc., one by one among the $2n$ candidates. Additionally, all mask initialization computations are linked by the index. For $|a_{(0)}|_{a_{(n+j)}}$, the same index $(n+j)$ is also used to access matrices A and A^{-1} for $|B_i|_{c_j}$, $|C_i^{-1}|_{c_i}$, and $|M_{\mathfrak{B}}^{-1}|_{c_j}$, while the index (0) addresses $|B_i^{-1}|_{b_i}$ and $|M_{\mathfrak{B}}^{-1}|_{c_j}$. Therefore, a single trace of the mask initialization will clearly provide sufficient interesting points to attack, and the leakage will disarm the LRA protection. Hence a second order attack on the max initialization face can be mounted, similar to the second order attacks known on masked AES.

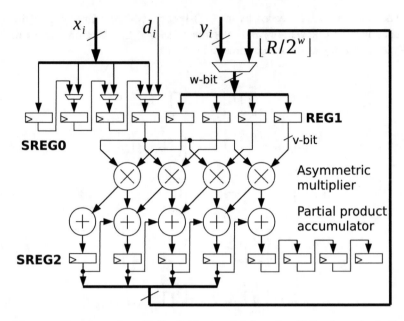

Fig. 6.4 $v \times w$-bit asymmetric multiplier architecture for channel multiplication and reduction

6.3.5 Channel Reduction Leakage

A typical processing cell design is depicted in Fig. 6.4. It utilizes a $v \times w$-bit asymmetric multiplier. One $w \times w$-bit multiplication takes $\lceil w/v \rceil$ cycles on this multiplier architecture using the multiply accumulate method. This asymmetric multiplication can also perform the channel reduction in Algorithm 1, where the multiplication $d_i \times \lfloor R/2^w \rfloor$ is carried out as in Fig. 6.4.

When addressing the architecture in Fig. 6.4, one can target REG1 for leakage extraction. In step 2 of Algorithm 1, REG1 stores $\lfloor x_i y_i / 2^w \rfloor$ and in step 3 it stores one of the following values:

$$\lfloor R/2^w \rfloor = \begin{cases} \lfloor \lfloor x_i y_i / 2^w \rfloor \times d_i / 2^w \rfloor, & t < 2^w \\ \lfloor \lfloor x_i y_i / 2^w \rfloor \times d_i / 2^w \rfloor + 1, & \text{otherwise} \end{cases} \qquad (6.21)$$

where $t = |x_i y_i|_{2^w} + |\lfloor x_i y_i / 2^w \rfloor \times d_i|_{2^w}$. Therefore, the leakage at REG1 is $\text{HD}(\lfloor x_i y_i / 2^w \rfloor, \lfloor R/2^w \rfloor)$. There are 2^w possible $\lfloor x_i y_i / 2^w \rfloor$ and $2n$ possible d_i values, so one may exploit $n2^{w+1}$ hypotheses to mount a DPA attack.

Furthermore, d_i is usually a small integer with a low Hamming weight so that its representation may be tailored efficiently as detailed below. Such a small d_i can considerably simplify the DPA attack. In steps 3 of Algorithm 1, $\lfloor R/2^w \rfloor$ is only of $\lceil \log_2(d_i + 1) \rceil$ bits length. Therefore, all the $(w - \lceil \log_2(d_i + 1) \rceil)$ leading bits are zeros. Note that before $\lfloor R/2^w \rfloor$ is loaded, $\lfloor x_i y_i / 2^w \rfloor$ is stored in REG1 and the Hamming weight of the leading $(w - \lceil \log_2(d_i + 1) \rceil)$ bits are exposed.

Another way to utilize the property of the small d_i is to reveal the Hamming weight of the values stored in SREG0. Small d_i implies that its binary representation consists mostly of 0's, and the values before and after d_i loaded will expose their Hamming weight at SREG0. Therefore, the time points before and after loading d_i are of interest, and we may input bit value of either all 1's or 0's to reveal the Hamming weight of d_i.

6.4 Countermeasures

In this section, we propose some dedicated countermeasures to protect the RNS modular multiplier from the outlined attacks. The functions of these methods are summarized in Table 6.2.

6.4.1 Enlarged Coprime Pool

Instead of only selecting n coprimes out of $2n$, we extend the selecting pool to m, where $m \geq 2n$ holds. Therefore, the equivalent randomness is $\binom{m}{n}$ and one can just select a proper m value to meet the required randomness degree. The overhead is an enlarged storage for A and A^{-1}.

However, the overhead might be undesirable when n is small. For $n = 4$ and $w = 64$ one requires to achieve a 33-bit randomness with $m = 768$, which means that one needs $2 \times 768^2 \times 64$ bits = 9 MB to store A and A^{-1}. It is quite a large overhead considering the operand is of only 256 bits, i.e., 32 bytes size. Additionally, to select 768 coprimes near 2^{64} is not a trivial work.

To avoid to oversize m, one way is to employ a larger n value. For $n = 8, w = 32$, which can also handle 256-bit operands, m only needs to be 72 for a 33-bit

Table 6.2 Countermeasures for RNS modular multiplier and their functions against various attacks

Counter-measures	Attacks			
	Limited random	Zero collision	Mask init.	Channel reduction
Enlarged pool	✓			
Plus-N	✓	✓		✓
Init. shuff.			✓	
Random pad.		✓		✓
Channel shuff.	✓	✓	✓	✓

Symbol ✓ means that the countermeasure is effectively against the attack

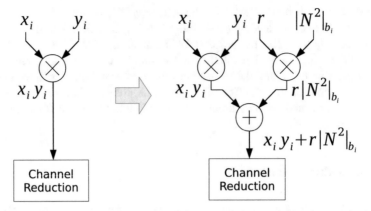

Fig. 6.5 Result of attack using the channel reduction leakage

randomness and the storage of A and A^{-1} decreases to just $2 \times 72^2 \times 32$ bits = 40.5 KB. We propose additional methods to increase the randomness in the following subsections.

6.4.2 Plus-N Randomness

Note that $x \equiv x + rN \pmod{N}$, where r is a random number, and one can use $x + rN$ to represent x in order to randomize the intermediate results. In the Montgomery domain, this is equivalent to add a multiple of N^2, i.e., $XY + rN^2$, after the product computation, and by Montgomery reduction, the result is given by $(XY+rN^2)N^{-1} = XYN^{-1} + rN$. In a rower the plus-$N$ is performed after the multiplication but before the channel reduction. The result for the ith channel is randomized from x_iy_i to $x_iy_i + r|N^2|_{b_i}$. The resulting data flow is as depicted in Fig. 6.5. The multiplication $r|N^2|_{b_i}$ can be performed either by the same multiplier for x_iy_i or by a dedicated multiplier to accelerate computation.

The randomness is determined from the bitlength of r. As in a channel, $r < 2^w$ and the equivalent randomness is up to w bits. Note that a small n implies a large w and this method compensates the limitation of the enlarged coprime pool quite well. For $n = 4$, one can add a 64-bit randomness for example. The overhead is that the product x_iy_i changes to $x_iy_i + r|N^2|_{b_i}$, which has one more bit, so the multiplier width is increased by 1-bit. As the overall dynamic range is enhanced to $(r + 3)N^2$, we need to add an additional channel or to enlarge w.

Besides providing a superior randomness, the plus-N countermeasure is also resistant to the zero-collision attack and to attacks on channel reduction. For zero collision attack, the intended intermediate value is $r_z b_i$, so that after the channel reduction, the result is zero. With plus-N redundancy, the value changes to $r_z b_i + r|N^2|_{b_i}$ and the zeros after the channel reduction disappear. For the channel reduction attacks, one cannot control the value $x_iy_i + r|N^2|_{b_i}$ by feeding certain input values and the size of the hypothesis space is maximized to 2^w.

6.4.3 Initialization Shuffling

In order to obstruct the attacks on the mask initialization, we propose to shuffle the computation. There are $2n^2 + 5n$ parameters for Eqs. (6.14)–(6.20). These computations are in general independent on each other, so that one can randomize the order of execution. The shuffle method can be applied at different levels.

1. All the rowers are synchronized to compute one set of the parameters simultaneously whereas the order for different sets is randomized. There are 7 sets of parameters and $|B_i|_{c_j}$ is calculated after the computation of $|M_\mathfrak{B}|_{c_j}$, as well as for $|C_i|_{b_j}$ and $|M_\mathfrak{C}|_{b_j}$. Therefore, there are in total $\binom{7}{2} \cdot \binom{5}{2} \cdot 3! = 1260$ possible permutations.
2. Consider $|B_i|_{c_j}$ or $|C_i|_{b_j}$ with the same j as one parameter. Then one rower has 7 parameters to compute and there are n rowers available. If a rower computes one parameter randomly at a time and the rowers are independent, there will be 1260^n permutations.
3. Each parameter needs n consecutive multiplications, and the order of the multiplicand can be randomized, too. There are $n!$ possible permutations.
4. Point 3 can be combined with Points 1 and 2 and the number of overall permutation results from a multiplication of the individual permutations.
5. One does not necessarily need to finish the computation of one parameter before starting the computation of the next parameter. One may suspend one computation, compute another parameter, and resume the previous one. For one rower, there are $7n$ multiplications, and these multiplications can be shuffled with the support of a RAM. In the case that the rowers are synchronized, the number of possible permutations is

$$\binom{7n}{2n} \cdot \binom{5n}{2n} \cdot (n!)^4 \cdot (3n)!$$

Otherwise,

$$\left(\binom{7n}{2n} \cdot \binom{5n}{2n} \cdot (n!)^4 \cdot (3n)! \right)^n$$

The variants of initialization shuffling in the above are listed in ascending order according to their overheads. For the synchronized rower, one control signal can handle all the rowers and this is easy to implement. For the asynchronized rowers, each rower should have an independent control logic. Note that the randomness provided by the initialization shuffling should be of the same order as that from other countermeasures: on one hand, the adversary cannot gain advantage by attacking mask initialization; on the other hand, the overhead from initialization shuffling should be minimal as long as the security specification is achieved.

6.4.4 Random Padding

One may argue to choose d_i to be around $2^{\lfloor w/2 \rfloor}$, so that $\lfloor R/2^w \rfloor$ in step 3 of Algorithm 1 is of approximately w bits. Thus, the Hamming weight of $\lfloor x_i y_i/2^w \rfloor$ is not directly leaked at REG1 in Fig. 6.4. Indeed, the leakage is not the Hamming weight of the values at SREG0 before and after d_i loading, but it is still linear to these values since d_i is fixed. Also, as shown in Eq. (6.21), we can still perform the DPA, although the computation of the hypothesis $\mathrm{HD}(\lfloor x_i y_i/2^w \rfloor, \lfloor R/2^w \rfloor)$ is more complex. Furthermore, since $d_i \simeq 2^{\lfloor w/2 \rfloor}$, the size of d_i is usually greater than v bits and the multiplication by d_i takes more than one cycle on the architecture depicted in Fig. 6.4.

We still suggest to use a small d_i, but the higher zero bits should be padded with random values, so that the leakage at SREG0 is minimized. In step 3 of Algorithm 1, the higher bits of input $\lfloor R/2^w \rfloor$ are also padded, so that the leakage at REG1 changes to $\mathrm{HD}(\lfloor x_i y_i/2^w \rfloor, r || \lfloor R/2^w \rfloor)$, where r is a random number of bit-length $(w - \lceil \log_2(d_i + 1) \rceil)$, and $||$ denotes the concatenation operator. Therefore, the attack targeting channel reduction is annihilated. The random padding can also obstruct the zero collision attack, since the higher bits feature not a zero but a random value.

6.4.5 Channel Task Shuffling

In most previous work, the independency of operations within RNS is exploited by a parallel architecture, i.e., in spacial domain. Instead of assigning one rower to perform one channel in $\mathfrak{B}$ and one channel in $\mathfrak{C}$, we propose to randomize the computation between the rowers, i.e., shuffle in the spacial domain. The overhead, however, results in more storage for the parameters. Originally, one rower needs to store $(2n + 5)w$ bits for parameters and several intermediate values. With spacial shuffling, one rower needs to store all the $(2n^2 + 5n)w$ parameters. This method is intended to cope with the attacks using spacial information, such as EM attack. However, for the power analysis, its advantage is limited since the overall power consumption remains unchanged. Besides the parallelism in the spacial domain, we can also aim at the independency in time domain. One can employ k rowers, and each rower is responsible for $2n/k$ channels. Then these n/k channel operations are independent to each other and thus the order to perform these computation may be randomized. Although the channel shuffling works for all the outlined attacks, its randomness is limited. For the time shuffling, the randomness depends on $2n/k$, which is usually a small number so that the computation can be accelerated by a parallel architecture. Therefore, we propose to combine channel shuffling with other countermeasures in order to enhance the resistivity against SCA.

6.5   Implementation

We exploited SASEBO-GII platform, which embeds a Virtex-5 FPGA, to test the efficiency of the proposed methods. Synthesis was performed on the top of Xilinx ISE 12.4 tool set. The countermeasures were implemented on a design for a 256-bit RNS multiplier with $n = 4, w = 67$. The logic utilization and the timing for the different countermeasures are listed in Tables 6.3 and 6.4.

The basic deign is without countermeasures enabled. It consists four Processing Elements (PEs), a microcoded controller and a data buffer, which utilizes 2498 slices, 16 DSPs and 11 18 Kb Block RAMs (BRAMs) in total. For each PE, 4 DSPs forms a 69×18 signed 2's complement multiplier, and 2 BRAMs serve as the simple dual port memory for the local storage. One BRAM is used for instruction ROM, and two are used as the interface buffer to communicate with the host. One 68 $\times$ 68-bit multiplication takes four cycles, and the channel reduction takes two. One BE takes five cycles. The coprocessor can achieve the frequency of over 165 MHz after placement and routing, and therefore, one modular multiplication takes 238.1 µs. Although it is not designed for speed or logic utilization performance, this work is comparable to the state-of-the-art 256-bit modular multiplier design.

Table 6.3 Utilization of the logic fabric for the countermeasures for RNS modular multiplier on XC5VLX50 FPGA

Designs	Logic utilization		
	#slices	#BRAM	# DSP
Basic design	2498 (34.7%)	11 (11.5%)	16 (33.3%)
CRT-LRA [9]	2504 (34.8%)	12 (12.5%)	16 (33.3%)
Enlarged pool	2502 (34.8%)	14 (14.6%)	16 (33.3%)
Plus-N	2498 (34.7%)	12 (12.5%)	16 (33.3%)
Init. shuff.	2510 (34.9%)	12 (12.5%)	16 (33.3%)
Random pad	2538 (35.2%)	12 (12.5%)	16 (33.3%)
Channel shuff.	2508 (34.9%)	12 (12.5%)	16 (33.3%)
All enabled	2543 (34.9%)	15 (15.6%)	16 (33.3%)

Table 6.4 Timing of mask initialization and one modular multiplication for the countermeasures for RNS modular multiplier

Designs	Frequency (MHz)	# Cycles		Time (µs)	
		Mask init.	Mod. mul.	Mask init.	Mod. mul.
Basic design	168	–	40	–	238.1
CRT-LRA [9]	165	210	40	1273	242.4
Enlarged pool	165	858	40	5200	242.4
Plus-N	168	–	47	–	279.8
Init. shuffling	165	210	40	1273	242.4
Random padding	165	–	40	–	242.4
Channel shuffling	165	–	40	–	242.4
All enabled	163	858	47	5264	288.3

Table 6.5 Performance comparison of hardware implementations of one 256-bit elliptic curve scalar multiplication

	ECC p	Arithmetic	n	w	Platform	Area		Freq.	Cycle	Delay
						# Logic	# DSPs	(MHz)	($\times 10^3$)	(ms)
Ours	Any p-256	Novel RNS	4	67	Virtex-5	2498 Slices	16	168	240.1	1.43
[10]	NIST p-256	Pseudo-Mersenne	–	–	Virtex-4	1715 Slices	32	490	303.4	0.620
[8]	Any p-256	RNS	8	33	Stratix II	9177 ALM	96[a]	157.2	106	0.68
[24]	Any p-256	RNS	15	35	Virtex	32,716 LUTs	–	39.7	157	3.95

[a] Altera gives the DSP occupation in number of 9-bit multipliers, 96 9-bit multipliers equivalents to 24 18-bit multipliers

The comparison of the ECC implementation is as shown in Table 6.5. Note that this design folds one 68×68 multiplications into four cycles, and we intentionally reduce the number of pipeline stages to suppress the noise for the SCA sake. A fully unrolled and pipelined design can achieve much less cycle count and faster frequency, and hence, have better timing performance [27].

With CRT-LRA enabled as [9] , the overhead is the mask initialization and more memory for the storage of matrix A and A^{-1}. For mask initialization, it consists of 140 modular multiplications and costs 210 cycles on four PEs. The storage overhead is 96 more entries, which is equivalent to 6.5 Kbit.

The enlarged pool design extends the number of coprimes from 8 to 16, so the possible number of mask changes from $\binom{8}{4} = 70$ to $\binom{16}{4} = 1820$. The storage overhead is 544 new entries, which is equivalent to 36.1 Kbit. The mask initialization also costs more cycles due to the additional channels [1]. It consists of 572 modular multiplications, and by distributing them to the four PEs, it costs 858 cycles.

The $r \times N$ or $r \times N^2$ computation in the plus-N technique uses the same 18×68-bit multiplier, so that there is no area overhead but it costs one cycle. The size of r for $r \times N^2$ is up to 17 bits, so that the computation of $r \times N^2$ only takes one cycle. The r for $r \times N$ cannot be beyond 10-bit, otherwise the dynamic range provided by $w = 67$, $n = 4$ is not sufficient. For a larger r, one needs to add more channels or enlarge w to extend the dynamic range. Also, due to one more cycle for plus-N operation, one modular multiplication costs seven more cycles.

The initial shuffling, channel shuffling basically are overhead free. The design is achieved by adding additional logic to the instruction pointer to shuffle the execution. The output is several cycles delayed to synchronize the signals, but the throughput maintains the same.

The overhead for random padding is that multiplexers is added to before SREG0 and REG1 and after SREG2, so that random values are padded to the higher bits when they are zeros, and discarded to yield the correct results.

6.6 Discussion

Although the outlined attacks and proposed countermeasures were implemented on the CRT-LRA design, they also work for MRS-LRA implementation. The only difference between the CRT-LRA and MRS-LRA is the load of mask initialization. For MRS-LRA, it only needs to compute $|M_{\mathfrak{B}}^{-1}|_{c_j}$ as given in Eq. (6.16), and one can still attack this step and use shuffling to counter the attack. Nevertheless, the MRS-based RNS modular multiplier is not suitable for parallel operation, while the proposed methods address the much faster CRT-based implementations.

As stated in Sect. 6.2, the advocated method is located at arithmetic level, and it can well cooperate with additional countermeasures implemented at both logic and algorithm level.

6.7 Conclusion

In this work, we first examine the effectiveness of the existing RNS-based SCA countermeasures and expose their vulnerabilities. Then, we elaborate on dedicated countermeasures and demonstrate their suitability and efficiency. The proposed methods do not compromise the parallelism and the speed of the RNS modular multiplier. At the same time, the logic overhead is negligible. They can work independently or in a combined manner, so that the randomness level is fully customizable according to the working specification and/or the budget. Furthermore, the proposed countermeasures are located at arithmetic level and are thus compatible to additional countermeasures at the algorithm and/or the logic level. Extensive experiments were detailed and the results confirm the advantages of the proposed methods.

References

1. J. Bajard, L. Imbert, P. Liardet, Y. Teglia, Leak resistant arithmetic, in *Cryptographic Hardware and Embedded Systems - CHES 2004*. Lecture Notes in Computer Science, vol. 3156 (Springer, Heidelberg, 2004), pp. 62–75
2. R. Cheung, S. Duquesne, J. Fan, N. Guillermin, I. Verbauwhede, G. Yao, FPGA implementation of pairings using residue number system and lazy reduction, in *Cryptographic Hardware and Embedded Systems – CHES 2011* Lecture Notes in Computer Science, vol. 6917 (Springer, Heidelberg, 2011), pp. 421–441
3. J. Coron, Resistance against differential power analysis for elliptic curve cryptosystems, in *Cryptographic Hardware and Embedded Systems – CHES 2011*. Lecture Notes in Computer Science, vol. 1717 (Springer, Heidelberg, 1999), pp. 292–302
4. J. Coron, P. Kocher, D. Naccache, Statistics and secret leakage, in *Financial Cryptography*. Lecture Notes in Computer Science, vol. 1962 (Springer, Heidelberg, 2001), pp. 157–173
5. P. Fouque, G. Leurent, D. Réal, F. Valette, Practical electromagnetic template attack on HMAC, in *Cryptographic Hardware and Embedded Systems – CHES 2009*. Lecture Notes in Computer Science, vol. 5747 (Springer, Heidelberg, 2009), pp. 66–80

6. F. Gandino, F. Lamberti, P. Montuschi, J. Bajard, A general approach for improving RNS Montgomery exponentiation using pre-processing, in *20th IEEE Symposium on Computer Arithmetic – ARITH 2011* (2011), pp. 195–204
7. D. Genkin, A. Shamir, E. Tromer, RSA key extraction via low-bandwidth acoustic cryptanalysis (2013), http://www.tau.ac.il/~tromer/papers/acoustic-20131218.pdf
8. N. Guillermin, A high speed coprocessor for elliptic curve scalar multiplications over $\mathbb{F}_p$, in *Cryptographic Hardware and Embedded Systems – CHES 2010*. Lecture Notes in Computer Science, vol. 6225 (Springer, Heidelberg, 2010), pp. 48–64
9. N. Guillermin, A coprocessor for secure and high speed modular arithmetic. Cryptology ePrint Archive, Report 2011/354 (2011), http://eprint.iacr.org/
10. T. Güneysu, Utilizing hard cores of modern FPGA devices for high-performance cryptography. J. Cryptogr. Eng. **1**, 37–55 (2011)
11. A. Joux, A one round protocol for tripartite Diffie-Hellman, in *Algorithmic Number Theory*. Lecture Notes in Computer Science, vol. 1838 (Springer, Heidelberg, 2000), pp. 385–393
12. A. Joux, A one round protocol for tripartite Diffie-Hellman. J. Cryptol. **17**(4), 263–276 (2004)
13. S. Kawamura, M. Koike, F. Sano, A. Shimbo, Cox-rower architecture for fast parallel Montgomery multiplication, in *Advances in Cryptology – EUROCRYPT 2000*. Lecture Notes in Computer Science, vol. 1807 (Springer, Heidelberg, 2000), pp. 523–538
14. N. Koblitz, Elliptic curve cryptosystem. Math. Comput. **48**, 203–209 (1987)
15. P. Kocher, Timing attacks on implementations of Diffie-Hellman, RSA, DSS, and other systems, in *Advances in Cryptology – CRYPTO 1996*. Lecture Notes in Computer Science, vol. 1109 (Springer, Heidelberg, 1996), pp. 104–113
16. P. Kocher, J. Jaffe, B. Jun, Differential power analysis, in *Advances in Cryptology – CRYPTO 1999*. Lecture Notes in Computer Science, vol. 1666 (Springer, Heidelberg, 1999), pp. 388–397
17. S. Mangard, E. Oswald, T. Popp, Power Analysis Attacks: Revealing the Secrets of Smart Cards, vol. 31 (Springer, New York, 2007)
18. V. Miller, Uses of elliptic curves in cryptography, in *Advances in Cryptology – CRYPTO 1985*. Lecture Notes in Computer Science, vol. 218 (Springer, Heidelberg, 1985), pp. 417–426
19. P. Montgomery, Modular multiplication without trial division. Math. Comput. **44**(170), 519–521 (1985)
20. H. Nozaki, M. Motoyama, A. Shimbo, S. Kawamura, Implementation of RSA algorithm based on RNS Montgomery multiplication, in *Cryptographic Hardware and Embedded Systems – CHES 2001*. Lecture Notes in Computer Science, vol. 2162 (Springer, Heidelberg, 2001), pp. 364–376
21. K. Posch, R. Posch, Base extension using a convolution sum in residue number systems. Computing **50**, 93–104 (1993)
22. K. Posch, R. Posch, Modulo reduction in residue number systems. IEEE Trans. Parallel Distrib. Syst. **6**(5), 449–454 (1995)
23. R. Rivest, A. Shamir, L. Adleman, A method for obtaining digital signatures and public-key cryptosystems. Commun. ACM **21**(2), 120–126 (1978)
24. D. Schinianakis, A. Fournaris, H. Michail, A. Kakarountas, T. Stouraitis, An RNS implementation of an elliptic curve point multiplier. IEEE Trans Circuits Syst. I Regul. Pap. **56**(6), 1202–1213 (2009)
25. A. Schlösser, D. Nedospasov, J. Krämer, S. Orlic, J. Seifert, Simple photonic emission analysis of AES, in *Cryptographic Hardware and Embedded Systems – CHES 2012*. Lecture Notes in Computer Science, vol. 7428 (Springer, Heidelberg, 2012), pp. 41–57
26. K. Tiri, I. Verbauwhede, A logic level design methodology for a secure DPA resistant ASIC or FPGA implementation, in *Proceedings of Design, Automation and Test in Europe Conference and Exhibition – DATE 2004*, vol. 1 (2004), pp. 246–251
27. G. Yao, J. Fan, R. Cheung, I. Verbauwhede, Faster pairing coprocessor architecture, in *Pairing-Based Cryptography – Pairing 2012*. Lecture Notes in Computer Science, vol. 7708 (Springer, Heidelberg, 2013), pp. 160–176

Chapter 7
Ultra-Low-Power Biomedical Circuit Design and Optimization: Catching the Don't Cares

Xin Li, Ronald D. (Shawn) Blanton, Pulkit Grover, and Donald E. Thomas

7.1 Introduction

As reported by United Nations and US Census Bureau, the US population has enormously grown during the past several decades, climbing from 209 million in 1970 to 310 million in 2010. Most importantly, the percentage of senior citizens (more than 65-year old) is expected to reach 21.28% in 2050. With the rapid booming of senior citizen population, the expenditure of healthcare continuously increases at a rate of 5–10% per year in the USA. Such a trend is also observed worldwide over a large number of other countries.

To reduce healthcare cost while simultaneously delivering high-quality health services, developing new portable and/or implantable biomedical devices is of great importance. Billions of US dollars could be saved by reforming today's healthcare infrastructure with these biomedical devices for various medical applications [8, 24]:

- *Health monitoring*: Health condition should be reliably monitored for each person to predict and diagnose chronic diseases at the very early stage. For instance, ECG signals can be continuously measured and automatically classified by a portable biomedical device to diagnose arrhythmia [3, 19, 29].
- *Clinical treatment*: Clinical therapy should be reliably delivered for each patient for both preventative care and disease treatment. Taking neuroprosthesis as an example, brain signals are sensed and decoded by an implantable device to control the prosthesis of a patient with neurological disorder [5, 11, 13, 21, 22, 26].

X. Li (✉) • R.D. (Shawn) Blanton • P. Grover • D.E. Thomas
Electrical and Computer Engineering Department, Carnegie Mellon University,
Pittsburgh, PA 15213, USA
e-mail: xinli@cmu.edu

© Springer International Publishing AG 2017
A. Chattopadhyay et al. (eds.), *Emerging Technology and Architecture for Big-data Analytics*, DOI 10.1007/978-3-319-54840-1_7

Towards these goals, miniaturized portable and/or implantable biomedical circuits must be designed and deployed to reliably sense, process, and transmit a large amount of physiological data with extremely low-power consumption. These circuits must carry several important "features":

- *High accuracy*: A biomedical device must accurately generate the desired output, such as diagnosis result for arrhythmia [3, 19, 29], and movement direction and velocity for neuroprosthesis [5, 11, 13, 21, 22, 26], that is not contaminated by artifacts, errors, and noises originated from human body and/or external environment [6, 17, 25].
- *Small latency*: The response of a biomedical device must be sufficiently fast for a number of real-time applications such as vital sign monitoring [7, 31] and deep brain stimulation [9, 18]. In these cases, physiological data must be locally processed within the biomedical device to ensure fast response time, especially when a reliable wired or wireless communication channel is not available to transmit the data to an external device (e.g., smart phone, cloud server, etc.) for remote processing. Even in the cases where data transmission is possible such as neuroprosthesis control [5, 11, 13, 21, 22, 26], the raw data must be locally processed and compressed before transmission in order to minimize the communication energy.
- *Low power*: To facilitate a portable and/or implantable device to continuously operate over a long time without recharging the battery, its power consumption must be minimized. Especially for the implantable applications where power consumption is highly constrained (e.g., less than 100 μW), it is necessary to design an application-specific circuit, instead of relying on general-purpose microprocessors, to meet the tight power budget [12, 15, 16, 27, 28, 30, 32].
- *Flexible reconfigurability*: Reconfigurability is needed to customize a biomedical device for different patients and/or different usage scenarios. For instance, the movement decoder of neuroprosthesis should be retrained every day to accommodate the time-varying characteristics of neural sources, recording electrodes and environmental conditions [26]. It, in turn, requires a reconfigurable circuit implementation that can be customized every day.

The aforementioned features, however, are considered to be mutually exclusive today. Taking neuroprosthesis as an example, executing a sophisticated movement decoding algorithm is overly power hungry for portable and/or implantable applications. For this reason, renovating the healthcare infrastructure with portable and/or implantable biomedical devices requires an even higher standard of performance than what can be offered by today's circuit technology.

In this chapter, we discuss a radically new design framework to seamlessly integrate data processing algorithms and their customized circuit implementations for co-optimization. The proposed framework could bring about numerous opportunities to substantially improve the performance of biomedical circuits. From this point of view, it offers a fundamental infrastructure that enables next-generation biomedical circuit design and optimization for many emerging applications.

7.2 How Can We Beat the State of the Art?

In this chapter, we attempt to address the following fundamental question: How can we further push the limit of accuracy, latency, power, and reconfigurability to meet the challenging performance required for portable and/or implantable biomedical applications? Historically, algorithm and circuit designs have been considered as two separate steps. Namely, a biomedical data processing algorithm is first developed and validated by its software implementation (e.g., MATLAB, C++, etc.). Next, a circuit is designed to implement the given algorithm. Such a two-step strategy suffers from several major limitations that motivate us to fundamentally rethink the conventional wisdom in this area.

First, since the biomedical data processing algorithms are particularly developed and tuned for their software implementations, they are *not* fully optimized for circuit implementations. Ideally, data processing algorithms should be customized to mitigate the non-idealities induced by circuit implementations (e.g., nonlinear distortion of analog front-end, quantization error of digital computing, etc.). Second, while a circuit implementation inevitably introduces various non-idealities, these non-idealities can be classified into two broad categories: (1) *critical non-idealities* that may significantly distort the output of a biomedical circuit, and (2) *non-critical non-idealities* that can be effectively mitigated or even completely eliminated by the data processing algorithm. A good circuit implementation should optimally budget the available resources (e.g., power) to maximally reduce the critical non-idealities rather than the non-critical ones.

Motivated by these observations, we propose to develop a radically new design framework to seamlessly *integrate* data processing algorithms and their customized circuit implementations for co-optimization, as shown in Fig. 7.1. Our core idea is to view a biomedical circuit, along with the data processing algorithm implemented by the circuit, as an information processing system. We develop an information-theoretic metric, referred to as *information processing capacity* (IPC) that extends the conventional communication notion of channel capacity to our application of biomedical data sensing, processing, and transmission. IPC quantitatively measures the amount of information that can be *processed* by the circuit. Intuitively, IPC is directly correlated to the accuracy of the circuit implementation. If a circuit can accurately process the input data and generate the desired output, its IPC is high. Otherwise, its IPC is low. In the extreme case, if a circuit cannot generate any meaningful output due to large errors, its IPC reaches the lowest value zero.

IPC can efficiently distinguish critical vs. non-critical non-idealities. It is strongly dependent on the critical non-idealities that distort the output, and is independent of the non-critical non-idealities that can be eliminated by the data processing algorithm. Hence, it serves as an excellent "quality" metric that we should maximize in order to determine the optimal data processing algorithm and the corresponding circuit implementation subject to a set of design constraints (e.g., latency, power, reconfigurability, etc.).

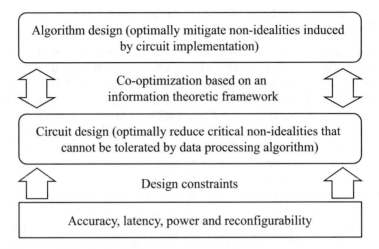

Fig. 7.1 An information-theoretic framework is proposed to co-optimize data processing algorithms and their customized circuit implementations for higher accuracy, smaller latency, lower power, and better reconfigurability of biomedical devices

It is important to note that our proposed design framework is *not* simply to combine algorithm and circuit designs. Instead, we aim to develop new methodologies that would profoundly revise today's data processing algorithms and integrated circuit designs for biomedical applications. In particular, our proposed information-theoretic framework can optimally explore the tradeoffs between accuracy, latency, power, and reconfigurability over all hierarchical levels from algorithm design to circuit implementation. From this point of view, the proposed framework based on IPC offers a fundamental infrastructure that enables next-generation biomedical circuit design and optimization for numerous emerging applications.

7.3 Information Processing Capacity

In this section, we describe an information-theoretic metric, IPC, to quantitatively measure the amount of information that can be processed by a biomedical circuit. It serves as a "quality" measure, when we co-optimize the algorithms and circuits for data sensing, processing, and transmission of biomedical devices. It, in turn, facilitates us to achieve superior accuracy, latency, power, and reconfigurability over the conventional design strategies.

7.3.1 Information-Theoretic Modeling

The IPC of a biomedical circuit can be mathematically modeled based on information theory. Without loss of generality, we consider a biomedical circuit, including

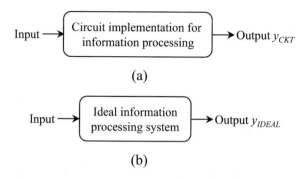

Fig. 7.2 An information-theoretic framework is proposed to co-optimize data processing algorithms and their customized circuit implementations for higher accuracy, smaller latency, lower power, and better reconfigurability of biomedical devices. (a) circuit implementation with non-idealities, and (b) ideal implementation

data sensing, processing, and transmission modules in general, as an information processing system shown in Fig. 7.2a. Since the circuit implementation is *not* perfect due to non-idealities, the aforementioned information processing system may generate errors where its output y_{CKT} deviates from the desired value.

To accurately characterize the "error" of the biomedical circuit, we further consider an ideal information processing system that guarantees to provide the correct output y_{IDEAL}, as shown in Fig. 7.2b. In other words, the ideal system is "conceptually" implemented with a circuit with "infinite" precision. It does not carry any non-ideality and, hence, is error-free. The "difference" between y_{CKT} and y_{IDEAL} indicates the non-idealities caused by the circuit implementation. However, quantitatively measuring such a difference is non-trivial, since a biomedical circuit can be applied to a broad range of usage scenarios (e.g., various physiological signals, various users, various environmental conditions, etc.). The comparison between y_{CKT} and y_{IDEAL} must cover all these scenarios where y_{CKT} and y_{IDEAL} are not just two numerical numbers and, hence, we cannot compare their difference by simply subtracting y_{CKT} from y_{IDEAL}. The information-theoretic metric, IPC, quantitatively measures the "quality" of approximating y_{IDEAL} by y_{CKT}. To derive the mathematical representation of IPC, we consider two different cases: (1) discrete output and (2) continuous output.

First, if the outputs y_{CKT} and y_{IDEAL} are discrete values (e.g., the diagnosis result of arrhythmia may be positive or negative), y_{CKT} and y_{IDEAL} can be modeled as two discrete random variables to cover the uncertainties over all usage scenarios. In general, we assume that y_{CKT} and y_{IDEAL} take M possible values $\{y_1, y_2, \cdots, y_M\}$. The statistics of these two random variables can be described by using their joint probability mass function (PMF) pmf(y_{CKT}, y_{IDEAL}). Table 7.1 shows a simple example for the binary random variables y_{CKT} and y_{IDEAL} (i.e., either TRUE or FALSE) where their statistics are fully described by four probabilities: true positive rate P_{TP}, false negative rate P_{FN}, false positive rate P_{FP}, and true negative rate P_{TN}.

Table 7.1 Confusion matrix
of a binary classifier

		y_{CKT}	
		True	False
y_{IDEAL}	True	P_{TP}	P_{FN}
	False	P_{FP}	P_{TN}

IPC is defined as the mutual information $I(y_{CKT}, y_{IDEAL})$ between y_{CKT} and y_{IDEAL} [2, 4]:

$$I(y_{CKT}, y_{IDEAL}) = \sum_{y_{CKT}} \sum_{y_{IDEAL}} \text{pmf}(y_{CKT}, y_{IDEAL}) \cdot \log\left(\frac{\text{pmf}(y_{CKT}, y_{IDEAL})}{\text{pmf}(y_{CKT}) \cdot \text{pmf}(y_{IDEAL})}\right).$$

(7.1)

Intuitively, the IPC metric in (7.1) measures the amount of information carried by y_{IDEAL} that can be learned from y_{CKT}. In one extreme case, if the circuit implementation is perfect, y_{CKT} is identical to y_{IDEAL} and, hence, IPC reaches its maximum. In the other extreme case, if y_{CKT} does not follow y_{IDEAL} at all due to large errors, there is no information about y_{IDEAL} that can be learned from y_{CKT} and, hence, IPC reaches its minimum (i.e., zero).

There are two important clarifications that should be made for IPC. First, instead of directly measuring the information carried by the circuit output y_{CKT}, we take the ideal output y_{IDEAL} as the "reference" and measure the information related to y_{IDEAL}. Since y_{IDEAL} represents all the important information of interest, IPC accurately captures our "goal" and ignores the "don't cares." This is the reason why IPC can serve as an excellent quality metric to guide our proposed algorithm/circuit co-optimization.

Second, IPC is different from other simple accuracy metrics that directly measure the difference between y_{CKT} and y_{IDEAL} based on statistical expectations. To understand the reason, we consider the example in Fig. 7.3a for which we may simply define the accuracy as the summation of the true positive rate P_{TP} and the true negative rate P_{TN}. Fig. 7.3b shows how this accuracy metric varies as a function of the false positive rate P_{FP} and the false negative rate P_{FN} where the probabilities for y_{IDEAL} to be TRUE and FALSE are set to pmf(y_{IDEAL} = TRUE) = 0.01 and pmf(y_{IDEAL} = FALSE) = 0.99, respectively. In this example, we set pmf(y_{IDEAL} = TRUE) to be much less than pmf(y_{IDEAL} = FALSE) to mimic the practical scenarios where pmf(y_{IDEAL} = TRUE) and pmf(y_{IDEAL} = FALSE) are highly unbalanced. For example, in the application of arrhythmia diagnosis [3, 19, 29], the probably of being positive (i.e., with arrhythmia) should be much less than the probability of being negative (i.e., without arrhythmia), since arrhythmia is only carried by a small group of unhealthy patients over the entire population.

Studying Fig. 7.3a, we observe that the simple accuracy metric heavily depends on the false positive rate P_{FP}, but weakly depends on the false negative rate P_{FN}, because the probability pmf(y_{IDEAL} = TRUE) is extremely small. If we maximize

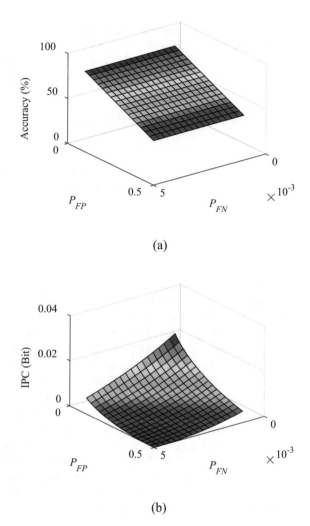

Fig. 7.3 (**a**) The conventional accuracy metric is not appropriately influenced by the false negative rate P_{FN}, if pmf(y_{IDEAL} = TRUE) is much less than pmf(y_{IDEAL} = FALSE). (**b**) The proposed IPC metric is appropriately influenced by both the false positive rate P_{FP} and the false negative rate P_{FN}, even if pmf(y_{IDEAL} = TRUE) and pmf(y_{IDEAL} = FALSE) are highly unbalanced

the aforementioned accuracy metric for algorithm/circuit co-optimization, it would aggressively minimize the false positive rate P_{FP}, thereby resulting in a large false negative rate P_{FN}. Consequently, a large portion of the unhealthy patients with arrhythmia may be mistakenly diagnosed as healthy ones.

On the other hand, Fig. 7.3b shows the relation between our proposed IPC and P_{FP} and P_{FN}. It can be observed that IPC is influenced by both P_{FP} and P_{FN}. Hence, by maximizing IPC, we take both P_{FP} and P_{FN} into account. This simple example demonstrates that when pmf(y_{IDEAL} = TRUE) and pmf(y_{IDEAL} = FALSE) are highly unbalanced, IPC can appropriately guide our proposed algorithm/circuit co-optimization, while the simple accuracy metric fails to work.

Finally, it is worth mentioning that if the outputs y_{CKT} and y_{IDEAL} are continuous values (e.g., movement decoding for neuroprosthesis results in the velocity value that is continuous), y_{CKT} and y_{IDEAL} can be modeled as two continuous random variables and their statistics can be described by the joint probability density function pdf(y_{CKT}, y_{IDEAL}). In this case, IPC can again be defined as the mutual information $I(y_{CKT}, y_{IDEAL})$ between y_{CKT} and y_{IDEAL} [2, 4]:

$$I(y_{CKT}, y_{IDEAL}) = \int_{-\infty}^{+\infty} \int_{-\infty}^{-\infty} \text{pdf}(y_{CKT}, y_{IDEAL}) \cdot \log\left(\frac{\text{pdf}(y_{CKT}, y_{IDEAL})}{\text{pdf}(y_{CKT}) \cdot \text{pdf}(y_{IDEAL})}\right).$$
(7.2)

In the following sub-sections, we will further discuss how IPC can be used for design and optimization of biomedical circuits.

7.3.2 Soft Channel Selection

Soft channel selection is an important task that is facilitated by our proposed algorithm/circuit co-optimization based on IPC. We consider the multi-channel biomedical device in Fig. 7.4, where channel selection is one of the most important tasks [1, 10, 23]. Appropriately selecting the important channels and removing the unimportant ones can efficiently minimize the amount of data for sensing, processing, and transmission, thereby substantially reducing the power consumption.

Today's channel selection is typically considered as a binary decision: a channel is either selected or not selected for recording. With the proposed information-theoretic framework based on IPC, we are now able to make a "soft" decision for each channel, referred to as soft channel selection. Namely, instead of simply including or excluding a given channel, we can finely tune the resolution of the channel (e.g., the number of bits required to represent the signal from the channel). Intuitively, an important channel should be designed with high resolution, while an unimportant channel can be designed with low resolution. The channel resolution is directly correlated to the power consumption of both analog front-end (e.g.,

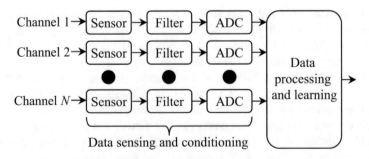

Fig. 7.4 A multi-channel biomedical device is shown to illustrate the application of soft channel selection

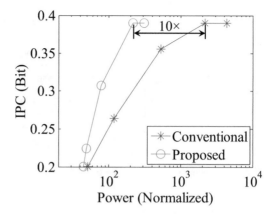

Fig. 7.5 The proposed soft channel selection reduces the power of the analog front-end by up to 10× compared to the conventional binary channel selection

sensors, analog filters, ADCs, etc.) and digital computing (e.g., digital filters, data processors, etc.). It, in turn, facilitates us to optimally explore the tradeoff between accuracy and power. In the extreme case, if the resolution of a channel is set to 0-bit, the channel is completely removed and it is equivalent to the conventional binary channel selection in the literature.

To demonstrate the efficacy of our proposed soft channel selection, we consider a preliminary example of movement decoding for neuroprosthesis, where our objective is to decode the movement direction from electrocorticography (ECoG) [26]. Fig. 7.5 compares the optimal IPC for both the conventional binary channel selection and the proposed soft channel selection. Note that the proposed approach successfully reduces the power of the analog front-end by up to 10×. It is important to mention that the proposed idea of soft channel selection can be further extended to other important applications such as data compression and transmission.

7.3.3 Robust Data Processing

To maximally reduce the power consumption for portable and/or implantable applications, fixed-point arithmetic, instead of floating-point arithmetic, is often adopted to implement data processing algorithms and the word length for fixed-point computing must be aggressively minimized. While fixed-point arithmetic has been extensively studied for digital signal processing during the past several decades [14, 20], it is rarely explored for many emerging data processing tasks that involve sophisticated learning algorithms (e.g., movement decoding for neuroprosthesis). It remains an open question how to revise these algorithms to maximally tolerate the quantization error posed by finite word length. Based upon IPC, data processing algorithms can be completely redesigned to mitigate the quantization error so that these algorithms can be mapped to fixed-point implementations with extremely low resolution.

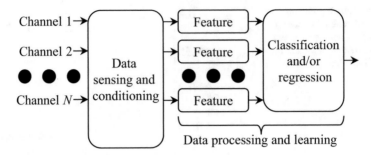

Fig. 7.6 A data learning algorithm typically consists of two steps: feature extraction and classification/regression

Fig. 7.7 The proposed data learning algorithm reduces the required word length by 2-bit, compared to the conventional approach

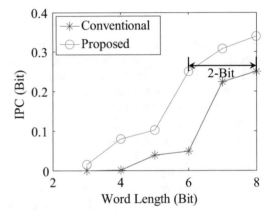

As shown in Fig. 7.6, a learning algorithm typically consists of two steps: (1) feature extraction and (2) classification (e.g., to determine movement direction for neuroprosthesis) and/or regression (e.g., to determine movement velocity for neuroprosthesis). We propose to maximize the IPC metric of a classification or regression engine subject to the constraint that all arithmetic operations for both feature extraction and classification/regression are quantized. Our reformulated learning algorithm solves a "robust" optimization problem to find the optimal, quantized classifier or regressor that is least sensitive to quantization error. It, in turn, offers superior performance over other conventional approaches where quantization error is not explicitly considered within the learning process.

As an example for illustration purpose, we consider the classification problem of decoding the movement direction from electrocorticography (ECoG) for neuroprosthesis. Fig. 7.7 shows the optimized IPC metric as a function of word length. To achieve the same IPC, our proposed approach can reduce the word length by 2-bit compared to the conventional classifier. Note that the word length of fixed-point arithmetic is directly correlated to the power consumption of its circuit implementation. Hence, reducing word length is of great importance for low-power portable and/or implantable biomedical devices.

7.4   Case Study: Brain–Computer Interface

Brain–computer interface (BCI) has been considered as a promising communication technique for patients with neuromuscular impairments. For instance, neural prosthesis provides a direct control pathway from brain to external prosthesis for paralyzed patients. It can offer substantially improved quality of life to these patients. To create a neural prosthesis, we must appropriately measure the brain signals and then accurately decode the movement information from the measured signals [5, 11, 13, 21, 22, 26].

A variety of signal processing algorithms have been proposed for movement decoding in the literature. Most of these algorithms first extract the important features to compactly represent the information carried by the brain signals. Next, the extracted features are provided to a classification and/or regression engine to decode the movement information of interest. While most movement decoding algorithms in the literature are implemented with software on microprocessors, there is a strong need to migrate these algorithms to hardware in order to reduce the power consumption for practical BCI applications.

7.4.1   System Design

Fig. 7.8 shows a simplified block diagram for the proposed hardware implementation of BCI. It consists of three major components:

- *Signal normalization*: The magnitude of brain signals varies from subject to subject and from channel to channel. Hence, representing brain signals by fixed-point arithmetic requires a large word length (i.e., a large number of bits). In

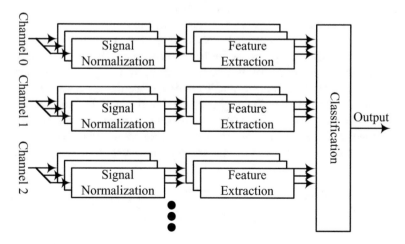

Fig. 7.8   A simplified block diagram is shown for the proposed hardware implementation of BCI

order to minimize the word length and, consequently, the power consumption for fixed-point computation, we must appropriately normalize the brain signal from each channel.

• *Feature extraction*: There are many different feature extraction approaches for movement decoding of BCI. For instance, given the brain signal recorded from a particular channel, we can apply discrete cosine transform (DCT) and consider the DCT coefficients as the features for decoding [28].

• *Classification*: Once all features are extracted for multiple channels, they are further combined to decode the movement information. For instance, all features can be linearly combined by a linear classifier to determine the movement direction of interest. Here, a variety of linear classification algorithms (e.g., linear discriminant analysis, support vector machine, etc.) can be used, where the classifier training is performed offline. The on-chip classification engine performs the multiply-and-accumulate operations to determine the final output (i.e., the movement direction) from the features.

7.4.2 Experimental Results

We consider the ECoG data set collected from a human subject with tetraplegia due to spinal cord injury [26]. The ECoG signals are recorded with a high-density 32-electrode grid over the hand and arm area of the left sensorimotor cortex. The sampling frequency is 1.2 kHz. The human subject is able to voluntarily activate his sensorimotor cortex using attempted movements.

Our objective is to decode the binary movement direction (i.e., left or right) from a single trial that is 300 ms in length. The ECoG data set contains 70 trials for each movement direction (i.e., 140 trials in total). For movement decoding, 7 important channels with 6 features per channel (i.e., 42 features in total) are selected based on the Fisher criterion. A linear classifier is trained and implemented with 8-bit fixed-point arithmetic to decode the movement direction.

The BCI system is implemented with a Xilinx FPGA Zynq-7000 board. For testing and comparison purposes, we further implement a reference design based on the conventional technique [30]. In this sub-section, we compare the performance between our proposed hardware implementation and the reference design.

We estimate the power and energy consumption for both the proposed and the reference designs by using Xilinx Power Analyzer, where the clock frequency is set to 0.5 MHz. Table 7.2 compares the power consumption for these two different designs. Note that the proposed design achieves more than 56× energy reduction

Table 7.2 Power and energy consumption per decoding operation

	Proposed design	Reference design
Power (mW)	0.72	3.8
Runtime (ms)	1.094	11.71
Energy (μJ)	0.787	44.5

Table 7.3 Power consumption of different functional blocks for the proposed design

Signal normalization (μW)	25.2
Feature extraction (μW)	690.2
Classification (μW)	2.6

Fig. 7.9 A Xilinx FPGA Zynq-7000 board is used to validate the proposed hardware design for movement decoding of BCI

over the reference design. Table 7.3 further shows the power consumption for different functional blocks of the proposed design. Note that feature extraction dominates the overall power consumption for our proposed hardware implementation. Hence, additional efforts should be pursued to further reduce the power consumption of feature extraction in our future research.

To validate the proposed design on the Xilinx Zynq-7000 board, we first load our hardware design to the FPGA chip through the programming interface. Next, the ECoG data set is copied to an SD card that is connected to the Zynq-7000 board. When running the movement decoding flow, a single trial of the ECoG signals is first loaded to the SRAM block inside the FPGA chip. Next, these signals are passed to the functional blocks of signal normalization, feature extraction and classification for decoding. The decoding results are read back to an external computer through an RS-232 serial port on the Zynq-7000 board so that we can verify the decoding accuracy. Fig. 7.9 shows a photograph of the Xilinx FPGA Zynq-7000 board where the RS-232 port and the programming interface are both highlighted.

7.5 Summary

In this chapter, we describe a new design framework for ultra-low-power biomedical circuits. The key idea is to co-optimize data processing algorithms and their circuit implementations based on an information-theoretic metric: IPC. The proposed design framework has been demonstrated by a case study of BCI. Our experimental results show that the proposed design achieves more than 56× energy reduction over a reference design. As an important aspect of our future research, we will further apply the proposed design framework to other emerging biomedical applications.

References

1. M. Arvaneh, C. Guan, K.K. Ang, C. Quek, Optimizing the channel selection and classification accuracy in EEG-based BCI. IEEE Trans. Biomed. Eng. **58**, 1865–1873 (2011). doi:10.1109/TBME.2011.2131142
2. C. Bishop, *Pattern Recognition and Machine Learning* (Springer, Berlin, 2006)
3. P. de Chazal, R.B. Reilly, A patient-adapting heartbeat classifier using ECG morphology and heartbeat interval features. IEEE Trans. Biomed. Eng. **53**, 2535–2543 (2006). doi:10.1109/TBME.2006.883802
4. T. Cover, J. Thomas, *Elements of Information Theory* (Interscience, Wiley, 2006)
5. J.P. Donoghue, Bridging the brain to the world: a perspective on neural interface systems. Neuron **60**, 511–521 (2008). doi:10.1016/j.neuron.2008.10.037
6. C. Guo, X. Li, S. Taulu, W. Wang, D.J. Weber, Real-time robust signal space separation for magnetoencephalography. IEEE Trans. Biomed. Eng. **57**, 1856–1866 (2010). doi:10.1109/TBME.2010.2043358
7. S. Khalid, D. Clifton, L. Clifton, L. Tarassenko, A two-class approach to the detection of physiological deterioration in patient vital signs with clinical label refinement. IEEE Trans. Inf. Technol. Biomed. **16**, 1231–1238 (2012). doi:10.1109/TITB.2012.2212202
8. J. Ko, C. Lu, M.B. Srivastava, J.A. Stankovic, A. Terzis, M. Welsh, Wireless sensor networks for healthcare. Proc IEEE **98**, 1947–1960 (2010). doi:10.1109/JPROC.2010.2065210
9. M.L. Kringelbach, N. Jenkinson, S.L. Owen, T.Z. Aziz, Translational principles of deep brain stimulation. Nature **8**, 523–635 (2007). doi:10.1038/nrn2196
10. T.N. Lal, M. Schroder, T. Hinterberger, J. Weston, M. Bogdan, N. Birbaumer, B. Scholkopf, Support vector channel selection in BCI. IEEE Trans. Biomed. Eng. **51**, 1003–1010 (2004). doi:10.1109/TBME.2004.827827
11. M.A. Lebedev, M.A. Nicolelis, Brain-machine-interface: past, present and future. Trends Neurosci. **29**, 536–546 (2006). doi:10.1016/j.tins.2006.07.004
12. K.H. Lee, S.Y. Kung, N. Verma, Low-energy formulations of support vector machine kernel functions for biomedical sensor applications. J Signal Process Syst **69**, 339–349 (2012). doi:10.1007/s11265-012-0672-8
13. E.C. Leuthardt, G. Schalk, J.R. Wolpaw, J.G. Ojemann, D.W. Moran, A brain-computer interface using electrocorticographic signals in humans. J. Neural Eng. **1**, 63–71 (2004). doi:10.1088/1741-2560/1/2/001
14. W. Meyer-Baese, *Digital Signal Processing with Field Programmable Gate Arrays* (Springer, Berlin, 2007)
15. C. Mora Lopez, D. Prodanov, D. Braeken, I. Gligorijevic, W. Eberle, C. Bartic, R. Puers, G. Gielen, A multichannel integrated circuit for electrical recording of neural activity, with independent channel programmability. IEEE Trans Biomed Circuits Syst **6**, 101–110 (2012). doi:10.1109/TBCAS.2011.2181842

16. R. Muller, S. Gambini, J.M. Rabaey, A 0.013 mm², 5 μW, DC-coupled neural signal acquisition IC with 0.5 V supply. J Solid-State Circuits 47, 232–243 (2012). doi:10.1109/JSSC.2011.2163552
17. K. Nazarpour, Y. Wongsawat, S. Sanei, J.A. Chambers, S. Oraintara, Removal of the eye-blink artifacts from EEGs via STF-TS modeling and robust minimum variance beamforming. IEEE Trans. Biomed. Eng. 55, 2221–2231 (2008). doi:10.1109/TBME.2008.919847
18. C.O. Oluigbo, A. Salma, A.R. Rezai, C. Oluigbo, Deep brain stimulation for neurological disorders. IEEE Rev. Biomed. Eng. 5, 88–99 (2012). doi:10.1109/RBME.2012.2197745
19. S. Osowski, L.T. Hoai, T. Markiewicz, Support vector machine-based expert system for reliable heartbeat recognition. IEEE Trans. Biomed. Eng. 51, 582–589 (2004). doi:10.1109/TBME.2004.824138
20. W. Padgett, D. Anderson, *Fixed-point Signal Processing* (Morgan and Claypool Publishers, Williston, 2009)
21. G. Pfurtscheller, C. Neuper, Motor imagery and direct brain-computer communication. Proc EEE 89, 1123–1134 (2001). doi:10.1109/5.939829
22. A.B. Schwartz, X.T. Cui, D.J. Weber, D.W. Moran, Brain-controlled interfaces: movement restoration with neural prosthetics. Neuron 52, 205–220 (2006). doi:10.1016/j.neuron.2006.09.019
23. P. Shenoy, K.J. Miller, J.G. Ojemann, R.P. Rao, Generalized features for electrocorticographic BCIs. IEEE Trans. Biomed. Eng. 55, 273–280 (2008). doi:10.1109/TBME.2007.903528
24. X.F. Teng, Y.T. Zhang, C.C. Poon, P. Bonato, Wearable medical systems for p-health. IEEE Rev. Biomed. Eng. 1, 62–74 (2008). doi:10.1109/RBME.2008.2008248
25. N.V. Thakor, Y. Zhu, Applications of adaptive filtering to ECG analysis: noise cancellation and arrhythmia detection. IEEE Trans. Biomed. Eng. 38, 785–794 (1991). doi:10.1109/10.83591
26. W. Wang, J.L. Collinger, A.D. Degenhart, E.C. Tyler-Kabara, A.B. Schwartz, D.W. Moran, D.J. Weber, B. Wodlinger, R.K. Vinjamuri, R.C. Ashmore, J.W. Kelly, M.L. Boninger, An electrocorticographic brain interface in an individual with tetraplegia. PLoS One 8, e55344 (2013). doi:10.1371/journal.pone.0055344
27. E.S. Winokur, M.K. Delano, C.G. Sodini, A wearable cardiac monitor for long-term data acquisition and analysis. IEEE Trans. Biomed. Eng. 60, 189–192 (2013). doi:10.1109/TBME.2012.2217958
28. M. Won, H. Albalawi, L. Li, D.E. Thomas, Low-power hardware implementation of movement decoding for brain computer interface with reduced-resolution discrete cosine transform. Conf. Proc. IEEE Eng. Med. Biol. Soc. 2014, 1626–1629 (2014). doi:10.1109/EMBC.2014.6943916
29. C. Ye, B.V. Kumar, M.T. Coimbra, Heartbeat classification using morphological and dynamic features of ECG signals. IEEE Trans. Biomed. Eng. 59, 2930–2941 (2012). doi:10.1109/TBME.2012.2213253
30. J. Yoo, L. Yan, D. El-Damak, M. Altaf, A.H. Shoeb, A.P. Chandrakasan, An 8-channel scalable EEG acquisition SoC with patient-specific seizure classification and recording processor. J Solid-State Circuits 48, 214–228 (2013). doi:10.1109/JSSC.2012.2221220
31. K. Yousef, M.R. Pinsky, M.A. DeVita, S. Sereika, M. Hravnak, Characteristics of patients with cardiorespiratory instability in a step-down unit. Am. J. Crit. Care 21, 344–350 (2012). doi:10.4037/ajcc2012797
32. X. Zou, X. Xu, L. Yao, Y. Lian, A 1-V 450-nW fully integrated programmable biomedical sensor interface chip. J Solid-State Circuits 44, 1067–1077 (2009). doi:10.1109/JSSC.2009.2014707

Chapter 8
Acceleration of MapReduce Framework on a Multicore Processor

Lijun Zhou and Zhiyi Yu

8.1 Introduction

In recent years, with the increasing development of integrated circuit technology, the society surrounded by great variety of electronic terminals is undergoing deep change in the era of big data. Processors, the core components of the era of big data, need to meet increasing computing requirements and energy challenges. Multi-core processors with parallel processing ability and energy efficiency are favored [1], and they can improve the energy efficiency further with hardware accelerations.

The famous parallel framework, MapReduce [2], is commonly used for high throughput big data applications. For example, one of the MapReduce platforms Hadoop provides strong support to deal with big data. How to implement MapReduce on multicore processors to fully utilizing the parallelism capability is becoming a critical issue. What's more, we also consider how to design a reconfigurable hardware acceleration solution based on multicore processors to further improve system efficiency.

The main work and contributions of this work include:

- *Feature extraction and acceleration of MapReduce Applications*: We analyze, implement, and accelerate several important high throughput algorithms of the Cloud Suite benchmark. For the PageRank algorithm which is commonly used in Web search, we propose a software–hardware co-design acceleration. It splits

L. Zhou
State Key Lab of ASIC & System, Fudan University, Shanghai, China
e-mail: 13210720082@fudan.edu.cn

Z. Yu (✉)
SYSU-CMU Joint Institute of Engineering, Sun Yat-sen University, Guangzhou, China

SYSU-CMU Shunde International Joint Research Institute, Shunde, China
e-mail: yuzhiyi@mail.sysu.edu.cn

© Springer International Publishing AG 2017
A. Chattopadhyay et al. (eds.), *Emerging Technology and Architecture for Big-data Analytics*, DOI 10.1007/978-3-319-54840-1_8

the matrix in software to improve parallelism, and it adds multiply adders in hardware to reduce write back operations. For text mining algorithm Naive-Bayes, we propose a method called topo-MapReduce [3], which significantly reduces the amount of inter-core communication.

• *A Scalable and Reconfigurable MapReduce Acceleration Framework*: We propose and design a multi-layer reconfigurable MapReduce acceleration framework based on a multi-core processor. It supports automatic configuration of MapReduce framework. A flexible packet/circuit switched inter-core communication method is designed to obtain flexible and high efficient network-on-chip (NoC). The multicore processors contain multi-port memory, supporting multiple operations in acceleration arrays to improve the performance. Experimental results show that the system can accelerate nearly 40 times than pure software approach.

The rest of the chapter is organized as follows. Section 8.2 describes MapReduce framework and related research on multicore platform. Section 8.3 proposes two optimized implementations of big data algorithms on multicore processors. Section 8.4 presents the Configurable MapReduce Acceleration Framework for MapReduce applications. Section 8.5 analyzes the experimental results. Finally, Sect. 8.6 concludes this chapter.

8.2 MapReduce Framework on Multicore Processors

8.2.1 Introduction to MapReduce

Many parallel programming models are proposed for big data analysis. Google proposed MapReduce [2] for large-scale data in 2006, which is particularly suitable for massive data search, mining, analysis, and machine learning. Figure 8.1 shows the process of text classification in MapReduce, and Fig. 8.2 is an overview of the framework of the operation. However, this method usually distributes the work across many servers, and neglects to exploit the parallel computation capability of one node/processor.

8.2.2 Related Work

MapReduce is developed by Google in order to deal with large scale of data. Yahoo supports a better implementation of Hadoop in open source version. In the Benchmark Competition Sort (data sorting contest), Yahoo completed 100 TB data sorting in 72 min, by using 2100 machines installed of Hadoop [4].

The Phoenix project from Stanford University implemented MapReduce in multicore processors [5]. They proposed a real-time system Phoenix based on MapReduce framework. It can automatically manage threads, dynamic task scheduling, data segmentation and fault tolerance and many other functions.

Zhou Haijie in Fudan University implemented a simple MapReduce framework in a 16-core processor. The parallel programming model of large data correlation

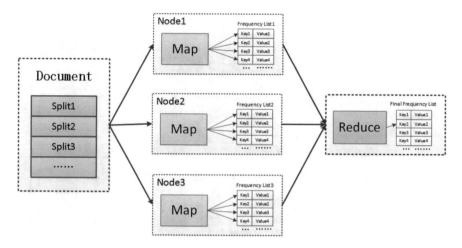

Fig. 8.1 Text classification in MapReduce framework

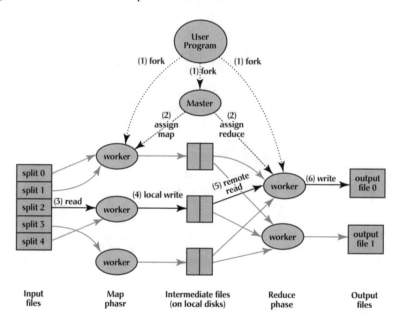

Fig. 8.2 Overview of MapReduce framework

is realized, and the data management of general multicore programming model is optimized. Xiao Zhiwei in Fudan University analyzed the performance bottleneck of multi-core cluster, and proposed a hierarchical MapReduce Model to optimize Hadoop. The model can make full use of the data locality and task parallelism of the system. Experimental data shows that the system can speed up 1.4–3.5 times compared to the original Hadoop.

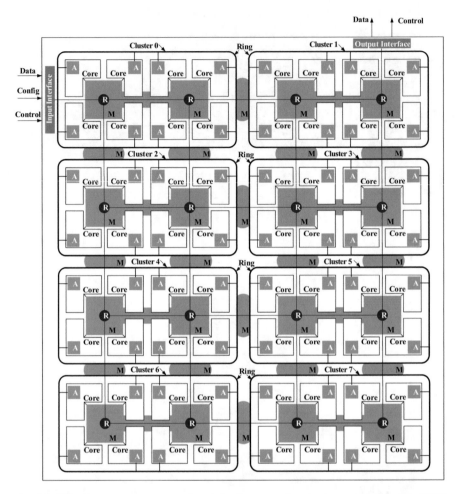

Fig. 8.3 This 64-core processor contains eight clusters, and each cluster contains eight RISC processors. In each cluster, a circuit-switching ring is implemented to achieve one-cycle communication within the cluster. Processors in different clusters communicate with each other using packet-switching routers

8.2.3 Experimental Platform

We have designed and implemented 16-core [6], 24-core [7] and 64-core processors in 2012, 2013, 2015, respectively. The 16-core processor supports both shared memory and message passing as communication mechanisms. The 24-core processor combines circuit switching and packet switching network-on-chip to obtain both efficiency and flexibility, and designed an application specific acceleration array to further improve system efficiency. The structure diagram of the 64-core processor is shown in Fig. 8.3. It has the following characteristics: (1) application level parallelism by partitioning the chip into multiple domains for different applications such as multimedia, communication, and big data applications, (2) efficient accelerators in each domain for different applications, (3) various hierarchical interconnection schemes to support communication between processor–processor, processor–accelerator, and accelerator–accelerator. The experiments of this chapter is based on this 64-core processor platform.

8.3 Accelerating Algorithms Based on MapReduce in Multicore Processors

This chapter selects two important algorithms: graph analysis (PageRank) and data analysis (Bayes classification) to show feature analysis and their accelerations.

8.3.1 Acceleration of PageRank Algorithm

8.3.1.1 Math Model of PageRank

PageRank is proposed to improve the network search efficiency. The main idea is to calculate the importance of a page by the quality of pages it is linked. An intuitive formula can be obtained as Eq. (8.1):

$$R(i) = C \sum_{j \in B(i)} \frac{R(j)}{N(j)} \qquad (8.1)$$

8.3.1.2 Hardware Accelerator for Pagerank

Figure 8.4 is the flow of Pagerank algorithm, and we can see that most of computation of PageRank is multiply. In addition, temporary variants G matrix and V vector need to be written back to the shared memory after every iteration because cache cannot fit the massive data, which wastes time and energy significantly. We design an acceleration unit which consists of 32-bit multiplier and 32-bit accumulator. The 32 bits input contains R_1 and L_1. Output $O = R_{1H} \times L_{1H} + R_{1L} \times L_{1L}$ (H means high part and L means low part). For the N bit, it will accelerate $N/2$ times in theory. The implementation of this accelerator is shown in Fig. 8.5. There are four pairs of accelerators in each cluster, each containing a 32-bit multiplier and 32-bit accumulator, which means four dimension of matrix can be calculated at the same time.

Fig. 8.4 The flow of PageRank algorithm. V_old is the temporary variable for V, which is the importance of one page, and G is the transferring matrix

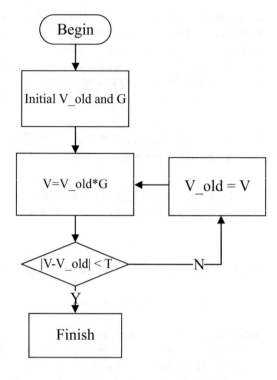

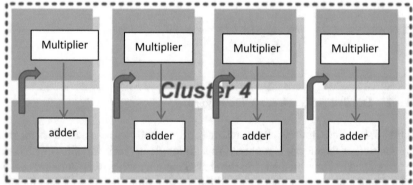

Fig. 8.5 Implementation of accelerators for Pagerank in the 64-core processor

8.3.2 *Acceleration of Naive-Bayes Algorithm*

Massive data mining has become an important area in big data era. A number of algorithms such as Bayes classification are proposed, and are widely used in business, life, and scientific research and other fields. The principle of Bayes classification algorithm is to classify text according to the frequency of emergence.

8.3.2.1 Math Model of Naive-Bayes Algorithm

The Naive-Bayes algorithm [8] is based on the probabilistic models. It calculates the score of a class c_i as the probability of a document $d_t(a_1, a_2, \ldots, a_j)$ to be assigned to c_i, as shown in Eq. (8.2), and classifies d_t in the class with highest score. Here $(a_1, a_2, \ldots, a_j)$ is a binary or weighted vector representing the terms in document d_t.

$$
\begin{aligned}
P(c_i|d_t) &= \frac{P(c_i)P(d_t|c_i)}{P(d_t)} \\
&= \frac{P(c_i)\Pi_{\forall j \in d_t}P(a_j|c_i)}{P(d_t)}
\end{aligned}
\tag{8.2}
$$

$$
P(a_j|c_i) = \frac{Tc_i(a_j)}{\sum_{v \in V} Tc_i(v)}
\tag{8.3}
$$

In Eq. (8.3), $Tc_i(a_j)$ is the number of occurrences of term a_j in class c_i and $\sum_{v \in V} Tc_i(v)$ is the summation of the numbers of occurrences of all terms in documents of class c_i, and V is the term vocabulary. Thus, it can conclude that $P(a_j|c_i)$ defines the representativity of a term a_j in class c_i as being the ratio of the frequency of term a_j in class c_i and the total frequency of all terms in class c_i.

8.3.2.2 Hardware Accelerator for Naive-Bayes

The existing technology for Naive-Bayes algorithm is based on the Mahout machine learning platform of the distributed cluster, which partitioning tasks into multiple machines to count word frequency. Each machine uses double hash linear detection to match the specific words and calculate word frequency statistics. This method can be divided into two steps. The first step is the segmentation of data and word frequency statistics. It sends data and operation rules to distributed node and calculates the results. The second step is to find the text word frequency using double hash matching.

The most computing cost of the Naive-Bayes is the indexing of the term and counting the frequency of every term in one class. So a highly efficient matching discipline between data and its storage location will accelerate the process of term indexing.

The Hash operation is divided into different functional modules to analyze its bottlenecks, as shown in Fig. 8.6. We can see that HashCode occupies the most time. Therefore, it can improve efficiency significantly by using hardware circuit to accelerate HashCode. Algorithm 1 shows the pseudo-code of HashCode, and its accelerator circuit is shown in Fig. 8.7.

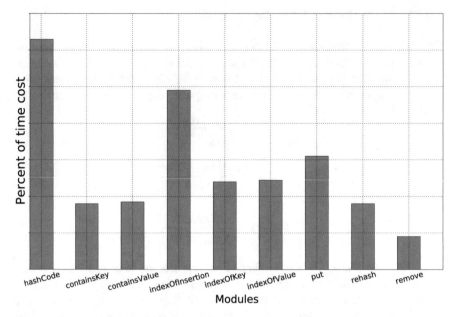

Fig. 8.6 Percent of time cost by modules of Hash operation

Algorithm 1 HashCode Algorithm

Require: *cha* is the element to be hashed, *i* is the index of character in *cha*, *EOF* is the end
    character of one word, *SEED* is the constant factor.
Ensure: $i > -1$ and is an integer.
 1: **function** HASHCODE (*cha*)
 2:     $i \leftarrow 0$
 3:     **while** $cha[i]! = EOF$ **do**
 4:         $temp \leftarrow cha[i]$
 5:         $Hash \leftarrow SEED * cha[i] + Hash$
 6:         $i \leftarrow i + 1$
 7:         $temp \leftarrow Hash$
 8:     **end while**
 9:     Return *Hash*
10: **end function**

8.3.2.3   Task Mapping Scheme: Topo-MapReduce

How to map applications efficiently on multi-core resources is also a key issue which
can impact the results significantly. The optimization idea for the Hash table is that
the huge hash table can be divided into a number of blocks, where the hash value in
conflict can be divided into the same block (Map). And then the statistical frequency
only needs to be completed (Reduce) in the same processor nodes. Looking up one
word's frequency only needs to match the hash value of the word to determine where
it is, avoiding cross-clusters access for accelerating frequency statistics.

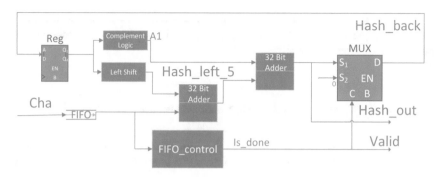

Fig. 8.7 Hardware accelerator for HashCode algorithm

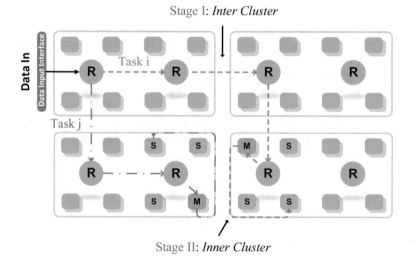

Fig. 8.8 Dataflow of task mapping. Stage I represents inter-cluster finding, and stage II represents inner-cluster finding

As shown in Fig. 8.8, we propose a new mapping model, Topo-MapReduce, as follows:

- *Avoid cross-cluster access*: Each cluster possesses one part of hash table, and avoids cross-cluster access.
- *Inter-cluster searching*: Each task is to be processed by one core. If the hash mapping entrance is outside of this core's range, the task will be sent to the adjacent cluster.
- *Inner-cluster searching*: When the task reaches the right cluster, it will be analyzed and find the memory by the master core, then sent to the destination processor.

Different from MapReduce, Topo-MapReduce divides tasks into several smaller independent parts. Those independent parts do not need to reduce together and save much inter-core communication.

8.4 Configurable MapReduce Acceleration Framework

Figure 8.9 is the overall architecture of the configurable MapReduce acceleration framework. This system is divided into four parts: controller, mapping and sorting, address allocation, and merge.

- Controller: it is responsible for data exchange between processors and accelerators, as shown in Fig. 8.10.
- Mapping and sorting: mapping is responsible for partitioning and transferring the received data to the address allocation. Sorting is necessary because the output from mapping needs to be sorted in MapReduce. Figures 8.11 and 8.12 shows the diagram for sorting.
- Address allocation: it receives data from mapping and sorting unit, and schedules address to merge part.
- Merge: it receives data from address allocation block and mapping block, merges data, and transfers the results to shared storage in clusters.

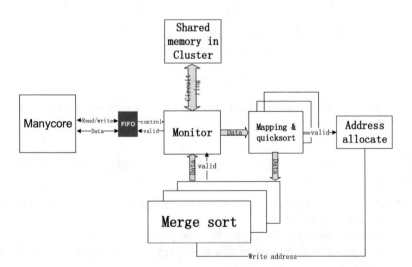

Fig. 8.9 Architecture of the configurable MapReduce acceleration framework

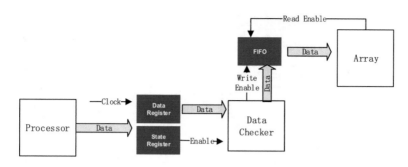

Fig. 8.10 Controller module

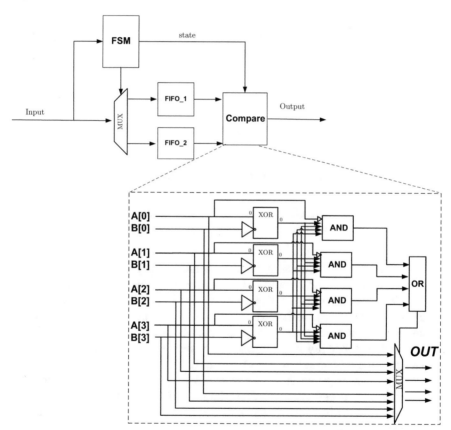

Fig. 8.11 Base comparing unit in sorting

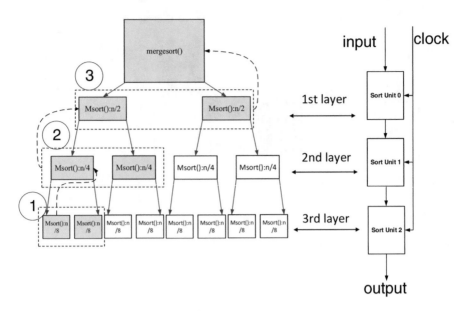

Fig. 8.12 Sorting unit combined with base comparing unit

8.4.1 High Throughput Data Transferring

We designed parallel and efficient data transferring solution, by taking full advantage of high communication bandwidth of internal NoC network, with two key features:

- Data transferring between processors/memories and acceleration execution array: The 64-core processors support hierarchical multi-port shared memory. The data from the memory of processor cores can be poured into acceleration array simultaneously. As shown in Fig. 8.13a, the processor p in bottom left corner configures the ring node, and arranges other processors to move data in the shared memory to the accelerators (shown as a) simultaneously, which increases the input speed of data significantly.
- Data transferring between accelerators: circuit switching is designed for data transferring between accelerators, to improve the data transfer capability. As shown in Fig. 8.13b, four different data transferring between accelerators can be done simultaneously.

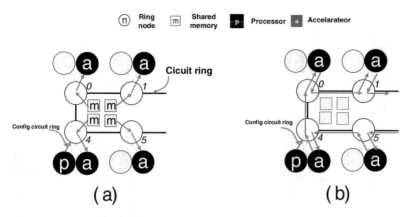

Fig. 8.13 Circuits for high throughput data transferring: (**a**) data transferring between processors and accelerators, and (**b**) data transferring between accelerators

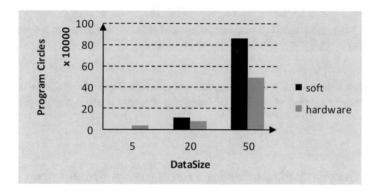

Fig. 8.14 Software and hardware accelerations for pagerank algorithm

8.5   Experiment Result Analysis

8.5.1   Pagerank with Hardware Accelerations

For pagerank algorithm, the hardware acceleration gains two times speedup at maximum when data size is big, as shown in Fig. 8.14.

8.5.2   Topo-Mapreduce

As shown in Fig. 8.15, Topo-MapReduce increases the performance by 17% at 8-core situation, and by 29% at 40-core situation, compared to the original MapReduce. In MapReduce model, more cores means more inter-core communication,

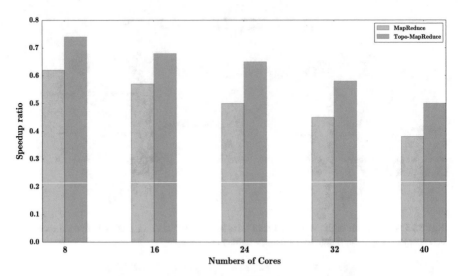

Fig. 8.15 The speedup ratio (speedup divided by number of cores) comparison between MapReduce and Topo-MapReduce with different numbers of cores

which slows down the speedup significantly, while Topo-MapReduce constrains the conflict items and saves lots of inter-core communication and improves performance.

8.5.3 Configurable Mapreduce Acceleration Framework

TeraSort sort is used as a benchmark to verify the efficiency of the Configurable MapReduce Acceleration Framework. 1 TB input data is generated by Hadoop's official test data generation tool, and takes 3 h and 40 min in a nine node cluster server, and the throughput $T = 1\,\text{TB/3\,h40\,min} = 69\,\text{Mbps}$. Using the multi-core platform we designed, we obtain a throughput of 333 Mbps under the amount of test data 1 GB. In the case of larger data and larger number of Reducer, the throughput is estimated to be 2630 Mbps, which is roughly 40 times than existing cluster.

Table 8.1 shows the results of several algorithms implemented in the Configurable MapReduce Framework Acceleration Array platform, using Xilinx FPGA VC707. Table 8.2 shows that the speedup of the framework with different number of accelerators. These comparisons show that the speedup of the framework increases as the application data volume increases. This is mainly due to the efficient inter-core communication of multi-core chip.

Table 8.1 Speedup of configurable mapreduce framework for different applications, system is implemented in Xilinx FPGA VC707

Algorithms	Area	Frequency	Data volume	Throughput	Speed up
Terasort	6%	333	1 GB	1.4	20.3
K-means	10%	244	3200 points	1.9	7.9
MatrixMul	11%	166	1000 dime	3.4	6.3

The area is the utilization percentage of FPGA

Table 8.2 Speedup of configurable mapreduce framework with different data

	Cluster	Multicore (10 accelerators)	Multicore (10,000 accelerators)
Throughput	69	333	2630
Compared with cluster	1	4.8	38.9 time

8.6 Conclusion

A Configurable MapReduce framework Accelerating Array including its algorithm, hardware accelerations, and platform is presented in this chapter. High performance is achieved by specific algorithm accelerators and a Configurable MapReduce Acceleration framework, based on a multicore processor equipped with circuit switching and packet switching. Some benchmark algorithms are accelerated by specific algorithm accelerators. An optimized mapping scheme, Topo-MapReduce, is adopted to avoid communication congestion and reduce communication latency. The Configurable MapReduce acceleration framework is designed to reduce the bottleneck of communication and computing. Experimental results show that the improved MapReduce framework with hardware acceleration can speed up by 40 times at maximum compared to the software solution, and the proposed Topo-MapReduce can further speed up by 29% at maximum compared to the original MapReduce.

Acknowledgements This work was supported by grants from Huawei Corporation, and SYSU-CMU Shunde International Joint Research Institute.

References

1. M. Saecker, V. Markl, Big data analytics on modern hardware architectures: a technology survey, in *Business Intelligence* (Springer, Berlin/Heidelberg, 2013), pp. 125–149
2. J. Dean, S. Ghemawat, MapReduce: simplified data processing on large clusters, in *Proceedings of Operating Systems Design and Implementation*, San Francisco, CA (2004), pp. 137–150
3. L. Zhou, Z. Yu, J. Lin, S. Zhu, W. Shi, H. Zhou, K. Song, X. Zeng, Acceleration of Naive-Bayes algorithm on multicore processor for massive text classification, in *International Symposium on Integrated Circuits (ISIC)* (2014), pp. 244–247

4. Spark the fastest open source engine for sorting a petabyte, https://databricks.com/blog/2014/10/10/spark-petabyte-sort.html
5. C. Ranger, R. Raghuraman, A. Penmetsa, et al., Evaluating MapReduce for multi-core and multiprocessor systems, in *IEEE 13th International Symposium on High Performance Computer Architecture, 2007. HPCA 2007* (IEEE, Washington, 2007), pp. 13–24
6. Z. Yu, K. You, R. Xiao, H. Quan, P. Ou, Y. Ying, H. Yang, M. Jing, X. Zeng, An 800MHz 320mW 16-core processor with message-passing and shared-memory inter-core communication mechanisms, in *IEEE International Solid-State Circuits Conference (ISSCC)* (2012), pp. 64–65
7. P. Ou, J. Zhang, H. Quan, Y. Li, M. He, Z. Yu, X. Yu, S. Cui, J. Feng, S. Zhu, J. Lin, M. Jing, X. Zeng, Z. Yu, A 65nm 39GOPS/W 24-core processor with 11Tb/s/W packet controlled circuit-switched double-layer network-on-chip and heterogeneous execution array, in *IEEE International Solid-State Circuits Conference (ISSCC)* (2013), pp. 56–57
8. V. Felipe, et al., GPU-NB: a fast CUDA-based implementation of Naive Bayes, in *Proceedings - Symposium on Computer Architecture and High Performance Computing* (2013), pp. 168–175

Chapter 9
Adaptive Dynamic Range Compression for Improving Envelope-Based Speech Perception: Implications for Cochlear Implants

Ying-Hui Lai, Fei Chen, and Yu Tsao

9.1 Introduction

A cochlear implant (CI) is currently the only electronic device that can facilitate hearing in people with profound-to-severe sensorineural hearing loss (SNHL). In a report published in December 2012, the US Food and Drug Administration estimated that approximately 324,200 people worldwide have received cochlear implants, and predicted that this number will continue to increase in the near future [1]. Although CI devices can considerably improve the hearing capabilities of individuals with profound-to-severe SNHL in quiet environments, their efficacy markedly degrades in a noisy and/or reverberating environment [2, 3]. In a CI device, the input sound signal is received via a microphone and fed into a speech processor. The speech processor captures the multi-channel temporal envelopes of the input signal, and then generates electric stimulations that directly excite the residual auditory nerves [4, 5]. Due to biological constraints, the dynamic range (DR) of stimulation generated by a speech processor in a CI is much smaller than that of a real speech signal. Hence, a compression scheme is required to compress the DR of the input signal to a desirable level.

Y.-H. Lai
Department of Electrical Engineering, Yuan Ze University, Chung Li, Taiwan
e-mail: yhlai@ee.yzu.edu.tw

F. Chen (✉)
Department of Electrical and Electronic Engineering, Southern University of Science and Technology, Shenzhen, China
e-mail: fchen@sustc.edu.cn

Y. Tsao
Research Center for Information Technology Innovation, Academia Sinica, Taipei, Taiwan
e-mail: yu.tsao@citi.sinica.edu.tw

© Springer International Publishing AG 2017
A. Chattopadhyay et al. (eds.), *Emerging Technology and Architecture for Big-data Analytics*, DOI 10.1007/978-3-319-54840-1_9

In the past, several compression strategies have been proposed. Among them, static envelope compression (SEC) is a popular method that is widely adopted in current CI devices. The SEC strategy uses a fixed compression ratio (CR) [6] to convert the acoustic amplitude envelope to an electric current signal [7, 8]. Although the SEC strategy can confine the overall electrical current signal applied to the CI to within a preset DR, the CR is not optimized to make the best use of the DR for speech perception. Yet, the DR of a temporal envelope is an important factor in speech intelligibility for CI users [9–11], particularly under noisy conditions. Therefore, designing a satisfactory, adaptive compression strategy that can perform compression effectively while maintaining a satisfactory DR is important.

More recently, a novel adaptive envelope compression (AEC) strategy has been proposed [12–14]. The AEC strategy aims to provide deeper modulation than the SEC strategy to endow CI recipients with improved speech intelligibility. The AEC strategy shares a concept with the adaptive wide-dynamic range compression (AWDRC) amplification scheme, which is designed for hearing aids [15]. The AEC strategy aims to optimize the CR while confining the compressed amplitude of the speech envelope within a preset DR. To this end, the AEC strategy dynamically updates the CR in real time, so that the local DR of the output envelope waveform can be effectively increased, resulting in a larger modulation depth and improved intelligibility.

In this chapter, we present four sets of experiments to show the effects of the AEC strategy. We first evaluated the performance of AEC under challenging listening conditions. Then, we investigated the effects of the adaptation rate in the AEC strategy on the intelligibility of envelope-compressed speech. Finally, we investigated the compatibility of the AEC strategy with noise reduction (NR) methods. We show that the AEC strategy has better speech perception performance than the SEC strategy, and can be suitably adopted in a CI speech processor.

The remainder of this chapter is organized as follows. Section 9.2 reviews speech processing in a CI device. Section 9.3 introduces vocoder-based speech synthesis, which is an important tool and is popularly used in the field of CI research. Section 9.4 introduces the compression scheme, including the conventional SEC strategy and the proposed AEC strategy. Section 9.5 presents the experimental setup and four sets of experimental results. Finally, Sect. 9.6 provides the concluding remarks and summarizes this chapter.

9.2 Speech Processor in CI Devices

A CI device consists of four fundamental units: (1) a microphone that picks up the sound, (2) a signal processor that converts the sound signals into electrical signals, (3) a transmission system that transmits the electrical signals to implanted electrodes, and (4) an electrode array that is implanted into the cochlea [16, 17]. The combination of the first two units comprises the speech processor in a CI device. Figure 9.1 shows the design of a speech processor with eight channels. As shown

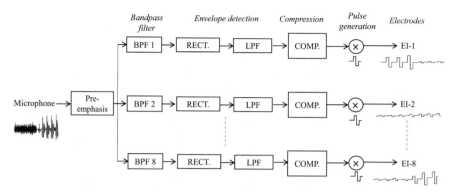

Fig. 9.1 Block diagram of an eight-channel speech processor [4]

in the figure, sound signals are first picked up by a microphone and then passed through a pre-emphasis filter to amplify high frequency components and attenuate low frequency components. The emphasized signals are then processed through a set of bandpass filters (eight filters in Fig. 9.1) to generate a set of bandpass signals. These bandpass signals are then processed by rectifiers and low-pass filters to create a set of temporal envelopes, each representing a specific frequency band. In general, the electrical DR of a CI recipient, between the threshold current (T-level) and the maximum comfortable current (C-level), is 5–20 dB [18, 19]. This DR is markedly less than that of sound levels encountered in real-world environments. For example, the DR of speech signals for a single speaker is 40–50 dB [9, 20]. Therefore, the speech processor of a CI device requires a compression (or automatic gain-control [17, 21]) strategy. The compression stage (i.e., COMP. in Fig. 9.1) is used to compress the wide temporal envelope to output magnitudes within the ranges of 0–1 [17]. The base level is the envelope level that produces a magnitude of 0, which yields a current at T-level. The saturation level is the envelope level that produces a magnitude of 1, which yields a current at C-level. Finally, the temporal envelopes are multiplied by non-simultaneous, biphasic pulse trains delivered as an electrical current through the cochlea via the electrode array.

Speech perception by CI users is primarily facilitated via temporal envelopes. The compression stage narrows down the envelope DR and, accordingly, may reduce speech comprehension by CI recipients, especially under challenging listening conditions (e.g., in noise and/or reverberation). Therefore, a suitable compression scheme is requisite so that CI users can obtain additional temporal envelope information.

9.3 Vocoder-Based Speech Synthesis

Although the number of CI recipients has greatly increased in recent years, one key challenge remains in the CI research field: it is difficult to conduct experiments on real CI recipients. To address this challenge, vocoder simulations derived from CI speech processing strategies have been presented to normal-hearing (NH) listeners in an attempt to predict the intelligibility of CI speech processing [12, 22]. Many studies have shown that vocoder simulations could predict the pattern of the performance observed by CI users, including the effects of background noise [23], the type of speech masker [24], and the number of electrodes [2, 25, 26].

Figure 9.2 shows the block diagram of an eight-channel tone-vocoder. The signal processing units of the tone-vocoder are similar to those of the CI speech processor (refer to Fig. 9.1). The input signals are first processed through the pre-emphasis filter (with a 3-dB/octave roll-off and 2000-Hz cutoff frequency). The bandpass filters (sixth-order Butterworth filters) are then used to filter the emphasized signal into eight frequency bands between 80 and 6000 Hz (with cutoff frequencies of 80, 221, 426, 724, 1158, 1790, 2710, 4050, and 6000 Hz). The temporal envelope of each spectral channel is extracted by a full-wave rectifier, followed by a low-pass filter. The envelope of each band is then compressed using a compression strategy (COMP in Fig. 9.3). Here, we implemented both the SEC and AEC strategies for COMP (refer to Figs. 9.3 and 9.4). The SEC strategy uses a fixed CR to confine the DR of the amplitude of the entire envelope to within a preset value; the AEC strategy, on the other hand, dynamically varies the CR value in a frame-by-frame manner (e.g., 2.5 ms in this study), with the maximum and minimum values of the compressed amplitude limited to a preset range. The compressed envelopes are then modulated using a set of sine waves (i.e., tone i in Fig. 9.2), where the frequencies for these sine waves are equal to the center frequencies of the bandpass filters. Finally, the amplitude-modulated sine waves of the eight bands are summed, and

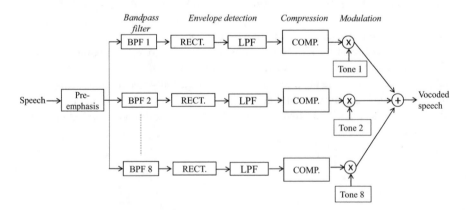

Fig. 9.2 Block diagram of an eight-channel tone-vocoder

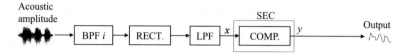

Fig. 9.3 Block diagram of the *i* channel speech processor using the SEC strategy

the level of the summed signal is adjusted to yield a root-mean-square (RMS) value equal to that of the original input signal. Notably, the vocoder simulations are not expected to predict the absolute results, but rather the performance trends noted by CI users. In the present chapter, we adopted the tone-vocoder, as shown in Fig. 9.2 to conduct four speech recognition tests on NH subjects.

9.4 Compression Scheme

The goal of the compression scheme in a CI device is to convert the sound signal into an electrical signal within the preset DR. In this section, we first review a popular compression scheme, namely the SEC strategy. Then, we present the recently proposed AEC strategy.

9.4.1 The Static Envelope Compression Strategy

The SEC strategy uses a linear transformation to compress the DR of the input signal to a desirable level [6]. Figure 9.3 shows the block diagram of the SEC-based speech processor in one channel. In Fig. 9.3, x and y denote the envelopes of input and output envelope signals, respectively. The output-compressed amplitude envelope signal y is computed as:

$$y = \alpha \times (x - \bar{x}) + \bar{x}, \tag{9.1}$$

where $\bar{x}$ is the mean of the input amplitude envelope x, and α is a scaling factor (SF), which is determined in order to ensure that the output amplitude envelope falls within a desirable DR:

$$UB = LB \times 10^{\frac{DR}{20}}. \tag{9.2}$$

In Eq. (9.2), UB and LB denote the upper bound (i.e., the maximum) and lower bound (i.e., the minimum) of the output amplitude values, respectively. From Eq. (9.2), it can be seen that the mean value of the output amplitude envelope equals that of the input amplitude envelope (i.e., $\bar{y} = \bar{x}$).

Notably, using a small SF in Eq. (9.1) induces a large CR, and vice versa. When α equals 0, the compressed amplitude envelope becomes a direct current (DC) signal with a constant value of $\bar{y}$ (i.e., $\bar{y} = \bar{x}$), and the DR is 0 dB; when α equals 1, the output amplitude envelope has the same DR as the input amplitude envelope. For the SEC strategy, α is set by audiology, and a fixed value is applied to the whole amplitude envelope so as to confine its DR to a preset value. A previous study [11] showed that by setting α to 1/3, 1/5, and 1/13 in Eq. (9.1), the DR of multi-channel amplitude envelopes of the Mandarin version of sentences for the Hearing in Noise Test (MHINT) [27] were adjusted to 15, 10, and 5 dB, respectively.

9.4.2   The Adaptive Envelope Compression Strategy

For the SEC strategy as shown in Fig. 9.3, a fixed α is applied to the whole amplitude envelope to confine its DR to a preset value. Although fixed mapping can effectively confine the DR, the SEC strategy does not make optimal use of the DR for speech perception. Recently, the AEC strategy was proposed to confine the amplitude envelope of speech signal within a fixed DR while continuously adjusting the SF for short-term amplitude [12–14]. Using the AEC strategy, the local DR approaches that of the uncompressed amplitude envelope, and thus the AEC-processed amplitude envelopes yield higher intelligibility than do the SEC-processed envelopes [12–14]. Figure 9.4 demonstrates the block diagram of the AEC-based speech processor in one channel. Compared with the SEC strategy (in Fig. 9.3), the AEC strategy uses a feedback unit to calculate the bounds and to apply the AEC rules and peak clipping unit in each channel. In Fig. 9.4, x, z, and y denote the input, compressed, and output envelopes, respectively, and $\Delta\alpha$ is the step size. Instead of using a fixed SF, the AEC strategy adjusts the SF value for each speech frame to compute the compressed amplitude envelope:

$$z_t = \alpha_t \times (x_t - \bar{x}) + \bar{x}, \tag{9.3}$$

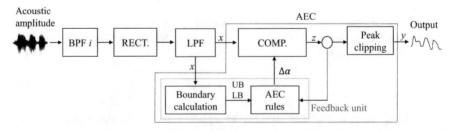

Fig. 9.4 The block diagram of the ith channel speech processor with the AEC strategy

where z_t and x_t are the compressed and original envelopes, respectively, at the tth speech frame, and α_t is the adaptive SF that is updated with $\Delta\alpha$:

$$\alpha_{t+1} = \alpha_t + \Delta\alpha. \tag{9.4}$$

In Fig. 9.4, the boundary calculation and the AEC rules units are used to determine $\Delta\alpha$ in Eq. (9.4). Given an input envelope, the boundary calculation unit computes the DR of the compressed envelope by estimating two bounds: the upper bound, UB; and the lower bound, LB, represented as:

$$\begin{aligned} UB &= \bar{x} + \alpha_0 \times (\max(x) - \bar{x}) \\ LB &= \bar{x} + \alpha_0 \times (\min(x) - \bar{x}) \end{aligned}, \tag{9.5}$$

where $\max(x)$ and $\min(x)$ are the maximum and minimum values of input amplitude envelope x, and α_0 is an initial SF for the AEC strategy. Generally speaking, the fixed compression rate used for the SEC strategy, α in Eq. (9.1), can be used as the initial SF α_0 for the AEC strategy in Eqs. (9.3) and (9.5).

With the estimated UB and LB based on Eq. (9.5), $\Delta\alpha$ in Eq. (9.4) is determined by two AEC rules, as:

1. *Increasing-envelope rule*: This rule aims to keep the compression process as close to linear as possible, such that $\alpha = 1$. By doing so, fewer input signals will be perturbed by compression when producing the output signal. When z_t lies between UB and LB, the AEC strategy will increase α_t by using a positive $\Delta\alpha$ in Eq. (9.4), accordingly increasing the SF value. This increasing-envelope rule stops when α_{t+1} reaches 1, which corresponds to the absence of compression application. In this scenario, the original signal is used as the output signal.
2. *Decreasing-envelope rule*: This rule aims to ensure that the amplitude of the output envelope will not fall outside the preset DR. When z_t in Eq. (9.3) becomes larger than UB, or lower than LB, the AEC strategy will decrease α_t by using a negative $\Delta\alpha$ in Eq. (9.4), which accordingly reduces the SF value. This decreasing-envelope rule stops when α_{t+1} reaches the initial value, α_0.

The AEC strategy follows the above two rules to dynamically adjust the SF values so as to compress the input amplitudes to fit the DR of the electrical current applied to the CI. However, when sudden changes in the input envelope happen, overshooting or undershooting may occur. To overcome such issues, the peak clipping unit is applied, as shown in Fig. 9.4, to ensure that the output envelope falls between the maximum and minimum levels, where UB and LB in Eq. (9.5) can be used as maximum and minimum values, respectively. Finally, the compressed amplitude envelope is computed as:

$$\begin{cases} y_t = z_t, & \text{if } LB < z_t < UB \\ y_t = UB, & \text{if } z_t \geq UB \\ y_t = LB, & \text{if } z_t \leq LB \end{cases}. \tag{9.6}$$

Figures 9.5a, b show examples of the SEC- and AEC-processed amplitude envelopes of the 6th channel (i.e., $f_{low} = 1790\,\text{Hz}$, $f_{high} = 2710\,\text{Hz}$) in a noisy

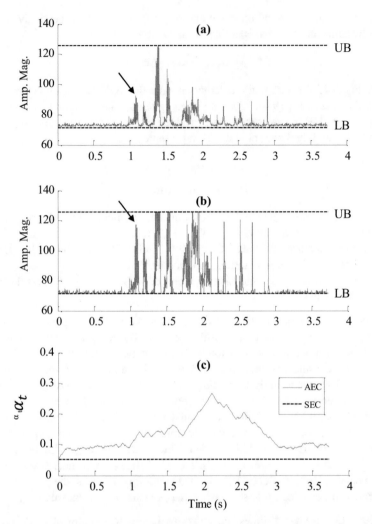

Fig. 9.5 Examples of the amplitude envelope processed using the (**a**) SEC and (**b**) AEC (i.e., $\Delta\alpha$ was 0.001 per 2.5 ms). The envelope waveforms were extracted from the 6th channel (i.e., $f_{\text{low}} = 1790\,\text{Hz}, f_{\text{high}} = 2710\,\text{Hz}$) of a testing sentence masked by an SSN masker at 0 dB, and compressed to a DR of 5 dB. In (**c**), the *solid line* shows the SF α_t used in the AEC strategy for the compressed amplitude envelope in (**b**); the *dashed line* indicates the fixed SF used in the SEC strategy in (**a**)

environment, when masked by speech-shape noise (SSN) at 0 dB signal-to-noise ratio (SNR). The UB and LB in this example are 125.3 and 70.6, respectively, yielding a DR of approximately 5 dB. As seen in Fig. 9.5a, b, both the SEC and AEC strategies effectively compress the DR of the amplitude envelope within the preset DR; however, the local (e.g., around 1.1 s) DR is larger for the envelope processed using the AEC strategy, which had a DR of 4.3 dB in Fig. 9.5b, than for that processed by the SEC strategy, which had a DR of 2.3 dB in Fig. 9.5a. Figure 9.5c shows the SF values used in the SEC and AEC strategies for compression of the sentences in Fig. 9.5a, b, respectively. From Fig. 9.5c, the SEC strategy applies a fixed SF of $\alpha = 1/13$, while the AEC strategy continuously adjusts its SF α_t. It has been noted that the SF α_t for the AEC strategy is generally larger than the fixed SF of $\alpha = 1/13$ employed in the SEC strategy. These results indicate that the AEC strategy can modulate the SF, based on the characteristics of the input signals so as to utilize the usable DR optimally. Moreover, the AEC strategy generates an amplitude envelope with a larger DR and, consequently, a larger modulation depth.

9.5 Experiments and Results

In this section, we present four sets of experiments used to verify the effects of the AEC strategy. In Experiment-1 and Experiment-2, we tested the performance of the AEC strategy in noise and in reverberation, respectively. In Experiment-3, we explored the effect of the adaptation rate in the AEC strategy. Experiment-4 was designed to investigate whether the advantage of a front-end noise reduction scheme could be preserved upon integration with a subsequent AEC strategy, and how this advantage would be influenced by the factors of input SNR, type of NR, and type of noise.

9.5.1 Experiment-1: The Speech Perception Performance of AEC in Noise

CI recipients have limited hearing DR for speech perception. This may partially account for their poor speech comprehension, particularly under noisy conditions. The proposed AEC strategy aims to maximize the modulation depth for CI recipients, while confining the compressed amplitude envelope to the preset DR. The purpose of Experiment-1 was to compare speech recognition synthesized by the proposed AEC strategy and by the SEC strategy under noisy conditions.

9.5.1.1 Subjects and Materials

Eleven (age range: 18–24 years; six females and five males) NH native-Mandarin speakers were recruited to participate in the listening tests. Sentence lists from the MHINT database were used to prepare the testing materials [27]. All sentences were pronounced by a male native-Mandarin speaker, with a fundamental frequency ranging from 75 to 180 Hz, and recorded at a sampling rate of 16 kHz. Two types of maskers: an SSN and two equal-level interfering male talkers (2T) were used to corrupt the testing sentences at two SNR levels (5 and 10 dB), which were chosen to avoid the ceiling/floor effects.

9.5.1.2 Procedure

The listening tests were conducted in a soundproof booth. The stimuli were played to listeners through a set of Sennheiser HD headphones at a comfortable listening level. The speech was compressed to an envelope DR of 5 dB in a vocoder simulation, which was done by using the SF $\alpha = 1/13$ and $\alpha_0 = 1/13$ in Eq. (9.1) and Eq. (9.4), respectively. Each subject participated in a total of eight [2 SNR levels × 2 types of maskers × 2 envelope compression strategies, i.e., the SEC and AEC strategies, respectively] testing tasks. Each task contained ten sentences, and the order of the eight tasks was randomized across subjects. None of the ten sentences were repeated across testing tasks. Subjects were instructed to repeat what they heard, and were allowed to listen to each stimulus twice. The sentence recognition score was used to evaluate the performance, which was calculated by dividing the number of words correctly identified by the total number of words in each testing task. During testing, each subject was given a 5-min break every 30 min during the test.

9.5.1.3 Results and Discussion

Figure 9.6 shows the listening test results in terms of mean sentence recognition scores at different SNR conditions. As shown in Fig. 9.6, the AEC strategy yielded higher speech recognition performance than did the SEC strategy in the three noisy testing tasks (5 dB and 10 dB for SSN, and 5 dB for 2T). To confirm the significance of the improvements further, two-way analysis of variance (ANOVA) and post-hoc comparisons were used to analyze the results of the two strategies under the four noisy conditions.

For the SSN results in Fig. 9.6, the ANOVA measures indicated significant effects of SNR level ($F[1, 10] = 36.03$, $p < 0.005$), compression strategy ($F[1, 10] = 33.41$, $p < 0.005$), and the interaction between SNR level and compression strategy ($F[1, 10] = 5.52$, $p = 0.041$). The post-hoc analyses further confirmed that the score differences between the SEC-processed sentences and AEC-processed sentences were significant ($p < 0.05$). For the 2T results, shown in Fig. 9.6b, the ANOVA measures indicated the significant effect ($F[1, 10] = 77.87$, $p < 0.005$) of SNR level, a non-significant effect of compression strategy ($F[1, 10] = 3.14$, $p = 0.107$), and a significant interaction between SNR level and compression strategy ($F[1, 10] = 12.63$, $p = 0.005$). The post-hoc analyses further showed that the score difference at 5 dB SNR was significant ($p < 0.05$) while that at 10 dB SNR was non-significant ($p = 0.46$) in Fig. 9.6b.

To analyze the advantage of the AEC strategy further, it is worthwhile to revisit the examples shown in Fig. 9.7. From this figure, it can be seen that the AEC-

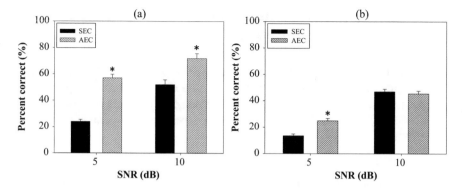

Fig. 9.6 The mean recognition scores of SEC and AEC in (**a**) SSN and (**b**) 2T maskers. The DR of the envelope amplitude is confined to 5 dB. The *error bars* denote the standard errors of the mean (SEM) values. An *asterisk* indicates a statistically significant ($p < 0.05$) difference between SEC and AEC scores

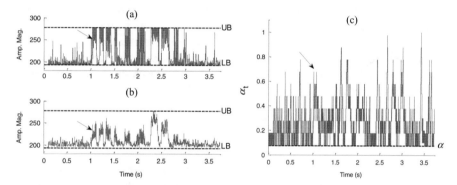

Fig. 9.7 Examples of amplitude envelope processed by (**a**) AEC (i.e., $\Delta\alpha = 0.1$) and (**b**) SEC strategies, and (**c**) the compression ratio α used in the AEC and SEC (*dashed line*). The envelope waveforms were extracted from the 6th channel of a testing sentence masked by SSN at 5 dB SNR, and compressed to 5 dB DR within [LB, UB]

processed envelope around 1 s and 2 s in Fig. 9.7a has a larger DR than that processed by the SEC strategy in Fig. 9.7b. Moreover, Fig. 9.7c demonstrates that the SEC strategy employs a fixed compression factor, where $\alpha = 1/13$, while the AEC strategy uses a small amplitude compression ratio, or a large SF α_t, for the frames around 1 s and 2 s, thus yielding a large modulation depth for the amplitude envelope and improved speech intelligibility.

In summary, the results of this experiment showed that the intelligibility of the AEC-processed sentences was significantly better than their SEC-processed counterparts under noisy conditions. This suggests that the proposed AEC strategy holds promise for improving speech perception performance under noisy listening conditions for patients with CI.

9.5.2 Experiment-2: The Speech Perception Performance of AEC in Reverberation

Reverberation, which results from multiple reflections of sounds from objects and surfaces in an acoustic enclosure, causes spectro-temporal smearing of speech [28]. Previous studies have indicated that a reverberating environment may reduce a CI recipient's ability to identify words [29, 30]. In Experiment-2, we intended to assess the effects of the AEC strategy in reverberation.

9.5.2.1 Subjects and Materials

Nine (age range: 19–27 years, four females and five males) NH native-Mandarin speakers were recruited to participate in the listening tests. As with Experiment-1, sentence lists from the MHINT [27] were used to prepare the testing materials. The reverberant conditions were simulated by head-related transfer functions recorded in a 5.5 m × 4.5 m × 3.1 m (length × width × height) room with a total volume of 76.8 m^3 [31]. The average reverberation time of the experimental room ($T_{60} = 1.0$ s) was reduced to $T_{60} = 0.6, 0.4$ and 0.2 s by adding floor carpeting, absorptive panels on the walls, and a ceiling, respectively. Additional details on simulating reverberant conditions can be found in previous studies [31, 32].

9.5.2.2 Procedure

The listening tests were conducted in a soundproof booth. The stimuli were played to listeners through a set of Sennheiser HD headphones at a comfortable listening level. The speech signals were compressed to the envelope DR of 5 dB in the vocoder simulation. This was done by using the SF $\alpha = 1/13$ and $\alpha_0 = 1/13$ in Eq. (9.1) and Eq. (9.3), respectively. Each subject participated in a total of eight [= 4 reverberant conditions (i.e., $T_{60} = 0, 200, 400,$ and 600 ms) × 2 envelope compression strategies (i.e., SEC and AEC)] testing tasks. Each task contained ten sentences. The order of the eight tasks was randomized across subjects, and none of the ten sentences were repeated across testing tasks. The subjects had repeated what they had heard during the experiments, and were allowed to listen to each stimulus twice. The sentence recognition score was used to compare speech recognition performance.

9.5.2.3 Results and Discussion

Figure 9.8 shows the listening test results in terms of mean sentence recognition scores for all testing tasks. The two-way ANOVA measures were computed by using the recognition score as the dependent variable. The reverberant condition, the

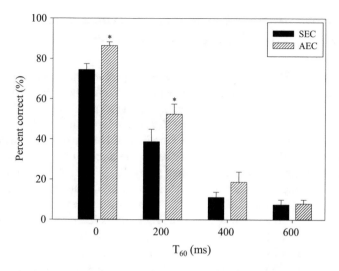

Fig. 9.8 The mean recognition scores of SEC and AEC. The DR of envelope amplitude is confined to 5 db. The *error bars* denote the SEM values. An *asterisk* indicates a statistically significant ($p<$ 0.05) difference between SEC and AEC scores

T_{60} value, and the compression strategy were considered as the two within-subject factors. ANOVA results indicated significant effects in the reverberant condition ($F[3, 24] = 208.62, p < 0.05$), the compression strategy ($F[1, 8] = 35.01, p < 0.05$), and in the interaction between the reverberant condition and compression strategy ($F[3, 24] = 8.05, p = 0.001$). Post-hoc analyses showed that the score differences between the SEC-processed sentences and AEC-processed sentences were significantly different ($p < 0.05$) under the reverberant conditions of $T_{60} = 0$ and 200 ms in Fig. 9.8.

In accordance with the intelligibility advantage observed in Experiment-1, the results of Experiment-2 showed that the amplitude envelope processed by the AEC strategy yielded higher intelligibility scores for vocoded sentences under reverberation compared to those processed using the SEC strategy. This makes the AEC strategy a highly promising way of enhancing speech comprehension in implanted listeners under reverberant conditions in the future. Interestingly, the above intelligibility advantage was not observed for all reverberant conditions. This indicated that there was no significant improvement in intelligibility found under the reverberant conditions of $T_{60} = 600$ ms in Fig. 9.8. This may be partially attributed to the usage of initial compression parameters (e.g., α_0 and $\Delta\alpha$) in the experiment. It is worthwhile to investigate optimal configuration of compression parameters so as to achieve the best performance for the AEC-based speech processing for CI recipients under reverberant listening conditions in future studies.

9.5.3   Experiment-3: The Effect of Adaptation Rate on the Intelligibility of AEC-Processed Speech

As mentioned in Sect. 9.4.2, the AEC strategy specifies the rate at which the SF value should be updated, using the adaptation rate outlined in Eq. (9.4), which is determined based on the two AEC rules. The adaptation rate used in the AEC strategy is similar to the attack time (AT) and release time (RT) of the wide-dynamic-range compression (WDRC) amplification scheme that is widely used in hearing aids [33]. The AT and RT describe the duration required for a hearing aid device to respond to a changing input signal [34]. When setting an inadequate time constant value for AT/RT, the gain will fluctuate rapidly and thus generate an undesirable pumping effect. Conversely, when setting an excessive time constant value, a lag in perception will be induced. Previous studies have explored the effects of AT/RT values on the intelligibility and satisfactory sound quality of a hearing aid for its users [35–38]. Their results indicated that AT and RT values should be carefully optimized in order to achieve satisfactory performances in speech intelligibility, listening comfort, and sound quality. For this reason, the adaptation rate is an important parameter in the AEC strategy. In this section, we investigate the effect of the adaptation rate in Eq. (9.4) on the intelligibility of the AEC-processed speech.

9.5.3.1   Subjects and Materials

Eight NH, native-Mandarin listeners (age range: 19–26 years, four females and four males) listeners were recruited to participate in the listening experiment. Sentences from the MHINT were used as the testing materials [27]. Two types of maskers, SSN and 2T, were used to prepare the noisy testing sentences at SNR levels of 5 and 10 db, which were chosen based on a pilot study to avoid ceiling and floor effects.

9.5.3.2   Procedure

The listening tests were conducted in a sound-proof room, and stimuli were played to listeners through a set of Sennheiser HD headphones at a comfortable listening level. The DR of the envelope waveforms were compressed to 5 db in tone-vocoder simulations, where the α values of SEC and α_0 of the AEC strategy were set to 1/13 [11]. Four different envelope compression methods were used in this experiment, i.e., SEC, AEC with $\Delta\alpha = 0.001$, AEC with $\Delta\alpha = 0.01$, and AEC with $\Delta\alpha = 0.1$. The last three are referred to as AEC.001, AEC.01, and AEC.1, respectively. Each subject participated in a total of 16 (2 SNR levels $\times$ 2 types of maskers $\times$ 4 envelope compression methods) listening tasks. Each task contained ten sentences, and the order of the 16 tasks was randomized across subjects. None of the ten sentences were repeated across the listening task. Subjects were instructed to repeat what they heard, and they were allowed to listen to each stimulus twice. The sentence recognition score was used to evaluate the performance.

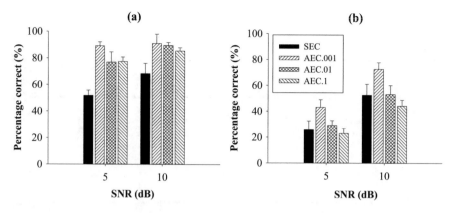

Fig. 9.9 The mean recognition scores for Mandarin sentences with (**a**) an SSN masker and (**b**) a 2T masker at SNR levels of 5 and 10 db. The *error bars* indicate SEM values

9.5.3.3   Results and Discussion

Figure 9.9 demonstrates the speech recognition scores in terms of the mean recognition rates for all testing tasks. From Fig. 9.10a, we noted that the three AEC strategies (AEC.001, AEC.01, and AEC.1) produced notably higher intelligibility scores than the SEC strategy in the SSN test condition. From Fig. 9.9b, we noted that AEC.001 yielded notably higher recognition scores than did the SEC strategy in the 2T masker condition. One-way ANOVA and Tukey post-hoc comparisons were conducted to analyze the results of the four compression strategies in the four testing conditions further. These analyses are summarized in Table 9.1. In the table, each mean score represents the corresponding recognition score in Fig. 9.10, and n denotes the sample size. For the results of the SSN masker, the ANOVA measures confirmed that the intelligibility scores differed significantly across the four groups, with ($F = 10.25$, $p < 0.001$) and ($F = 5.80$, $p = 0.003$) at SNR levels of 5 db and 10 db, respectively. The Tukey post-hoc results further verified the significant differences for the following group pairs at both SNR levels of 5 and 10 db: SEC with AEC.001; SEC with AEC.01; and SEC with AEC.1. Moreover, the ANOVA results for the 2T masker confirmed that the intelligibility scores differed significantly across the four groups, with ($F = 3.00$, $p = 0.048$) and ($F = 3.49$, $p = 0.029$) at SNR levels of 5 and 10 db, respectively. The Tukey post-hoc comparisons verified the significant differences between the group pairs of SEC and AEC.001 at both SNR levels (5 and 10 db).

These results confirm that the value of adaptation rate, $\Delta\alpha$, indeed affects the intelligibility of AEC-processed speech. The use of an inappropriate value of $\Delta\alpha$ may diminish the benefits affecting speech intelligibility, especially under testing conditions using an interfering masker. To demonstrate the effect of the adaptation rate further, Fig. 9.10 highlights the results obtained by the traditional SEC strategy; the AEC strategy with a very slow adaptation rate, where $\Delta\alpha = 0.0001$; and the AEC strategy with a fast adaptation rate, where $\Delta\alpha = 0.1$. The examples showed that, when using an inadequate $\Delta\alpha$ value, such as 0.0001 as shown in Fig. 9.10b, the

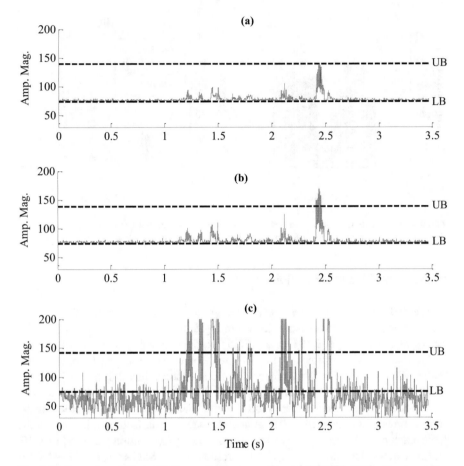

Fig. 9.10 Examples of the amplitude envelope processed by (**a**) SEC, (**b**) AEC with a very slow adaptation rate, $\Delta\alpha = 0.0001$, and (**c**) AEC with a fast adaptation rate, $\Delta\alpha = 0.1$. The envelope waveforms were extracted from the 6th channel of a testing sentence masked by SSN at 5 db, and compressed to 5 db DR

benefits of the AEC strategy are limited, as the available DR is not effectively used. Conversely, when using a similarly inadequate $\Delta\alpha$ value, such as 0.1, as shown in Fig. 9.10c, a pumping effect may occur. Consequently, some envelopes will fall within the range of peak clipping, accordingly causing speech signal distortions. In addition, when the envelope waveform varies too rapidly, speech intelligibility will also be decreased. From the results of Figs. 9.9 and 9.10, when $\Delta\alpha$ is equal to 0.001, the optimal balance between benefits, the pumping effect, and the distortion in the AEC strategy becomes evident. In this experiment, we only selected these three values of $\Delta\alpha$ (slow, moderate, and fast adaptation rate) to investigate the performance of the AEC strategy. Future studies should further investigate the effect of $\Delta\alpha$ while considering the wider characteristics of language, such as the band importance function or the tone of Mandarin, in addition to the noise types.

Table 9.1 The mean recognition scores for different strategies, where each factor was included in the one-way ANOVA and Tukey post-hoc testing

Test condition	Strategy	n	Mean score	F	p	Post-hoc comparison* (group$_i$, group$_j$)
SSN				10.25	<0.001	
(SNR = 5 db)	SEC	8	51.8			(SEC, AEC.001)
	AEC.001	8	89.1			(SEC, AEC.01)
	AEC.01	8	76.9			(SEC, AEC.1)
	AEC.1	8	77.4			
SSN				5.80	0.003	
(SNR = 10 dB)	SEC	8	68.3			(SEC, AEC.001)
	AEC.001	8	91.0			(SEC, AEC.01)
	AEC.01	8	89.5			(SEC, AEC.1)
	AEC.1	8	85.4			
2T				3.00	0.048	
(SNR = 5 db)	SEC	8	25.9			(SEC, AEC.001)
	AEC.001	8	43.2			
	AEC.01	8	29.2			
	AEC.1	8	23.0			
2T				3.49	0.029	
(SNR = 10 dB)	SEC	8	52.5			(SEC, AEC.001)
	AEC.001	8	72.5			
	AEC.01	8	53.1			
	AEC.1	8	52.5			

9.5.4 Experiment-4: The Effect of Joint Envelope Compression and Noise Reduction

An NR method plays a crucial role in the improvement of sound quality/intelligibility in noisy conditions [39–42]. Chung [43] found that NR methods greatly enhanced the modulation depth of noise-suppressed signals, but these benefits were somehow eliminated by the compression stage. On the other hand, the benefits of NR approaches can be maintained or even further improved by using a suitable compression strategy. Previous studies of NR have found that integrating the NR approach with a fixed compression ratio strategy benefits speech perception by CI recipients [44]. The aim of this experiment was to evaluate how the AEC strategy interacted with the NR approaches in the handling of noisy speech.

9.5.4.1 Subjects and Materials

Eight NH, native-Mandarin listeners (age range: 19–26 years, four females and four males) were recruited to participate in the listening test. The MHINT sentences were used to test the performance, with SSN and 2T maskers used to corrupt testing sentences. The test speeches included 0, 5, and 10 db SNR levels.

9.5.4.2   Signal Processing with NR and Envelope Compression

In this set of experiments, an NR process was implemented before the AEC strategy in an eight-channel tone-vocoder. Figure 9.11 shows the block diagram of the NR-and AEC-based tone-vocoder in one channel. In the figure, the noisy speech signal was first processed by an NR method and then fed into a standard tone-vocoder, where the SEC strategy or the AEC strategy could be used as the compression scheme. We adopted two types of NR approaches, namely the Wiener filtering approach [45] and the Karhunen Loeve theorem (KLT) algorithm [46] in this set of experiments. The Wiener filtering approach utilizes a priori SNR statistics to design a gain function to filter out noise components from the noise input. For the KLT method, the KLT algorithm is first applied to the noisy signal. The KLT components that represent the signal subspace were modified by a gain function, while the remaining KLT components that represent the noise subspace were nulled. An enhanced signal was obtained by applying the inverse KLT of the modified components. The techniques used in these two algorithms have been detailed in previous studies [45, 46].

Following the NR stage in Fig. 9.11, the envelope of the noise-suppressed signal was extracted via bandpass filtering and waveform rectification. Next, in the CR estimation (CRE) stage, the appropriate SF for the SEC strategy or the initial SF (i.e., α_0) for the AEC strategy was determined based on the input envelope (x in Fig. 9.11). This SF was then used to transform the output envelope to generate a compressed signal (z in Fig. 9.11). The final peak clipping stage was used to confine the compressed envelope to within the expected DR (y in Fig. 9.11). The compressed envelopes were then modulated by a set of sine waves (i.e., tone i) with frequencies equal to the center frequencies of the bandpass filters. Finally, the envelope-modulated sine waves of the eight bands were combined, and the level of the combined signal was adjusted to produce an RMS value equal to that of the original input signal.

Figure 9.12 shows examples of an amplitude envelope processed by the Wiener filtering and KLT methods, followed by the SEC and AEC strategies, under same

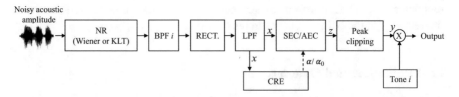

Fig. 9.11 Block diagram for obtaining the output tone-vocoded speech for speech enhancement methods followed by a compression strategy (either AEC or SEC) in the ith channel

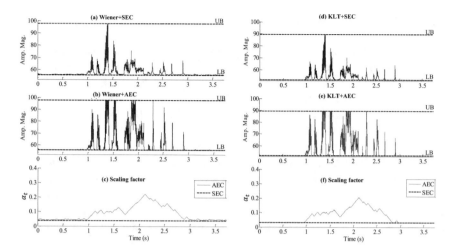

Fig. 9.12 Examples of the amplitude envelope for (**a**) Wiener+SEC, (**b**) Wiener+AEC, (**d**) KLT+SEC, and (**e**) KLT+AEC processing. The envelope waveforms were extracted from the 6th channel of a testing sentence masked by the SSN masker at SNR 5 db, and compressed to a 5-dB DR. In (**c**) and (**f**), the *solid lines* show the SF used in the AEC strategy for (**b**) and (**e**); the *dashed lines* show the fixed SF in the SEC strategy for (**a**) and (**d**)

testing conditions (SSN masker at SNR 5 db). The envelope was extracted from the sixth channel, and compressed to a 5-dB DR, where the initial SF was computed by the CRE stage. We noted two findings: (1) the KLT performance was similar to that of the Wiener filtering approach when integrated with the same compression strategy, and (2) the AEC strategy can provide better modulation depth than the SEC strategy when it is integrated with the same NR algorithm.

9.5.4.3 Procedure

As NR methods were used in this experiment, the SF should differ from the values used in Experiment-1, in order to ensure that the output envelope falls within the desirable DR, which was 5 db in this experiment. As shown in Fig. 9.11, the CRE stage computes an α value for the SEC strategy and an α_0 for the AEC strategy, based on the NR-processed signals. Four different signal processing methods were used: (1) Wiener+SEC, (2) Wiener+AEC, (3) KLT+SEC, and (4) KLT+AEC. The adaptation rate ($\Delta\alpha$) of the AEC strategy was 0.001 in this experiment.

The listening tests were conducted in a soundproof booth. Each subject participated in a total of 24 (3 SNR levels × 2 types of maskers × 4 types of signal processing) testing tasks. Each task contained ten sentences, and the order of these 24 tasks was randomized across subjects. None of the ten sentences were repeated across the testing conditions. Subjects were instructed to repeat what they heard, and they were allowed to listen to each stimulus twice.

9.5.4.4 Results and Discussion

Figure 9.13 shows the mean speech recognition scores for all of the testing tasks. For the SSN results indicated in Fig. 9.13a, the mean recognition rates for Wiener+SEC, Wiener+AEC, KLT+SEC, and KLT+AEC, respectively, were: 18.4%, 33.6%, 15.8%, and 41.1% for 0 db SNR; 23.3%, 53.9%, 14.1%, and 53.0% for 5 db SNR; and 15.7%, 52.6%, 24.1%, and 64.1% for 10 db SNR. For the 2T results in Fig. 9.13b, the mean recognition rates were: 2.3%, 3.1%, 1.1%, and 2.6% for 0 db SNR; 4.3%, 15.6%, 6.5%, and 14.4% for 5 db SNR; and 17.1%, 40.6%, 17.3%, and 41.3% for 10 db SNR. Three-way ANOVA measures was used to analyze these data for the following three factors: the type of masker (masker), SNR level (SNR), and processing method (F1). The results indicated that all of the main effects and second-order interaction were significant (refer to Table 9.2). Tukey's post-hoc analysis showed significant differences for the following group pairs: Wiener+SEC and Wiener+AEC; Wiener+SEC and KLT+AEC; Wiener+AEC and KLT+SEC; and KLT+SEC and KLT+AEC.

The results of the three-way ANOVA and Tukey post-hoc comparisons indicated that the AEC strategy can provide higher speech recognition scores than the SEC strategy when integrated with a Wiener filter or KLT. The reason for the poorer performance of the SEC strategy compared to the AEC strategy in noisy conditions may be similar to that applicable in hearing aids [47, 48], where the static compression processing will increase the low-level noise during the pauses in the speech signal and thereby decrease the SNR performance for the NR algorithm. Therefore, the SEC strategy did not benefit as much from the NR algorithm as the AEC strategy. Moreover, the results in Table 9.2 show that different NR algorithms, such as the Wiener filter and KLT, did not produce significant differences when integrated with the same compression strategy (i.e., SEC or AEC). In contrast, different compression strategies produced significantly different results when integrated with the same NR algorithm, where the AEC strategy consistently

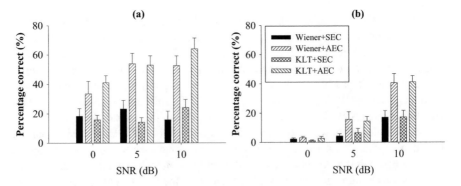

Fig. 9.13 Mean recognition scores for Mandarin sentences with (**a**) an SSN masker and (**b**) a 2T masker, at SNR levels of 0, 5, and 10 dB. The *error bars* indicate SEM values

Table 9.2 The mean recognition scores for different strategies, where three factors (types of masker [masker], SNR levels [SNR], and processing method [F1]) were included in three-way ANOVA and Tukey's post-hoc testing

Source of variance	Type III sum of squares	df	Mean square	F	p	Post-hoc comparison* (group$_i$, group$_j$)
Corrected model	$65,113.9^a$	17	3830.2	20.03	<0.001	
Intercept	$110,544.0$	1	110,544.0	578.15	<0.001	
SNR × F1	3800.57	6	633.43	3.31	0.004	
Masker × F1	4892.81	3	1630.94	8.53	<0.001	
Masker × SNR	2508.07	2	1254.04	6.56	0.002	
Masker	$19,784.38$	1	19,784.38	103.47	<0.001	(1,2), (1,4), (2,3), (3,4)
SNR	$12,065.76$	2	6032.88	31.55	<0.001	
F1	$22,062.35$	3	7354.12	38.46	<0.001	
Error	$33,269.05$	174	191.20			
Total	$208,927.00$	192				
Corrected total	$98,382.99$	191				

F1 group variable: 1, Wiener+SEC; 2, Wiener+AEC; 3, KLT+SEC; 4, KLT+AEC
Dependent variable: speech intelligibility scores
$^a R^2 = 0.669$ (adjusted $R^2 = 0.624$)

outperformed the SEC strategy. NR algorithms probably did not yield significant differences due to the very narrow electrical DR (5 db) used for CIs. Since noisy signals processed by an NR algorithm will generate speech with increased DR, a smaller SF has to be used to ensure that the sound signals remain within the audible range (i.e., between UB and LB). From the examples of Figs. 9.5c and 9.12c, the initial SF, α_0, was larger without than with NR, under the same testing conditions, and the speech intelligibility scores were higher for a larger SF. This provides further evidence that the compression strategy is more important than the NR method in noisy conditions for compressed speech perception.

In summary, this experiment investigated the performance of the AEC strategy when integrated with different NR algorithms. The results indicate that integration of NR algorithms with the AEC strategy provided better speech intelligibility than with the SEC strategy, implying that the AEC strategy, integrated with NR methods, is useful for further improving the speech perception performance in CI recipients.

9.6 Summary

A non-linear, compressive mapping function is normally used in CI devices to convert an acoustic amplitude envelope to an electric current signal (with a narrow DR). The present chapter assessed the performance of compression strategies (i.e., static vs. adaptive) relative to that of CI speech processing by vocoded simulation. More specifically, we used a simple compression function, as shown in Eqs. (9.3) and (9.6), called the AEC strategy, to compress the amplitude envelope into a preset DR. Note that most of the present acoustic-to-electric conversions in CI devices use a fixed mapping function. It is reasonable to foresee that the adaptive mapping (from acoustic to electric) function will improve the speech comprehension of implanted patients. In addition, the signal processing in the AEC strategy is similar to that used by the SEC strategy, but is characterized by additional boundary calculations (for UB and LB). Furthermore, the AEC rules optimally and continuously adjust for the compression ratio on a frame-by-frame basis. Since these two additional units are rather simple, the computation load for the AEC strategy is reasonable when compared with the conventional SEC strategy. This enables for the practical implementation of the AEC strategy by means of microprocessors.

The results of these experiments showed that the amplitude envelope processed by the AEC strategy yielded significantly higher intelligibility scores for vocoded sentences in noisy and in reverberation conditions than when it was processed by the SEC strategy. Moreover, integration of NR methods with the AEC strategy outperformed integration of NR methods with the SEC strategy under noisy conditions. This makes the proposed AEC strategy a highly promising approach for the enhancement of speech comprehension in noisy conditions for listeners with CIs.

Acknowledgements This work was supported by the Ministry of Science and Technology of Taiwan under Project MOST 104-2221-E-001-026-MY2 and MOST 105-2218-E-155-014-MY2. This work was also supported by the National Natural Science Foundation of China (Grant No. 61571213). We thank Dr. Dao-Peng Chen of the Institute of Biomedical Sciences, Academia Sinica, for help with the statistical analysis.

References

1. NIDCD, Cochlear implants, vol. 116. NIH Publication, no. 11–4798 (2013)
2. L.M. Friesen, R.V. Shannon, D. Baskent, X. Wang, Speech recognition in noise as a function of the number of spectral channels: comparison of acoustic hearing and cochlear implants. J. Acoust. Soc. Am. **110**(2), 1150–1163 (2001)
3. B.L. Fetterman, E.H. Domico, Speech recognition in background noise of cochlear implant patients. Otolaryngol. Head Neck Surg. **126**(3), 257–263 (2002)
4. P.C. Loizou, Introduction to cochlear implants. IEEE Eng. Med. Biol. Mag. **18**(1), 32–42 (1999)
5. F.G. Zeng, Trends in cochlear implants. Trends Amplif. **8**(1), 1–34 (2004)
6. P.C. Loizou, M. Dorman, J. Fitzke, The effect of reduced dynamic range on speech understanding: implications for patients with cochlear implants. Ear Hear. **21**(1), 25–31 (2000)
7. D.K. Eddington, W. Dobelle, D. Brackmann, M. Mladejovsky, J. Parkin, Auditory prostheses research with multiple channel intracochlear stimulation in man. Ann. Otol. Rhinol. Laryngol. **87**(6 Pt 2), 1–39 (1977)
8. F.G. Zeng, R.V. Shannon, Loudness balance between electric and acoustic stimulation. Hear. Res. **60**(2), 231–235 (1992)
9. F.G. Zeng, G. Grant, J. Niparko, J. Galvin, R. Shannon, J. Opie, P. Segel, Speech dynamic range and its effect on cochlear implant performance J. Acoust. Soc. Am. **111**(1), 377–386 (2002)
10. R. van Hoesel, M. Böhm, R.D. Battmer, J. Beckschebe, T. Lenarz, Amplitude-mapping effects on speech intelligibility with unilateral and bilateral cochlear implants. Ear Hear. **26**(4), 381–388 (2005)

11. F. Chen, L.L. Wong, J. Qiu, Y. Liu, B. Azimi, Y. Hu, The contribution of matched envelope dynamic range to the binaural benefits in simulated bilateral electric hearing. J. Speech Lang. Hear. Res. **56**(4), 1166–1174 (2013)
12. Y.H. Lai, Y. Tsao, F. Chen, Effects of adaptation rate and noise suppression on the intelligibility of compressed-envelope based speech. Plos One **10**(7), e0133519 (2015)
13. Y.H. Lai, F. Chen, Y. Tsao, Effect of adaptive envelope compression in simulated electric hearing in reverberation, in *2014 14th International Symposium on Integrated Circuits (ISIC)* (IEEE, Singapore, 2014), pp. 204–207
14. Y.H. Lai, F. Chen, Y. Tsao, An adaptive envelope compression strategy for speech processing in cochlear implants, in *Interspeech* (2014), pp. 481–484
15. Y.H. Lai, P.C. Li, K.S. Tsai, W.C. Chu, S.T. Young, Measuring the long-term snrs of static and adaptive compression amplification techniques for speech in noise. J. Am. Acad. Audiol. **24**(8), 671–683 (2013)
16. R.S. Tyler, S. Waltzman, S. Bankoski, *Cochlear Implants: Audiological Foundations* (Singular Publishing Group, San Diego, 1993)
17. P.P. Khing, B.A. Swanson, E. Ambikairajah, The effect of automatic gain control structure and release time on cochlear implant speech intelligibility. Plos One **8**(11), e82263 (2013)
18. F.G. Zeng, J.J. Galvin III, Amplitude mapping and phoneme recognition in cochlear implant listeners. Ear Hear. **20**(1), 60–74 (1999)
19. K. Kasturi, P.C. Loizou, Use of s-shaped input-output functions for noise suppression in cochlear implants. Ear Hear. **28**(3), 402–411 (2007)
20. A. Boothroyd, F.N. Erickson, L. Medwetsky, The hearing aid input: a phonemic approach to assessing the spectral distribution of speech. Ear Hear. **15**(6), 432–442 (1994)
21. C.J. James, P.J. Blamey, L. Martin, B. Swanson, Y. Just, D. Macfarlane, Adaptive dynamic range optimization for cochlear implants: a preliminary study. Ear Hear. **23**(1), 49S–58S (2002)
22. R.V. Shannon, F.G. Zeng, V. Kamath, J. Wygonski, M. Ekelid, Speech recognition with primarily temporal cues. Science **270**(5234), 303–304 (1995)
23. Q.J. Fu, R.V. Shannon, X. Wang, Effects of noise and spectral resolution on vowel and consonant recognition: acoustic and electric hearing. J. Acoust. Soc. Am. **104**(6), 3586–3596 (1998)
24. G.S. Stickney, F.G. Zeng, R. Litovsky, P. Assmann, Cochlear implant speech recognition with speech maskers. J. Acoust. Soc. Am. **116**(2), 1081–1091 (2004)
25. M.F. Dorman, P.C. Loizou, D. Rainey, Simulating the effect of cochlear-implant electrode insertion depth on speech understanding. J. Acoust. Soc. Am. **102**(5), 2993–2996 (1997)
26. M.F. Dorman, P.C. Loizou, D. Rainey, Speech intelligibility as a function of the number of channels of stimulation for signal processors using sine-wave and noise-band outputs. J. Acoust. Soc. Am. **102**(4), 2403–2411 (1997)
27. L.L. Wong, S.D. Soli, S. Liu, N. Han, M.-W. Huang, Development of the Mandarin hearing in noise test (MHINT). Ear Hear. **28**(2), 70S–74S (2007)
28. O. Hazrati, S.O. Sadjadi, P.C. Loizou, J.H. Hansen, Simultaneous suppression of noise and reverberation in cochlear implants using a ratio masking strategy. J. Acoust. Soc. Am. **134**(5), 3759–3765 (2013)
29. O. Hazrati, J. Lee, P.C. Loizou, Blind binary masking for reverberation suppression in cochlear implants. J. Acoust. Soc. Am. **133**(3), 1607–1614 (2013)
30. O. Hazrati, P.C. Loizou, Reverberation suppression in cochlear implants using a blind channel-selection strategy. J. Acoust. Soc. Am. **133**(6), 4188–4196 (2013)
31. T. Van den Bogaert, S. Doclo, J. Wouters, M. Moonen, Speech enhancement with multichannel wiener filter techniques in multimicrophone binaural hearing aids. J. Acoust. Soc. Am. **125**(1), 360–371 (2009)
32. F. Chen, O. Hazrati, P.C. Loizou, Predicting the intelligibility of reverberant speech for cochlear implant listeners with a non-intrusive intelligibility measure. Biomed. Signal Process. Control **8**(3), 311–314 (2013)
33. T. Venema, *Compression for Clinicians* (Delmar, Clifton Park, 2006)
34. P.E. Souza, Effects of compression on speech acoustics, intelligibility, and sound quality. Trends Amplif. **6**(4), 131–165 (2002)

35. A.C. Neuman, M.H. Bakke, C. Mackersie, S. Hellman, H. Levitt, Effect of release time in compression hearing aids: paired-comparison judgments of quality. J. Acoust. Soc. Am. **98**(6), 3182–3187 (1995)
36. A.C. Neuman, M.H. Bakke, C. Mackersie, S. Hellman, H. Levitt, The effect of compression ratio and release time on the categorical rating of sound quality. J. Acoust. Soc. Am. **103**(5), 2273–2281 (1998)
37. M. Hansen, Effects of multi-channel compression time constants on subjectively perceived sound quality and speech intelligibility. Ear Hear. **23**(4), 369–380 (2002)
38. S. Gatehouse, G. Naylor, C. Elberling, Linear and nonlinear hearing aid fittings–1. patterns of benefit. Int. J. Audiol. **45**(3), 130–152 (2006)
39. R. Van Hoesel, G.M. Clark, Evaluation of a portable two-microphone adaptive beamforming speech processor with cochlear implant patients. J. Acoust. Soc. Am. **97**(4), 2498–2503 (1995)
40. V. Hamacher, W. Doering, G. Mauer, H. Fleischmann, J. Hennecke, Evaluation of noise reduction systems for cochlear implant users in different acoustic environment. Otol. Neurotol. **18**(6), S46–S549 (1997)
41. J. Wouters, J.V. Berghe, Speech recognition in noise for cochlear implantees with a two-microphone monaural adaptive noise reduction system. Ear Hear. **22**(5), 420–430 (2001)
42. P.C. Loizou, A. Lobo, Y. Hu, Subspace algorithms for noise reduction in cochlear implants. J. Acoust. Soc. Am. **118**(5), 2791–2793 (2005)
43. K. Chung, Challenges and recent developments in hearing aids part i. speech understanding in noise, microphone technologies and noise reduction algorithms. Trends Amplif. **8**(3), 83–124 (2004)
44. F. Chen, Y. Hu, M. Yuan, Evaluation of noise reduction methods for sentence recognition by mandarin-speaking cochlear implant listeners. Ear Hear. **36**(1), 61–71 (2015)
45. P. Scalart, et al., Speech enhancement based on a priori signal to noise estimation, in *IEEE International Conference on Acoustics, Speech, and Signal Processing*, vol. 2 (IEEE, Atlanta, 1996), pp. 629–632
46. Y. Hu, P.C. Loizou, A generalized subspace approach for enhancing speech corrupted by colored noise. IEEE Trans. Speech and Audio Process. **11**(4), 334–341 (2003)
47. Y.H. Lai, Y. Tsao, F. Chen, A study of adaptive wdrc in hearing aids under noisy conditions. Int. J. Speech Lang. Pathol. Audiol. **1**(2), 43–51 (2013)
48. G. Naylor, R.B. Johannesson, Long-term signal-to-noise ratio at the input and output of amplitude-compression systems. J. Am. Acad. Audiol. **20**(3), 161–171 (2009)

Part III
Emerging Technology, Circuits and Systems for Data-Analytics

Chapter 10
Neuromorphic Hardware Acceleration Enabled by Emerging Technologies

Zheng Li, Chenchen Liu, Hai Li, and Yiran Chen

10.1 Introduction

As demand on high performance computation continuously increases, the traditional von Neumann computer architecture becomes less efficient as the appearance of "Memory Wall[1], which greatly hindered the overall performance of computing engines. In recent years, neuromorphic hardware systems have gained great attention. Under such a condition, many improved or alternative computing architectures were motivated. As an important instance, neuromorphic computing systems have emerged as a promising solution for "Big Data" applications. Neuromorphic computing systems can potentially provide the capabilities of biological perception and information processing within a compact and energy-efficient platform[2]. Many research activities have been carried out on algorithm enhancement [3] and/or system implementations built upon the conventional CPU, GPU, or FPGA [4].

As a highly generalized and simplified abstract of a neuromorphic system, an artificial neural network (NNW) usually uses a connection matrix to represent a set of synapse networks. Nowadays, many NNW models have been proposed, which can be generally classified as feedforward (FFW) and feedback or recurrent types. In the FFW class, the signals flow only in the forward direction, whereas brain state in a box (BSB), a representative of feedback NNW, in which the signals can flow in forward as well as backward direction.

BSB was used by Anderson er al. [5] as a fully connected neural network called to model psychological effects observed in probability learning. The BSB model is a simple, auto-associative,nonlinear, energy-minimizing neural network [6, 7]. In this network, each unit, which has no self-connection, is fully connected to every other

Z. Li • C. Liu • H. Li • Y. Chen (✉)
ECE Department, University of Pittsburgh, Pittsburgh, PA, USA
e-mail: zhl85@pitt.edu; chl192@pitt.edu; hal66@pitt.edu; yic52@pitt.edu

© Springer International Publishing AG 2017
A. Chattopadhyay et al. (eds.), *Emerging Technology and Architecture for Big-data Analytics*, DOI 10.1007/978-3-319-54840-1_10

217

unit in the network. A common application of the BSB model is optical character recognition for printed text [8]. Recently, a multi-answer character recognition method based on the BSB model has been developed to improve reliability and robustness for noisy or occluded text images [9]. An input character image is processed through the BSB models in parallel for the recall (pattern recognition) operation. When all recalls are completed, a set of candidates are selected based on the convergence speed.

Different from BSB, an FFW neural network feathering an open loop data path as its name implies. The FFW neural network was the first and simplest type of artificial neural network devised. Despite its simplicity, around 90/100 of NNW applications use FFW architecture[10]. According to the layers count of computational units, FNN can be sorted as single layer or multi-layer perceptron. A single-layer perceptron network consisting of a single layer of output nodes.

On the other hand, for all NNWs regardless of FFW or feedback, the net inputs of a group or groups of neurons can be transformed into matrix-vector multiplication(s). Similar to the biological systems, the neural network algorithms inherently are adaptive to the environment and resilience to random noise. As a consequence, hardware realizations of neural networks require a large volume of memory and are associated with high design complexity and hardware cost [11]. Algorithm enhancement can alleviate the situation but cannot fundamentally resolve it. More efficient hardware-level solutions become necessary.

Traditionally, Complementary Metal–Oxide–Semiconductor (CMOS) transistor-based Static Random-Access Memory (SRAM), Ternary Content Addressable Memory (TCAM) were employed to construct the weight matrix between two layers. Prohibitive area and power consumption prevent these designs from extensive application. Fortunately, the emerging memristor provides a promising solution for neural network implementations.

The existence of the memristor was predicted in circuit theory nearly 40 years ago [12]. However, it wasn't until 2008 that the first physical realization was demonstrated by HP Lab through a TiO_2 thin-film structure [13]. Afterward, many memristor materials and devices have been reported or rediscovered. The memristor has many promising features, such as non-volatility, low-power consumption, high integration density, and excellent scalability [14, 15]. More importantly, the unique property to record the historical profile of the excitations on the device makes it an ideal candidate to realize synapse behavior in electronic neural networks [16, 17].

In this chapter, we take BSB and single layer perceptron as examples of feedback and FFW to validate the neuromorphic hardware acceleration enabled by emerging memristors, we build a training as well as recall circuit for a BSB system and high reliable active function component for a single layer perceptron. Weight mapping and algorithm optimization for hardware oriented design was studied. Key design parameters and physical constraints have been extracted and analyzed. Effectiveness of our designs is demonstrated by performing comprehensive simulations w.r.t. quality of domain of attraction, distinguish margin, failure rate of recognition, etc.

10.2   Background

10.2.1   Neural Network

An NNW is constituted by a number of artificial neurons that are interconnected together. The structure of artificial neuron is inspired by the concept of biological neuron shown in Fig. 10.1a. Basically, it is the processing element in the nervous system of the brain that receives and combines signals from other similar neurons through thousands of input paths referred to as dendrites. Each input signal (electrical in nature), flowing through dendrite, passes through a synapse or synaptic junction. The accumulated signals are nonlinearly modified at the output before flowing to other neurons through the branches of axon [7].

The artificial neuron is modelled based on the biological neuron shown in Fig. 10.1b. Basically, it has an op-amp summer-like structure. Each input signal flows through a weighted path which can be positive (excitory) or negative (inhibitory). The summing node accumulates all the input-weighted signals, and then passes to the output through an activation or transfer function as shown in Fig. 10.1b. In other words, the neuron performs two types of operations, (1) a dot product of the inputs $x_1, \ldots, x_n$ and the weights $w_1, \ldots, w_n$, and (2) the evaluation of an activation function. The dot product operation can be seen in Eq. 10.1. The activation function of the neuron is shown in Eq. 10.2.

$$DP_j = \sum_{i=1}^{n} x_i w_{ij} \tag{10.1}$$

$$y_j = f(DP_j) \tag{10.2}$$

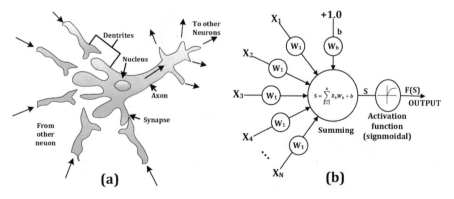

Fig. 10.1 (a) Sketch of a biological neuron (b) Model of artificial neuron

$$y_i = S(y_{in_j}) = \begin{cases} 1, & y_{in_j} > 1(TH) \\ y_{in_j}, & -1(-TH) \leq y_{in_j} \leq 1(TH) \\ -1, & y_{in_j} < -1(-TH) \end{cases} \tag{10.3}$$

In our FFW design,the sigmod function shown in Eg.1.3 is used as active function. On the other hand, the mathematical model of the BSB recall function will be represented as [18]

$$\mathbf{x}(t+1) = S(\alpha \cdot \mathbf{Ax}(t) + \beta \cdot \mathbf{x}(t)) . \tag{10.4}$$

10.2.2 Memristor Preliminaries

The memristor is a non-linear passive two-terminal device of which the resistance is determined by the historical profile of the applied electrical excitations. The existence of memristor devices was predicted as early as 1970s by Chua et. al [12] and firstly demonstrated in 2008 by HP labs [13]. Besides TiO_2 thin film developed by HP labs, many materials with various mechanisms were discovered to be memristive behavioral. Here we majorally focus on the TiO_2 memristor technology for its long-term validation and authority.

Figure 10.2 illustrates the cross-section of the TiO_2 thin-film memristor and the corresponding variable resistor model, which can be regarded as two serially connected resistors. Here, R_L and R_H , respectively, denote the low resistance state (LRS) and the high resistance state (HRS). The overall memristance can be expressed as $M(p) = p \cdot R_H + (1-p) \cdot R_L$, where $p(0 \leq p \leq 1)$ is the relative doping front location, which is the ratio of doping front position over the total thickness of the TiO_2 thin film. The velocity of doping front movement $v(t)$, driven by the voltage applied across the memristor $v(t)$, can be expressed as

$$v(t) = \frac{dp(t)}{dt} = \mu_v \cdot \frac{R_L}{h^2} \cdot \frac{V(t)}{M(p)} \tag{10.5}$$

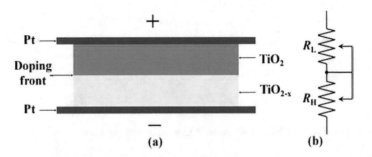

Fig. 10.2 (a) TiO_2 memristor structure. (b) Equivalent circuit

where μ_v is the equivalent mobility of dopants, h is the total thickness of the thin film, and $M(p)$ is the total memristance when the relative doping front position is p. In general, a certain energy (or threshold voltage) is required to enable the state change in a memristive device [19]. When the electrical excitation through a memristor is greater than the threshold voltage, i.e., $V(t) > Vth$, the memristance changes (in training). Otherwise, a memristor behaves like a resistor.

10.2.3 Memristor Array

There are two types of structures to organize the memristor-based memory matrix, crossbar and 1T1R. The crossbar employs a memristor device at each intersection of horizontal and vertical metal wires without any selectors [20], as shown in Fig. 10.3a. The crossbar structure is characterized by the high storage density by providing a large number of signal connections within a small footprint and conduct the weighted combination of input signals [21, 22]. This structure is usually applied in designs with analog input. However, sneakpath [23] in such selection free structures hinders reliability and efficiency of sensing and programming.

Practically, the one-transistor-one-resistive (1T1R) structure shown in Fig. 10.3b can safely solve this problem by adding an access transistor (or emerging selector [24]) for each memristor at a sacrifice of more area overhead. NNW designs using digital input will benefit from the 1T1R structures in R/W reliability. For instance, in the 1T1R array illustrated in Fig. 10.3b.

In this chapter, we use analog and digital computational scheme in BSB and FFW design, respectively. As a result, crossbar is employed by BSB design and 1T1R structure is applied in FFW design.

Fig. 10.3 (a) Memristor crossbar (b) 1T1R cell structure

10.3   Design Methodology

In this section, we will conceptually explain how to program a memristor crossbar to store the information of connection matrix. In other words, the mapping method. And the training method of BSB design, mimics the software training algorithm and adjusts the memristors iteratively to reach the required input/output function. The last but not the least is the recall component design for FFW design which applies the offline train scheme. A set of modifications to the algorithm are required since the hardware constrains which we will illustrate with details in the following sections.

10.3.1   Weight Mapping

10.3.1.1   Mapping Method for BSB System

Let us use the N-by-N memristor crossbar array shown in Fig. 10.3a to demonstrate its matrix computation functionality. Here, we apply a set of input voltages $\mathbf{V_I^T} = [V_{I,1}, V_{I,2}, \ldots, V_{I,N}]$ on the word-lines (WLs) of the array, and collect the current through each bit-line (BL) by measuring the voltage across a sensing resistor. The same sensing resistors are used on all BLs with resistance r_s or conductance $g_s = 1/r_s$. The output voltage vector $\mathbf{V_O^T} = [V_{O,1}, V_{O,2}, \ldots, V_{O,N}]$. Assume the memristor sitting on the connection between WL_i and BL_j has a memristance of $m_{i,j}$. The corresponding conductance $g_{i,j} = 1/m_{i,j}$. Then, the relation between the input and output voltages can be represented by

$$\mathbf{V_O} = \mathbf{C} \cdot \mathbf{V_I} \tag{10.6}$$

Here, matrix C can be represented by the memristors conductance and the load resistors as

$$\mathbf{C} = \mathbf{D} \cdot G^T = \mathrm{diag}(d_1, \ldots, d_N) \begin{bmatrix} g_{1,1} & \cdots & g_{1,N} \\ \vdots & \ddots & \vdots \\ g_{N,1} & \cdots & g_{N,N} \end{bmatrix}^T \tag{10.7}$$

where $d_i = 1/(g_s + \sum_{i=1}^{N} g_{i,j})$. To differentiate the mathematical connection matrix A in neural network, we use C to describe the physical relation between $\mathbf{V_I}$ and $\mathbf{V_O}$. Thus, all the terms in C must be positive values.

Please note that some noniterative neuromorphic hardware uses the output currents $\mathbf{I_O}$ as output signals. Since the BSB algorithm discussed in this chapter is an iterative network, we take $\mathbf{V_O}$ as output signals, which can be directly fed back to inputs for the next iteration without extra design cost.

Equation (10.6) indicates that a trained memristor crossbar array can be used to construct the positive matrix **C**, and transfer the input vector $\mathbf{V_I}$ to the output vector $\mathbf{V_O}$. However, **C** is not a direct one-to-one mapping of conductance matrix **G** as indicated in Eq. (10.7). Though a numerical iteration method can be used to obtain the exact mathematical solution of **G**, it is too complex and hence impractical when frequent updates are needed.

For simplification, assume $g_{i,j} \in \mathbf{G}$ satisfies $g_{min} \leq g_{i,j} \leq g_{max}$, where g_{min} and g_{max}, respectively, represent the minimum and maximum conductance of all the memristors in the crossbar array. Thus, a simpler and faster approximation solution to the mapping problem is defined as

$$g_{j,i} = c_{i,j} \cdot (g_{max} - g_{min}) + g_{min} \qquad (10.8)$$

A decayed version of **C**, referred to as $\hat{\mathbf{C}}$ can be approximately mapped to the conductance matrix G of the memristive array. Plugging Eq. (10.8) into Eq. (10.7), we have

$$\hat{c}_{i,j} = \frac{c_{i,j} \cdot (g_{max} - g_{min}) + g_{min}}{g_s + (g_{max} - g_{min}) \cdot \sum_{j=1}^{N} c_{i,j} + N \cdot g_{min}} \qquad (10.9)$$

Note that many memristive materials, such as TiO_2, demonstrate a large g_{max}/g_{min} ratio [13]. Thus, a memristor at the HRS under a low-voltage excitation can be regarded as an insulator, that is, $g_{min} \approx 0$. Moreover, the BSB connection matrix is a special matrix with a small $\sum_{j=1}^{N} c_{i,j}$. For example, all BSB models used for character recognition in our experiments show $\sum_{j=1}^{N} c_{i,j} < 5$ when $N = 256$. The term $\sum_{j=1}^{N} c_{i,j}$ can be further reduced by increasing the ratio g_s/g_{max}. As a result, the impact of $\sum_{j=1}^{N} c_{i,j}$ can be ignored. These two facts indicate that Eq. (10.9) can be further simplified as

$$\hat{c}_{i,j} = c_{i,j} \cdot g_{max}/g_s \qquad (10.10)$$

A memristor is a physical device with conductance $g > 0$. Therefore, all elements in matrix C must be positive as shown in Eq. (10.7). However, in the original BSB recall model, $a_{i,j} \in \mathbf{A}$ can be either positive or negative. An alternative solution is moving the whole **A** into the positive domain. Referring Eq. (10.4), since the output $\mathbf{x}(t + 1)$ will be used as input signal in the next iteration, a biasing scheme at $\mathbf{x}(t + 1)$ is needed to cancel out the shift induced by the modified **A**. The biasing scheme involves a vector operation since the shift is determined by $\mathbf{x}(t)$. To better maintaining the meaning of **A** in physical mapping and leveraging the high integration density of memristor crossbar, we propose to split the positive and negative elements of **A** into $\mathbf{A^+}$ and $\mathbf{A^-}$ as

$$a_{i,j}^+ = \begin{cases} a_{i,j}, & if \quad a_{i,j} > 0 \\ 0, & if \quad a_{i,j} \leq 0 \end{cases} \quad and \quad a_{i,j}^- = \begin{cases} 0, & if \quad a_{i,j} > 0 \\ -a_{i,j}, & if \quad a_{i,j} \leq 0 \end{cases} \qquad (10.11)$$

As such, (10.4) becomes

$$\mathbf{x}(t+1) = S(\mathbf{A}^{+}\mathbf{x}(t) - \mathbf{A}^{-}\mathbf{x}(t) + \mathbf{x}(t)) \tag{10.12}$$

where we set $\alpha = \beta = 1$. Thus, $\mathbf{A}^{+}$ or $\mathbf{A}^{-}$ can be mapped to two memristor crossbar arrays M1 and M2 in a decayed version $\hat{\mathbf{A}}^{+}$ and $\hat{\mathbf{A}}^{-}$, respectively, by following (10.9).

10.3.1.2 Mapping Method for Feedforward System

Instead of crossbar, 1T1R structure is employed in our FFW design because of the digital operational scheme. And we set the resistance range of Memristor devices from $50\,\text{K}\Omega$ to $1\,\text{M}\Omega$ [25, 26]. The according conductance $g \in (1\,\mu\text{S}, 20\,\mu\text{S})$.

Figure 10.4a compares the effective cell conductance $\tilde{g}$ and the Memristor conductance g. The result shows that they are very close when the select transistor is on. This is because the transistor's conductance g_{ON} (at the order of mS) is much higher than g. When the transistor is turned off, the extremely small g_{off} ($\approx nS$) dominates the cell conductance. So $\tilde{g}$ of an OFF cell has negligible impact on the computation. Therefore, a BL current virtually comes only from those ON cells enabled by WL pulses and the computing on every BL is nearly independent from others.

We also investigated the relationship between g and the BL voltage when the cell is on. Figure 10.4b presents the result when $g = 20\,\mu\text{S}$. As BL voltage increases, both V_{GS} and V_{DS} of the select transistor decrease, leading to the reduction of g_{ON}. Because the memristor and the transistor in a cell are connected in series and g is much smaller than gon, $\tilde{g}$ is primarily determined by g. The reduction of g_{ON} causes only 4.0% in the change of $\tilde{g}$. As g decreases, the variance of $\tilde{g}$ induced by the change of BL voltage becomes even more significant.

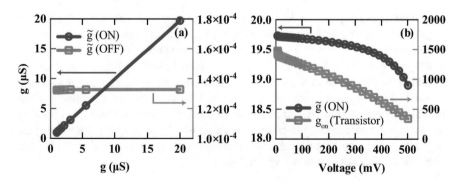

Fig. 10.4 (a) $\tilde{g}_{ij}$ vs. g_{ij} at ON/OFF states. (b) Change of $\tilde{g}_{ij}$ and g_{ij} as BL voltage varies

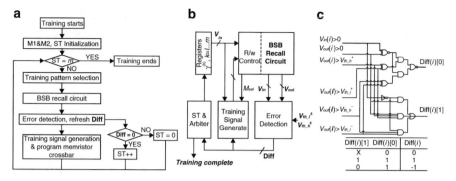

Fig. 10.5 (**a**) Training flow. (**b**) Conceptual circuit diagram. (**c**) Error detection circuit

10.3.2 Training Algorithm Optimization

Figure 10.5a shows the operational flow of the BSB training circuit, including two memristor crossbars, referred to as **M1** and **M2**. In addition, the corresponding circuit diagram is shown in Fig. 10.5b. Our goal is to develop a method to train the memristor crossbars as autoassociative memories for prototype patterns. The training scheme leverages the recall circuit to verify the training result and generate the control signals.

Step 1. Initializing the Crossbar Arrays: At the beginning of a training procedure, all memristance values in **M1** and **M2** are initialized to an intermediate value. The initialization does not have to be precisely accurate. Indeed, even when all of the memristors are all at either LRS or HRS, the crossbar arrays can still be successfully trained but it requires more time to reach convergence according to the simulation results.

Step 2. Selecting a Prototype Pattern $\gamma^{(k)} \in B^n (k = 1, \ldots, m)$: Here, B^n is the n-dimension binary space (1, 1). Assume a training set includes m prototype patterns and each pattern $\gamma^{(k)}$ has the same probability to be chosen every time. The counter ST is used to record in sequence the number of patterns that have been successfully trained. When ST>0, those patterns that have been trained are excluded from the selection.

Step 3. Sending $\gamma^{(k)}$ to the BSB Recall Circuit: We convert $\gamma^{(k)}$ in binary space $(-1,1)$ to a set of input voltages within the boundary $(-0.1\,\text{V}, 0.1\,\text{V})$. These input signals are supplied to the two memristor crossbars simultaneously. The resulting signals V_O can be obtained at the output of the BSB recall circuit.

Step 4. Error Detection: An error is defined as the difference between the prototype pattern and the recall result; that is, the difference between the input and output signals of the recall circuit. A piece of error detection circuitry for bit i is shown in Fig. 10.5c, which generates only the direction of the weight change based on the simplified algorithm[27]. In total, N pieces of error detection blocks are needed for an N×N crossbar array. Considering that the range of $V_{\text{out}}(i)$ could be different from that of $V_{in}(i)$, we apply a scalar γ to the input vector and take $\gamma \cdot V_{in}(i)$ as the target output signal. Rather than generating $\gamma \cdot V_{in}(i)$ in every training, we use the preset threshold voltages for error detection. Since $V_{in}(i)$ is either 0.1 or 0.1 V, four thresholds are needed, including

$$V_{th\_h}^+ = 0.1\lambda + \theta \quad V_{th\_l}^+ = 0.1\lambda - \theta$$
$$V_{th\_l}^- = -0.1\lambda - \theta \quad V_{th\_l}^- = -0.1\lambda + \theta$$
(10.13)

Here, θ represents the tolerable difference.

The error detection output Diff(i) could be 1, 0, or 1. When $|V_{out}(i) - \gamma \cdot V_{in}(i)| < \theta$, Diff($i$) = 0, meaning the difference between the normalized $V_{in}(i)$ and $V_{out}(i)$ is so small that we consider them logically identical. Otherwise, Diff(i) = +1 or −1, indicating the normalized $|V_{out}(i)|$ is greater or less than the normalized $|V_{in}(i)|$, respectively.

Step 5. Training Memristor Crossbar Arrays: If **Diff** is not a zero vector, which means some error has been detected, the crossbar arrays need to be further tuned. In order to control the training step with a finer granularity, we modify only one memristor crossbar each time. For example, one could train **M1** or **M2** when the iteration number is odd or even, respectively.

The weight updating of a memristor crossbar array is conducted by columns, during which constant voltage pulse signals are applied to **M1** or **M2**. Note that the real resistance change of a memristor is also determined by its array location and device characteristics. In the design, such difference can be compensated by properly controlling the amplitude/width/shape of training pulses and paying more training iterations.

The polarity of the training pulse for the jth column is determined by Diff(j). The design supplies the training pulses on all the rows of a memristor crossbar. The jth column is connected to ground and all the others are supplied with half of the training voltage. For **M1**, the training pattern is either the current selected prototype pattern $\gamma^{(k)}$ (if Diff(j) = 1) or its element-wise negated version (if Diff(j) = −1). The training signals to M1 and M2 have opposite polarities. That is, the training pattern of M2 uses the current prototype pattern when Diff(j) = 1 or its element-wise negated version when Diff(j) = −1.

Note that the mapping method uses M1 and M2 to represent the positive and negative terms of the BSB connection matrix, respectively. However, the proposed training scheme operated in real design circumstance cannot and does not have to guarantee an identical mapping to software generated matrix. In fact, what matters most is the overall effect of **M1** and **M2**, not exact memristance values in each individual crossbar array.

Step 6. If Training is Completed?: The counter ST increases by 1 if a prototype pattern goes through Step 25 and reports no error without further tuning **M1** and **M2**. Otherwise, the ST is reset to 0 whenever an error is detected and all of the patterns in B^n are available in Step 2. ST = m means the entire training set has been successfully learned and hence the training stops.

10.3.3 Recall Component Optimization

10.3.3.1 BSB Recall Implementation

To realize the BSB recall function at circuit level, we first convert the normalized input vector x(t) to a set of input voltage signals $\mathbf{V}(t)$. The corresponding functional description of the voltage feedback system can be expressed as

$$\mathbf{V}(t+1) = S'(G_1\mathbf{A}^+\mathbf{V}(t) - G_1\mathbf{A}^-\mathbf{V}(t) + G_2\mathbf{V}(t))$$
$$= S'(G_1\mathbf{V}_{\mathbf{A}+}(t) - G_1\mathbf{V}_{\mathbf{A}-}(t) + G_2\mathbf{V}(t)) \tag{10.14}$$

Here, G_1 and G_2 are the signal gain amplitudes resulted by peripheral circuitry, corresponding to α and β in (10.4).

We use V_{bn} to represent the boundary of the input voltage, that is, $V_{bn} V_i(t) V_{bn}$ for any $V_i(t) \in \mathbf{V}(t)$. The new saturation boundary function is modified accordingly. Note that V_{bn} must be smaller than Vth so that the memristances do not change during the recall process. Practically speaking, V_{bn} can be adjusted based on the requirement of convergence speed and accuracy.

Figure 10.6 shows the diagram of the BSB recall circuit built based on Eq. (10.4). The design is an analog system consisting of three major components.

1. Memristor Crossbar Arrays As the key component of the overall design, memristor crossbar arrays are used to realize the matrix–vector multiplication function in the BSB recall operation. Two memristor crossbar arrays $\mathbf{M_1}$ and $\mathbf{M_2}$ are required to represent the matrices $\mathbf{A}^+$ and $\mathbf{A}^-$, respectively. They both have the same dimension as the BSB connection matrix $\mathbf{A}$.

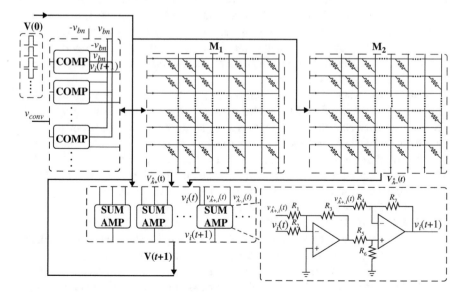

Fig. 10.6 Conceptual diagram of the BSB recall circuit

2. Summing Amplifier The conceptual structure of SUM-AMP is shown in the inner set of Fig. 10.6. In our design, the input signal $V_i(t)$ along with the voltage outputs of crossbar arrays $V_{\hat{A}+,i}(t)$ and $V_{\hat{A}-,i}(t)$ is fed into a summing amplifier (SUM-AMP). Assume $R1 = R4 = R6 = 1/g_s, R2 = R3 = R5 = R7 = 1/g_{min}$, and $G1 = G2 = 1$, the output of the SUM-AMP is

$$V_i(t+1) = \frac{g_s}{g_{min}} \cdot V_{\hat{A}+,i}(t) - \frac{g_s}{g_{min}} \cdot V_{\hat{A}-,i}(t) + V_i(t)$$

$$= V_{\hat{A}+,i}(t) - V_{\hat{A}-,i}(t) + V_i(t)$$

(10.15)

indicating that the decayed effect has been canceled out. The SUM-AMP naturally conducts $S()$ function by setting its output voltage boundary to V_{bn}. Moreover, the resistance values $R_1 R_7$ can be adjusted to match the required and in Eq. (10.12), if they are not the default value 1. For an N dimensional BSB model, N SUM-AMPs are required.

3. Comparator A new set of voltage signals $\mathbf{V}(t + 1)$ generated from the SUM-AMPs will be used as the input for the next iteration. Meanwhile, ever $V_i \in \mathbf{V}$ compares with $+V_{bn}$ and V_{bn} to determine if path i has converged. The recall operation stops when all N paths reach convergence. In total, N COMPs are needed to cover all of the paths.

10.3.3.2   FFW Active Function Implementation

Figure 10.7 depicts an overview of our proposed spiking computing architecture that leverages the compact resistive crossbar structure. The design adopts the rate coding model and represents data using the frequency of spikes (pulses) [28]. Through different bitlines (BLs) in a resistive crossbar array, the synaptic weighting functions of different entries are executed in parallel. The integrate and fire circuits (IFCs) as post-neurons generate output spikes based on the strength of the weighted pre-neuron signals from the crossbar.

A single-layer neural network with N pre-neurons and M post-neurons can be implemented using an $N \times M$ resistive crossbar array in the following approach: First, the activity pattern of pre-neurons $x_{N\times 1}$ is transferred into a set of pulses to wordlines (WLs). Here we assume the duration of an input pulse is t_m. The number of spikes on WL_i within a computation period T (that is, $n_{x,i}$) is determined by $x_i \in \mathbf{x}$. The synaptic weight between the jth pre-neuron and the ith post-neuron is mapped to conductance g_{ij} at the crosspoint of WL_i and BL_j. The total weighted signal to post-neuron j is transferred to the current flowing through BL_j and accumulated on a capacitor C_m in IFC. Once the voltage on Cm reaches to a predefined threshold Vth, the IFC fires an output spike and resets C_m. The activity function of postneurons $\mathbf{y}M1$ is represented by a set of spike numbers such as $[n_{y,0}, n_{y,1}, \ldots, n_{y,M1}]^T$.

We use $V_{x,i}(t)$ and $V_{y,j}(t)$ to denote the voltages on WL_i and BL_j at time t, respectively. The current flows through all the connected resistive devices contributes to

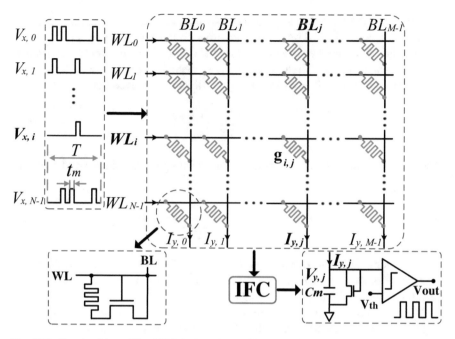

Fig. 10.7 Sketch of the spiking FFW design with a 1T1R cell array

the total current on BL_j, such as

$$I_{y,i}(t) = \sum_{i=0}^{N-1} g_{i,j}[V_{x,i}(t) - V_{y,i}(t)] \tag{10.16}$$

On the other hand, the voltage across C_m and the current flowing through it also follows

$$I_{y,i}(t) = C_m \frac{dV_{y,i}(t)}{dt} \tag{10.17}$$

By combining Eqs. (10.16) and (10.17), the increase of V_{yj} within a small epoch at time t can be derived as

$$\frac{dV_{y,j}(t)}{dt} = \frac{[1 - e^{-\frac{1}{C_m}\sum_{i=0}^{N-1} g_{ij}}] \sum_{i=0}^{N-1} g_{ij}V_{x,i}(t)}{\sum_{i=0}^{N-1} g_{ij}}. \tag{10.18}$$

indicating that the change of $Vy, j(t)$ is approximately proportional to the weighted pre-neuron signals $\sum_{i=0}^{N-1} g_{i,j}V_{x,i}(t)$. Moreover, the IFC fires a spike whenever V_{yj} reaches Vth. Thus, the spike number produced at post-neuron j is

$$n_{y,j}(t) \propto \int_{\tau=0}^{t} \sum_{i=0}^{N-1} g_{ij}V_{x,i}(\tau)d\tau. \tag{10.19}$$

Equation 10.19 implies that the computation of connection matrix in neural network can be performed by resistive crossbar array using spike signals.

In addition, we observed that the delay of IFC is a critical parameter determining the performance of the spiking neuromorphic system. Let's set $k = V$th$/V_I$. Then the time duration to switch Vy,j from 0 V to Vth can be derived by:

$$\Delta\tau = \frac{-C_m ln(1-k)}{\sum_{i=0}^{N-1} \tilde{g}_{ij}\delta_i}. \tag{10.20}$$

The number of output pulses generated during an input pulse duration of t_m can be calculated by

$$n_{y,j} = \frac{t_m}{\Delta\tau + t_0} = \frac{t_m}{\frac{\alpha}{\sum_{i=0}^{N-1} \tilde{g}_{ij}\delta_i} + t_0}, \tag{10.21}$$

where t_0 is the delay overhead of the IFC. $\alpha = -C_m \cdot ln(1-k)$ represents the BL charging efficiency, which is determined by the integration capacitor C_m and the threshold voltage Vth of the IFC design.

We proposed a new IFC design featuring high speed and low power consumption. Figure 10.8a depicts its schematic. During the operation, the BL voltage V_y continues increasing until it reaches Vth. Then the differential pair (M_1M_4) together with the following two cascaded inverters generates a high voltage at Vs, which in turn enables the discharging transistor M_{13}. Consequently, V_y decreases quickly and eventually turns off M_{13}. As such, the firing of one output spike at V_{out} is completed

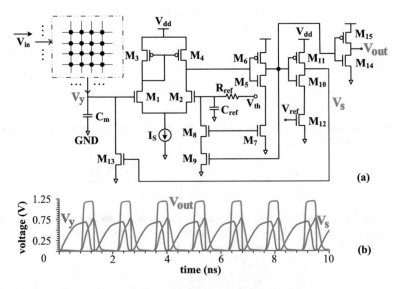

Fig. 10.8 The IFC circuit: (**a**) the schematic. (**b**) the simulation waveforms

and a new iteration of integrate-and-fire starts. To improve the IFC throughput, we tended to reduce its intrinsic operation delay and make it shorter than the integrating time in Eq. (10.20). A positive feedback loop ($M_7 - M_9$) was deployed based on the traditional comparator for this purpose. Another approach was to minimize the discharge time of C_m once a spike is fired out, i.e., using a large M_{13} to provide sufficient discharging current.

We implemented and simulated the IFC design with Globle Foundry 130 nm technology. VI is set to 1.2 V , and Vth is set to 0.5 V in which the system shows best computation accuracy. An MIM capacitor with a capacitance of 153fF (which is the minimum value offered by the PDK) is used as C_m. The design parameters were carefully selected so that the intrinsic delay of the integrate-and-fire is shorter than even the minimum BL integrating time. Also, it will achieve fast output spikes if the frequency of which is still within the range that can be reliably captured by the sensing circuit. The waveforms of V_y, V_s, and V_{out} under the fastest firing frequency (568.2 M spikes/s) are shown in Fig. 10.8b.

The area of the IFC design at Global Foundry 130 nm technology is 175.3 μm^2, which is compatible to that of traditional designs, e.g., 120 μm^2 at 65 nm technology in [22]. The energy consumption of our design is 0.48pJ-per-spike, which is about a quarter of the one in [22] (2pJ-per-spike).

10.4   Simulation and Evaluation

10.4.1   BSB System Evaluation

10.4.1.1   BSB training

The training method iteratively programs the memristor circuit until the required input–output function is achieved, therefore it can overcome most of the impact of process variations and signal fluctuations. In this section, simulation results are presented to demonstrate the training results for the proposed hardware training method and to compare with existing software synthesis methods. First, we compare the convergence speeds between prototype patterns and untrained patterns, essential for the realization of the racing the BSB recall function [17]. Second, the performance of a memristor crossbar as an autoassociative memory is analyzed.

In convergence speed analysis, we start with the simple linear memristor model and then employ the nonlinear TiO_2 memristor model based on the real device measurement [13] to demonstrate the training effect. Last, fabrication defects are considered by assuming that defected cells exist and are randomly distributed in the crossbar arrays. All input patterns are $(-1,+1)$ binary patterns with a length of n. The experimental setup is listed in Table 10.1.

Table 10.1 Simulation setup

Recall	Circuit	Memristor parameters					
		$R_H(\Omega)$	$R_L(\Omega)$	$h(nm)$	$\mu_v(m^2 \cdot s^{-1} \cdot V^{-1})$	Vth(V)	
		$10K\Omega$	$1K\Omega$	10	$1.00E^{-14}$	1.1	
		Summing amp parameters					
		G_1	G_2(Recall/Train)	V_{op+}(V)	V_{op-}(V)		
		30	0.6/0	1.05	-1.05		
		Comparater		Sensing res.	Recall voltage V_RV		
		V_{ref_h}(V)	V_{ref_l}(V)	$R_S(\Omega)$	For "1"		For "-1"
		1.0	-1.0	1000	0.1		-0.1
Training	Circuit	Memristance movement ΔM					
		Linear model		Nonlinear model			
		$\pm3\Omega$ each step		$\frac{(R_H - R_L \cdot \mu_v \cdot R_L \cdot V_T \cdot t)}{h^2 \cdot M}$			
		Comparator		Training time	Training voltage V_T(V)		
		V_{th_h}(V)	V_{th_l}(V)	$t(\mu s)$	For "1"		For "0"
		±0.125	±0.115	10	1.5		-1.5

Convergence Speed

In the BSB recall process, the learned prototype patterns should converge much
faster than the unlearned patterns. If this phenomenon appears, then the circuit
has remembered the prototype patterns and has the ability to classify whether an
input pattern is in the set of prototype patterns or not. We conduct the following
two experiments to analyze the BSB circuit performance based on the convergence
speeds.

Experiment 1 There are eight different randomly generated prototype patterns,
$N = 16$. The BSB system is trained to remember these patterns and all eight learned
prototype patterns and 100 unlearned random patterns are then recalled.

The results in Fig. 10.9 clearly show that there is a convergence speed gap
between the prototype patterns and the unlearned patterns. Especially, the eight
prototype patterns all converge to the magnitude boundary before the ninth iteration,
whereas the fastest convergence speed for unlearned patterns is the 12th iteration.
The larger the hamming distance between the input pattern and the prototype
patterns, the more iteration are required to converge, if convergence is even possible.

Experiment 2 The 26 lowercase characters from "a" to "z" are used as input
patterns. We use 20 patterns of lowercase character a representing different size-
font-style combinations as the prototype patterns for training. To compare their
convergence speeds, 500 patterns representing the other 25 lowercase characters
with different sizes/fonts/styles are also recalled. Figure 10.10 shows the result. It
can be clearly observed that the 20 prototype patterns of the lowercase character
"a" converge much faster than the other character patterns. Compare with result of
experiment 1 in Fig. 10.9, as the size of the BSB memory N increases (from 16

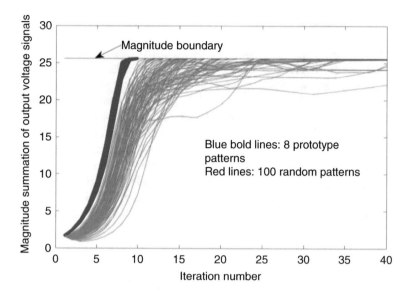

Fig. 10.9   Exp. 1: Iteration number vs. magnitude summation of output voltage signals

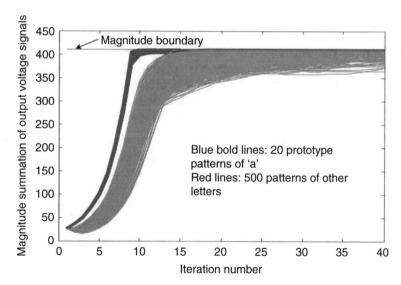

Fig. 10.10   Exp.2: Iteration number vs. magnitude summation of output signals

to 256), the convergence speed gap between prototype patterns and the unlearned patterns becomes more obvious. In conclusion, a simple but effective training method is realized by circuits and it can be used to construct the hardware architecture for the racing BSB algorithm proposed in [18].

BSB as Autoassociative Memory

Uniform Size of Domain of Attraction: An associative memory prefers a large overall domain of attraction, indicating that every input pattern eventually converges to a prototype pattern. When optimizing the training algorithm, it requires to uniformly increasing the domain of attraction for every prototype pattern rather than focusing only on a few of them. Thus, uniform size of domain of attraction is a useful measurement standard for the performance of associative memory. The number of $(1, 1)$ binary input patterns that are at the Hamming distance of l away from $\gamma^{(k)}$ and whose final states are $\gamma^{(k)}$ is defined as its domain of attraction, denoted by $\text{Doa}(\gamma^{(k)}, l)$. The uniform size of domain of attraction, denoted by $\text{UniDoa}(k)$, means the percentage for the $\sum_{l=0}^{p} \text{Doa}(\gamma^{(k)}, l)$ over the maximum of the $\sum_{l=0}^{p} \text{Doa}(\gamma^{(k)}, l)$ for all prototype patterns $\gamma^{(1)}, \ldots, \gamma^{(m)}$, which is defined as [29]

$$\text{Uni-Doa}(k) = \left\{ \frac{\sum_{l=0}^{p} \text{Doa}(\gamma^{(k)}, l)}{\max_{1 \le k \le m} \sum_{l=0}^{p} \text{Doa}(\gamma^{(k)}, l)} \right\} \times 100 \tag{10.22}$$

Quality of Domain of Attraction From testing all the possible input patterns, we can use the corresponding output patterns to evaluate the quality of domain of attraction, which reflects the overall performance of the BSB associative memory (not only for different prototype patterns). As we generate random binary patterns to test, we can calculate their Hamming distance with all prototype patterns. The prototype pattern with the least hamming distance to the input pattern is regarded as the most likely prototype pattern in the sense of Hamming distance. Then, we can divide the $(1, +1)$ binary input patterns into four classes based on their final state.

1. *Best:* among the nearest prototype patterns in the sense of Hamming distance.
2. *Good:* a prototype pattern that is not one of the nearest prototype patterns in the sense of Hamming distance, meaning it is not the most likely prototype pattern.
3. *Negative:* a spurious state (final state is none of the prototype patterns).
4. *Bad:* a state that is not convergent but trapped in a limit cycle.

The quality of the domain of attraction is represented by the number of $(-1, +1)$ binary input patterns in each class (Table 10.2).

Experiment 3 We compare the training effect of our embedded hardware circuit with the classic BSB training algorithms proposed by Lillo et al. [30] and Perfetti [31], and a more recent BSB training algorithm, Park [29]. The test case is taken from [19]. In this experiment, we consider the following five prototype patterns with n = 10:

Table 10.2 Specs of the fabricated STT-MRAM

	Best	Good	Negative	Bad
Hardware(linear)	419	6	465	134
Hardware(nonlinear)	465	0	473	86
Lillo et al. [7]	164	1	859	0
Perfetti [31]	164	1	859	0
Park [29]	164	1	859	0

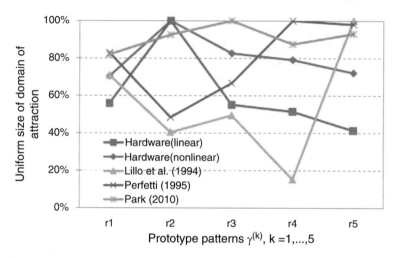

Fig. 10.11 Exp.3: Uniform size of domain of attraction

$$\gamma^{(1)} = [-1, +1, -1, +1, +1, +1, -1, +1, +1, +1]^T$$

$$\gamma^{(2)} = [+1, +1, -1, -1, +1, -1, +1, -1, +1, +1]^T$$

$$\gamma^{(3)} = [-1, +1, +1, +1, -1, -1, +1, -1, +1, -1]^T \qquad (10.23)$$

$$\gamma^{(4)} = [+1, +1, -1, +1, -1, +1, -1, +1, +1, +1]^T$$

$$\gamma^{(5)} = [+1, -1, -1, -1, +1, +1, +1, -1, -1, -1]^T$$

Figure 10.11 and Table 10.2 summarize the simulation results of the uniform size of domain of attraction, and the quality of domain of attraction, respectively. The results obtained from our proposed hardware design are labeled as Hardware.

The simulation results show that our hardware circuit performs better than Lillo et al. [7] in the class of Best in the quality of domains of attraction test. Compared with Perfetti [18], our scheme is competitive for the similar performance in uniform size and the quality of domains of attractions. However, it has a large drop in error correction rate when Hamming distance $l \geq 2$. Since our hardware circuit was built based on the fundamental training algorithm, it cannot be as good as Park [19], the state-of-the-art training algorithm developed based on constrained optimization. However, our scheme advances for its simple structure and low computation requirement. Moreover, it provides much faster training speed than the traditional software solutions since we utilize memristor crossbars embedded on-chip.

10.4.1.2   BSB Recall

The robustness of the BSB recall circuit was analyzed based on Monte Carlo simulations at the component level. The experimental setup is listed in Table 10.1. Memristor device parameters are taken from [6].

We tested 26 BSB circuits corresponding to the 26 lowercase letters from "a" to "z". The character imaging data was taken from [15]. Each 16×16 points character image can be converted to a $(1, +1)$ binary vector with $N = 256$. Accordingly, each BSB recall matrix has a dimension of 256 256. The training set of each character consists of 20 prototype patterns representing different size/font/style combinations.

In each test, we created 500 design samples for each BSB circuit and ran 13,000 Monte Carlo simulations. The defected input pattern in Fig. 10.12 has been considered in evaluation.

BSB Recall Circuit Under Ideal Condition

Sending an input pattern to different BSB circuits will result in different converging speeds. Figure 10.13 is the example when processing a perfect "a" image through BSB circuits trained for 26 lowercase letters. The BSB circuits for "a", "i", and "s" reach convergence with the least iteration numbers. The multianswer character recognition method considers the three letters as winners and takes them to context aware word recognition, such as perception-prediction model [18, 32]. Figure 10.14 shows the performance of the BSB circuit design under ideal condition without input defects, process variations, or signal fluctuations. The x-axis and y-axis represent input images and the BSB circuits, respectively. All of the winners are highlighted by the black blocks.

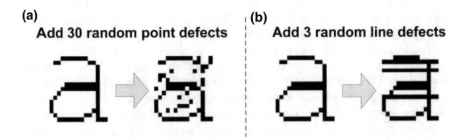

(a) Add 30 random point defects

(b) Add 3 random line defects

Fig. 10.12 (a) Random point defects. (b) Random line defects

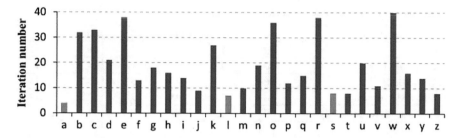

Fig. 10.13 Iterations of 26 BSB circuits for a perfect "a" image

Fig. 10.14 Performance of
26 BSB circuits under ideal
condition

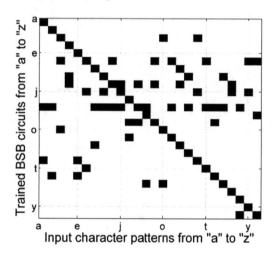

Process Variations and Signal Fluctuations

A BSB circuit corresponding to its trained input pattern always wins under the ideal condition. However, after injecting noise into the input pattern or circuit design, some BSB circuits might fail to recognize its trained input pattern. We use the probability of failed recognitions (PF) to measure the performance of a BSB circuit (Table 10.3).

1. Random Noises The random noise in the BSB circuit comes from process variations as well as the electrical signal fluctuations. The second to fifth rows in Table 10.3 summarize the impact of every single random noise contributor based on Monte Carlo simulations. Here, we assume two memristor crossbar arrays are fully correlated, i.e., $Corr_M = 1$. The simulation results show that the BSB circuit design has a high tolerance for random noise: compared with the ideal condition without any fluctuation (IDEAL), the random noise of circuits causes only slight performance degradation. This is because resilience to random noise is one of the most important inherent features for the BSB model as well as other neural networks.

Table 10.3 $P_F(\%)$ of 26 BSB circuits for 26 input patterns

Random point numbers	0	10	20	30	40	50
IDEAL	0	2.1	4.2	5.3	10.0	20.8
$M(\sigma_{sys} = 0.1 \& \sigma_{rdm} = 0.1)$	0	1.9	4.6	6.5	14.2	24.7
$R_s(\sigma = 0.1)$	0	1.8	4.3	6.2	13.7	24.1
SUM-AMP ($\sigma = 0.1$)	0	1.9	4.4	7.7	13.5	23.1
COMPARATOR ($\sigma = 0.1$)	0	2.3	5.5	5.4	11.1	22.0
$\text{Corr}_M = 0.6$	5.6	10.2	17.2	22.7	30.8	38.6
$\text{OVERALLCorr}_M = 0.6$	4.6	8.2	15.2	20.7	32.8	36.6
Random line numbers	0	1	2	3	4	5
IDEAL	0	7.3	13.8	21.5	35.8	50.2
$M(\sigma_{sys} = 0.1 \& \sigma_{rdm} = 0.1)$	0	7.4	14.8	25.5	38.8	53.6
$R_s(\sigma = 0.1)$	0	7.4	14.8	25.3	35.1	51.8
SUM-AMP ($\sigma = 0.1$) ($\sigma = 0.1$)	0	7.7	15.3	23.4	34.7	52.6
COMPARATOR	0	6.9	14.5	23.3	33.7	53.2
$\text{Corr}_M = 0.6$	5.1	14.4	24.7	34.6	44.2	55.1
$\text{OVERALLCorr}_M = 0.6$	6.3	15.4	24.2	34.1	44.0	58.2

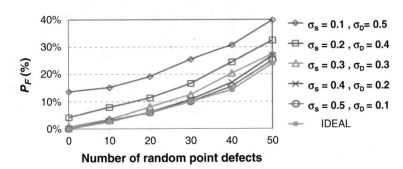

Fig. 10.15 Comparison of the impacts of static/dynamic noise

2. Static Noise Versus Dynamic Noise The noise matrices of $\mathbf{N_M}$ and $\mathbf{N_{Rs}}$ mainly affect the mapping between the connection matrix and memristor crossbar array. Physically, these noise elements come from process variations and remain unchanged. So, they can be regarded as static noise (N_S). In contrast, the noise of the SUM-AMPs and COMPs induced by electric fluctuations demonstrates a dynamic behavior during circuit operation. We classify them as dynamic noise (N_D). We can adjust N_S and N_D and observe the combined impact on the BSB circuit performance. For simplicity, we set $\sigma_{rdm}(M) = \sigma(R_S) = \sigma S, \sigma(\text{AMP}) = \sigma(\text{COMP}) = \sigma_D$, and $\text{Coor}_M = 1$ to exclude the impact of correlations between $\mathbf{M_1}$ and $\mathbf{M_2}$. The result in Fig. 10.15 shows that the dynamic noise dominates P_F. For example, when $\sigma_D = 0.5$ and $\sigma_S = 0.1$, PF is high even with a clean input image. Decreasing D but increasing S results in P_F reduction in all the regions. From BSB circuit design point of view,

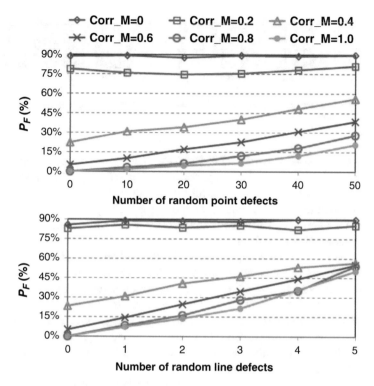

Fig. 10.16 Impact of Corr$_M$

the accuracy and the stability of the iteration circuit is more important than the programming accuracy of the memristor crossbar.

3. Impact of CorrM The BSB circuit uses two memristor crossbar arrays to split the positive and negative elements of A. Reducing Corr$_M$, that is, increasing the difference in the systematic noises of M_1 and M_2, can be regarded as A^+ and A^- having different overall shifts. This induces a directional noise in the recall function. As a consequence, Corr$_M$ demonstrates a higher impact. As shown in the sixth row in Table 10.3, decreasing the correlation between two crossbars from 1 to 0.6 demonstrates a large impact on the performance. Figure 10.16 shows that when decreasing Corr$_M$ from 1 to 0, the average P_F dramatically increases.

10.4.2 FFW System Evaluation

We evaluated the performance and robustness of the proposed spiking neuromorphic design by using the application of digital image recognition. Since the simulation time increases dramatically with the design scale, the crossbar with 32 rows was selected for training and recognition of images with 32 pixels. Six images corresponding to number $0 \sim 5$ in Fig. 10.17a were used as the standard training set.

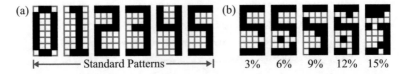

Fig. 10.17 (**a**)The standard training set of six patterns. (**b**) An example of noise pattern "5" under different single bit error rate

In the *feedforward* network ("**F**") implementations, the back-propagation and delta rule[humiao] were adopted to perform training and programmed Memristors to particular resistance states. Meanwhile, a double crossbar array is used in each system to obtain negative weights as we did in the BSB system design. The 1T1R structure in the design can effectively suppress the sneak path leakage and therefore make programming very efficient.

The system performance was measured by the probability of failed recognition (P_F). The analysis was conducted based on Monte-Carlo simulations (i.e., 10,000 simulations per configuration). Noise patterns with a certain bit error rate (BER) were generated and used as the testing images. The example in Fig. 10.17b shows that visually, it is already very difficult to identify an image with a BER >9%.

In this design, we are particularly interested in the impacts of physical constraints, including the limited available resistance state levels and output spike number. We evaluated and compared the designs of which the ReRAM devices provide the continuous analog resistance states ("**A**") or only 8 discrete resistance levels ("**D**"). The efficiency of the digitized output spikes was studied by comparing the result obtained directly from the analog BL current ("**C**") and that achieved based on real output pulses ("**P**").

Thus, the configuration "**AC**" performs closely to the mathematical neural network model and was taken as the baseline in the following evaluations. The configuration "**DP**" corresponds to our proposed spiking neuromorphic design.

Here, we assumed 8 discrete resistance levels of ReRAM devices and 20 mV sensing margin of BL output op-amp design.

The reliability analysis was conducted by assuming that the ReRAM resistances ("**PV**") and the IFC spike generation speed ("**IFC**") follow normal distributions with a standard deviation of 10% and 5%, respectively.

10.4.2.1   FFW Recall

We realized a 1-layer feedforward network based on the scheme in Fig. 10.7 for image recognition. It maps an input pattern to the output through a direct graph without iterations. A 32×6 crossbar array was trained so that output j has the strongest response to number j of the six training patterns. Figure 10.18 gives the simulation result when any standard pattern is used as the input. Output j generates the biggest spike number to pattern j but demonstrates much weaker responses to

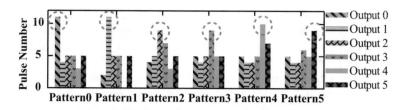

Fig. 10.18 Pattern recognition result in forward neural network in ideal condition

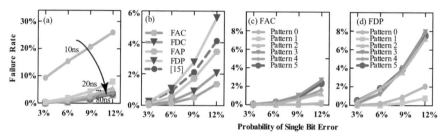

Fig. 10.19 The simulation results of the FFW implementation. (**a**) $P_{F,\text{all}}$ of FDP at various Ts; (**b**) $P_{F,\text{all}}$ under different configurations; (**c**) $P_{F,\text{ind}}$ of FAC; (**d**) $P_{F,\text{ind}}$ of FDP

other patterns. Here, we say the recognition of a noisy testing image is failed when the corresponding output doesn't produce the most spikes or another output has the same number of spikes.

The Computation Period T Selection The input pulse duration T determines the output spike granularity and hence greatly affects the system performance. Correspondingly, we investigated the performance of FDP configuration by varying T from 10 ns to 80 ns. Figure 10.19 presents the results expressed by the average failure rate of all the patterns ($P_{F,\text{all}}$) under different BERs.

The results show that when T is less than 20 ns, the design cannot produce enough output spikes to differentiate the top two strongest outputs, resulting in a lot of failures. $P_{F,\text{all}}$ quickly drops to 5.57% when increasing T to 30 ns. Further prolonging T demonstrates marginal improvement. To guarantee sufficient system performance, $T = 30$ ns was selected, corresponding up to 15 input pulses per computation period.

The Impact of Physical Constraints was evaluated by comparing the recognition qualities of different configurations shown in Fig. 10.19. FAC as the baseline has the least failures. Reducing the resistance states to 8 discrete levels inevitably results in quality loss when mapping the analog values of a connection matrix to the limited conductances in a crossbar. Compared to FAC, $P_{F,\text{all}}$ of FDC increases 0.71% at BER=12%. Changing from the analog BL current to the digitalized spikes (FAC vs. FAP) causes up to 2.08% more recognition failures, implying that this feedforward network implementation is more sensitive to the granularity of output signals. Overall, our proposed FDP obtains a $P_{F,\text{all}}$ of 5.57% at BER=12%, which is 4.21% higher than the baseline FAC and 1.46% worse than the computing engineer of [15].

Figure 10.19c, d shows the statistical results of each individual pattern ($P_{F,\text{ind}}$) of FAC and FDP, respectively. Numbers 2~5 with high similarity in training patterns are more sensitive to the input defects during testing. The probability of failures is much smaller for numbers 0 and 1.

The system reliability was conducted by including the variations in ReRAM resistances and the IFC spiking generation. Figure 10.19e shows the relative $P_{F,\text{all}}$ of FDP under different conditions, all of which are normalized to the ideal one without any variations. The variations in memresistances can barely affect the system performance because it is buried under the resistance offset caused by the mapping from connection matrix to crossbar array. The impact of the fluctuation in IFC spiking generation is more obvious. Even though, $P_{F,\text{all}}$ under the worst scenario is still $< 5.65\%$.

10.5 Conclusion

In this chapter, we proposed two neural network implementations including feed-forward and feedback (BSB) network based on the emerging memristor technology. We realize the transformation of the mathematical expression of BSB training and recall model to pure physical device relation and design the corresponding circuit architecture. The multi-answer character recognition algorithm is used in the experiments for robustness analysis of the proposed design. We also thoroughly study the impacts of various noises induced by process variations and electrical fluctuations and discuss the physical constraints in circuit implementation.

And we proposed a novel spiking based FFW design, which is a mixed-signal system that uses the digitalized spikes for data transferring and leverages the high density crossbar structure for parallel computation in analog format. Such a design naturally minimizes the use of analog components and therefore obtains significant savings in design area and energy consumption, compared with BSB crossbar-based computing engine with fully analog operation. We carefully studied the feasibility of the proposed spiking based design in terms of the computation accuracy, efficiency, and reliability. The realization of neural network models demonstrated that our design has a good tolerance in resistive device imperfection but more vulnerable to the fluctuations in output spike generation.

References

1. W.A. Wulf, S.A. McKee, Hitting the memory wall: implications of the obvious. ACM SIGARCH Comput. Archit. News **23**(1), 20–24 (1995)
2. P. Camilleri, M. Giulioni, V. Dante, D. Badoni, G. Indiveri, B. Michaelis, J. Braun, P. Del Giudice, A neuromorphic avlsi network chip with configurable plastic synapses, in *7th International Conference on Hybrid Intelligent Systems (HIS 2007)* (IEEE, 2007), pp. 296–301

3. M. Wang, B. Yan, J. Hu, P. Li, Simulation of large neuronal networks with biophysically accurate models on graphics processors, in *The 2011 International Joint Conference on Neural Networks (IJCNN)* (IEEE, 2011), pp. 3184–3193

4. H. Shayani, P.J. Bentley, A.M. Tyrrell, Hardware implementation of a bio-plausible neuron model for evolution and growth of spiking neural networks on FPGA, in *NASA/ESA Conference on Adaptive Hardware and Systems (AHS)* (IEEE, 2008), pp. 236–243

5. J.A. Anderson, J.W. Silverstein, S.A. Ritz, R.S. Jones, Distinctive features, categorical perception, and probability learning: Some applications of a neural model. Psychol. Rev. **84**(5), 413 (1977)

6. T. Kohonen, *Self-Organization and Associative Memory*, vol. 8 (Springer Science & Business Media, New York, 2012)

7. W.E. Lillo, D.C. Miller, S. Hui, S.H. Zak, Synthesis of brainstate-in-a-box (BSB) based associative memories. IEEE Trans. Neural Netw. **5**(5), 730–737 (1994)

8. A. Schultz, Collective recall via the brain-state-in-a-box network. IEEE Trans. Neural Netw. **4**(4), 580–587 (1993)

9. Q. Qiu, Q. Wu, M. Bishop, R.E. Pino, R.W. Linderman, A parallel neuromorphic text recognition system and its implementation on a heterogeneous high-performance computing cluster. IEEE Trans. Comput. **62**(5), 886–899 (2013)

10. M. Kubat, Neural networks: a comprehensive foundation by Simon Haykin, Macmillan, 1994. Knowl. Eng. Rev. **13**, 409–412 (1999). ISBN:0-02-352781-7

11. J. Partzsch, R. Schuffny, Analyzing the scaling of connectivity in neuromorphic hardware and in models of neural networks. IEEE Trans. Neural Netw. **22**(6), 919–935 (2011)

12. L. Chua, Memristor-the missing circuit element. IEEE Trans. Circuit Theory **18**(5), 507–519 (1971)

13. D.B. Strukov, G.S. Snider, D.R. Stewart, R.S. Williams, The missing memristor found. Nature **453**, 80–83 (2008)

14. Y. Ho, G.M. Huang, P. Li, Nonvolatile memristor memory: device characteristics and design implications, in *Proceedings of the 2009 International Conference on Computer-Aided Design* (ACM, 2009), pp. 485–490

15. D. Niu, Y. Chen, C. Xu, Y. Xie, Impact of process variations on emerging memristor, in *2010 47th ACM/IEEE Design Automation Conference (DAC)* (IEEE, 2010), pp. 877–882

16. D.B. Strukov, J.L. Borghetti, R.S. Williams, Coupled ionic and electronic transport model of thin-film semiconductor memristive behavior. Small **5**(9), 1058–1063 (2009)

17. Y.V. Pershin, M. Di Ventra, Experimental demonstration of associative memory with memristive neural networks. Neural Netw. **23**(7), 881–886 (2010)

18. Q. Qiu, Q. Wu, M. Bishop, R.E. Pino, R.W. Linderman, A parallel neuromorphic text recognition system and its implementation on a heterogeneous high-performance computing cluster. IEEE Trans. Comput. **62**(5), 886–899 (2013)

19. Y.V. Pershin, M. Di Ventra, Practical approach to programmable analog circuits with memristors. IEEE Trans. Circuits Syst. I Regul. Pap. **57**(8), 1857–1864 (2010)

20. A. Heittmann, T.G. Noll, Limits of writing multivalued resistances in passive nanoelectronic crossbars used in neuromorphic circuits, in *Proceedings of the great lakes symposium on VLSI* (ACM, 2012), pp. 227–232

21. U. Ramacher, C. von der Malsburg, *On the Construction of Artificial Brains* (Springer Science & Business Media, New York, 2010)

22. T. Hasegawa, T. Ohno, K. Terabe, T. Tsuruoka, T. Nakayama, J.K. Gimzewski, M. Aono, Learning abilities achieved by a single solid-state atomic switch. Adv. Mater. **22**(16), 1831–1834 (2010)

23. Y. Cassuto, S. Kvatinsky, E. Yaakobi, Sneak-path constraints in memristor crossbar arrays, in *2013 IEEE International Symposium on Information Theory Proceedings (ISIT)* (IEEE, 2013), pp. 156–160

24. J.-J. Huang, Y.-M. Tseng, C.-W. Hsu, T.-H. Hou, Bipolar nonlinear selector for 1s1r crossbar array applications. IEEE Electron Device Lett. **32**(10), 1427–1429 (2011)

25. S. Yu, Y. Wu, H.-S.P. Wong, Investigating the switching dynamics and multilevel capability of bipolar metal oxide resistive switching memory. Appl. Phys. Lett. **98**(10), 103514 (2011)
26. S.H. Jo, K.-H. Kim, W. Lu, High-density crossbar arrays based on a si memristive system. Nano Lett. **9**(2), 870–874 (2009)
27. M. Hu, H. Li, Y. Chen, Q. Wu, G.S. Rose, R.W. Linderman, Memristor crossbar-based neuromorphic computing system: a case study. IEEE Trans. Neural Netw. Learn. Syst. **25**(10), 1864–1878 (2014)
28. W. Gerstner, W.M. Kistler, *Spiking Neuron Models: Single Neurons, Populations, Plasticity* (Cambridge University Press, Cambridge, 2002)
29. Y. Park, Optimal and robust design of brain-state-in-a-box neural associative memories. Neural Netw. **23**(2), 210–218 (2010)
30. W.E. Lillo, D.C. Miller, S. Hui, S.H. Zak, Synthesis of brain-state-in-a-box (BSB) based associative memories. IEEE Trans. Neural Netw. **5**(5), 730–737 (1994)
31. R. Perfetti, A synthesis procedure for brain-state-in-a-box neural networks. IEEE Trans. Neural Netw. **6**(5), 1071–1080 (1995)
32. Q. Qiu, Q. Wu, R. Linderman, Unified perception-prediction model for context aware text recognition on a heterogeneous many-core platform, in *The 2011 International Joint Conference on Neural Networks (IJCNN)* (IEEE, 2011), pp. 1714–1721

Chapter 11
Energy Efficient Spiking Neural Network Design with RRAM Devices

Yu Wang, Tianqi Tang, Boxun Li, Lixue Xia, and Huazhong Yang

11.1 Introduction

The explosion of big data brings huge demands for higher processing speed, lower power consumption, and better scalability of computing systems. However, the traditional "scaling down" method is approaching its limit, making it more and more difficult for CMOS-based computing systems to achieve considerable performance improvements from device scaling [1]. Moreover, from the architecture level, the memory bandwidth required by high-performance CPUs has also increased beyond what conventional memory architectures can efficiently provide, leading to an ever-increasing memory wall [2] challenge to the efficiency of von Neumann architecture. In this way, new technologies, from both the device level and the architecture level, are required to overcome these challenges.

The spiking neural network (SNN) is an emerging computing model, as shown in Fig. 11.1, which encodes and processes information with time-encoded neural signals [3]. As a bio-inspired architecture abstracted from actual neural system, SNN not only provides a promising solution to deal with cognitive tasks, such as the object detection and speech recognition, but also inspires new computational paradigms beyond the von Neumann architecture and boolean logics, which can drastically promote the performance and efficiency of computing systems [4, 5]. However, an energy efficient hardware implementation and the difficulty of training the model remain as two important impediments that limit the application of SNN.

On the one hand, we need an applicable computing platform to utilize the potential ability of SNN. IBM [6] proposes a neurosynaptic core named TrueNorth. To mimic the ultra-low-power processing of brain, TrueNorth uses several approaches

Y. Wang (✉) • T. Tang • B. Li • L. Xia • H. Yang
Department of Electronic Engineering, Tsinghua University, Beijing, China
e-mail: yu-wang@mail.tsinghua.edu.cn; ttq14@mails.tsinghua.edu.cn;
lbx13@mails.tsinghua.edu.cn; xialx13@mails.tsinghua.edu.cn; yanghz@mails.tsinghua.edu.cn

© Springer International Publishing AG 2017
A. Chattopadhyay et al. (eds.), *Emerging Technology and Architecture for Big-data Analytics*, DOI 10.1007/978-3-319-54840-1_11

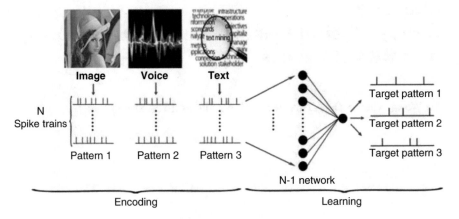

Fig. 11.1 Spiking neural network

to reduce the power consumption. Specifically, TrueNorth uses digital messages between neurons to reduce the communication overhead and event-driven strategy to further save the energy computation [5]. However, the CMOS based implementation still has some limitations that are hard to avoid, while some RRAMs' inherent advantages can overcome these difficulties. First, on-chip SRAM, where the synapse information is stored, is a kind of volatile memory with considerable leakage power, while RRAM is non-volatile with very low leakage power [7]. Another limitation is that TrueNorth may still need adders to provide the addition operation of neuron function, but RRAM crossbar can do the addition, or the matrix–vector multiplication, with ultra-high energy efficiency by naturally combining the computation and memory together [8–10]. Consequently, RRAM shows potential on implementing low-power spiking neural network.

On the other hand, from the perspective of algorithm, the efficient training of SNN and mapping a trained SNN onto neuromorphic hardware present unique challenges. Recent work of SNN mainly focuses on increasing the scalability and level of realism in neural simulation by modeling and simulating thousands to billions of neurons in biological real time [11, 12]. These techniques provide promising tools to study the brain but few of them support practical cognitive applications, such as the handwritten digit recognition. Even TrueNorth [13] uses seven kinds of applications to verify its performance, but the training and mapping methods for spike-oriented network are not discussed in detail. In other words, the mapping problem and efficient training method for SNN, especially for the real-world applications, to achieve an acceptable cognitive performance is severely demanded. Moreover, SNN can also be used for brain system simulation. For example, IBM made the cat cortex simulation (with $\sim 10^9$ neurons and $\sim 10^{13}$ synapses) on Blue Gene supercomputer cluster (with 147,456 CPUs and 144 TB memory) [14]. And such applications in the field of biological researches are out of discussion in this chapter.

These two problems are always coupled together and only by overcoming these two challenges can we actually utilize the full power of SNN for real-time data processing applications. In this chapter we discuss these two problems with the RRAM based system architecture and two different offline training algorithms of SNN. We use the MNIST digit recognition task [15] as an application example for the real-time classification. The goal of this chapter is to design an RRAM-based SNN system with higher classification accuracy and to analyze its strengths and weaknesses compared with other possible implementations.

The rest of this chapter is organized as follows:

- Section 11.2 introduces the background knowledge, including SNN and RRAM.
- Section 11.3 compares different models of spiking neural networks for practical cognitive tasks, including the Spike Timing Dependent Plasticity (STDP), the Remote Supervised Method (ReSuMe), and the latest Neural Sampling Learning Scheme. We show that the neural sampling method which transfers the ANN to SNN is promising for real-word applications while STDP and ReSuMe can **NOT** be used alone in the classification task since both of them are unsupervised learning method.
- Section 11.4 shows an RRAM-based implementation of SNN architecture. Two different specific networks, i.e. (1) STDP cascaded with three-layered ANN and (2) four-layered SNN transferred from full-connected ANN, are built and mapped to our system. The RRAM implementation mainly includes an RRAM crossbar array working as network synapses, an analog design of the spiking neuron, an input encoding scheme, and a mapping algorithm to configure the RRAM-based spiking neural network. And these elements will be described separately.
- In Sect. 11.5, a case study of digit recognition tasks is introduced to evaluate the performance of RRAM-based SNN. We compare the power efficiency and recognition performance of SNN and the RRAM-based artificial neural network (ANN). The experiment results show that ANN can beat SNN on the recognition accuracy, while SNN usually requires less power consumption. Based on these results, we discuss the possibility of using boosting methods, which combine some weak SNN learners together, to further enhance the recognition accuracy for real-world application.

11.2 Preliminaries

11.2.1 Spike Neurons

The neuron is the basic building block of SNN. Different mathematical models of spiking neurons have been explored with different levels of computational efficiency and biological plausibility [16]. The model of *Leaky Integrate and Fire (LIF)* [17] is one of the most widely used models for its computing efficiency. In this model, a one-order differential function determines the state variable $V(t)$ and a

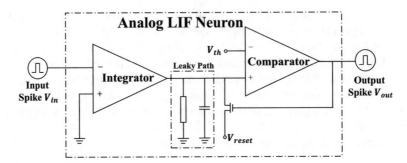

Fig. 11.2 Analog LIF neuron

threshold function determines whether the neuron spikes and then resets. And it is described as:

$$V(t) = \begin{cases} \beta \cdot V(t-1) + V_{in}(t) & \text{when } V < V_{th} \\ V_{reset} \text{and set as spike} & \text{when } V \geq V_{th} \end{cases} \tag{11.1}$$

where $V(t)$ is the state variable and β is the leaky parameter; V_{th} is the threshold state which the state variable makes comparison with and once exceeding, the state variable will reset to V_{reset}.

An analog *LIF* neuron implementation is shown in Fig. 11.2: the integrator calculates the state of the neuron $V(t)$ and the *RC* works as the leaky path. When $V(t) > V_{th}$, the transistor will be conducted and $V(t)$ will be reset.

11.2.2 RRAM Device Characteristics

Figure 11.3a shows a 2D filament model of HfO_x based RRAM device [18]. The model is a sandwich structure with a resistive layer between two metal electrodes. The conductance is exponentially dependent on the tunneling gap (d). Therefore, we will take advantage of the variable conductance of the RRAM device by setting the value of tunneling gap d. For the HfO_x based RRAM device, the I–V relationship can be empirically expressed as follows [18]:

$$I = I_0 \cdot \exp\left(-\frac{d}{d_0}\right) \cdot \sin h\left(\frac{V}{V_0}\right) \tag{11.2}$$

where d is the average tunneling gap distance. I_0 ($\sim$1 mA), d_0 ($\sim$0.25 nm) and V_0 ($\sim$0.25 V) are fitting parameters through experiments. When $V << V_0$, there exists the approximation that $\sinh(\frac{V}{V_0}) \approx \frac{V}{V_0}$. The I–V relationship is linear under this condition. In this work, we will scale down the RRAM voltage to under 0.1 V in order to take advantage of the approximately linear I–V relationship.

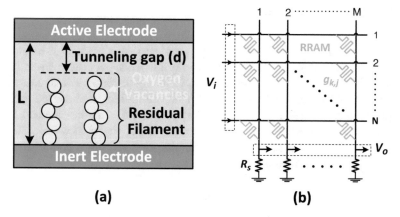

Fig. 11.3 (a) Physical model of the HfO$_x$ based RRAM. The resistance of the RRAM device is determined by the tunneling gap distance d, and d will evolve due to the filed and thermally driven oxygen ion migration. (b) Structure of the RRAM Crossbar Array

As shown in Fig. 11.3b, the relationship between the input voltage vector ($\mathbf{V_i}$) and output voltage vector ($\mathbf{V_o}$) can be expressed as follows [19]:

$$V_{o,j} = \sum_k c_{k,j} \cdot V_{i,k} \qquad (11.3)$$

where k ($k = 1, 2, \ldots, N$) and j ($j = 1, 2, \ldots, M$) are the index numbers of input and output ports, and the matrix parameter $c_{k,j}$ can be represented by the conductivity of the RRAM device ($g_{k,j}$) and the load resistors (g_s) as:

$$c_{k,j} = \frac{g_{k,j}}{g_s + \sum_{l=1}^{N} g_{k,l}} \qquad (11.4)$$

The continuous variable resistance states of RRAM devices enable a wide range of weight matrices that can be represented by the crossbar. The precision of RRAM crossbar based computation may be limited by non-ideal factors, such as process variations, IR drop [20], drifting of RRAM resistance [21], etc. However, SNN only requires low precision of single synaptic value, meanwhile the binary input and LIF operation also alleviate the precision requirement of matrix vector multiplication. Therefore, the RRAM crossbar array is a promising component to realize matrix–vector multiplication for synapse weight computation in neural networks.

11.3   Training Scheme of SNN

The spiking neural network faces a huge problem that it is difficult to train the synaptic weights when applied in the real-world applications. In this section, we compare different SNN training algorithms, including the Spike Timing Dependent Plasticity (STDP), Remote Supervision Method (ReSuMe), and the latest Neural Sampling learning scheme. We show that the neural sampling method which transfers the ANN to SNN is promising for real-word applications while STDP and ReSuMe can **NOT** be used alone in the classification task since both of them are unsupervised learning method.

11.3.1   Spike Timing Dependent Plasticity (STDP)

Synapses connect neurons to each other and transmit signals between them. The synaptic weights, which determine the connecting strength of neurons, are learnable. Spike Timing Dependent Plasticity (STDP) [22] is an unsupervised learning rule that updates the synaptic weights as a function of the relative spiking time of pre- and post-synaptic neurons and the exponential window form of STDP is shown as:

$$\Delta w = \begin{cases} a^+ \cdot w_{ij}(1 - w_{ij}) \cdot \exp\left(-\frac{|t_j - t_i|}{\tau}\right) & \text{if } t_j \geq t_i \\ a^- \cdot w_{ij}(1 - w_{ij}) \cdot \exp\left(-\frac{|t_j - t_i|}{\tau}\right) & \text{if } t_j < t_i \end{cases} \qquad (11.5)$$

where w_{ij} is the synaptic weight between pre- and post-synapse neuron n_i, n_j; t_i, t_j are the spiking time of neuron n_i, n_j; a is the maximum learning rate and τ is the time constant of the learning window. According to Eq. (11.5), the synaptic weight is limited in the interval of $[0, 1]$. The learning rate is decided by the time interval of n_i, n_j spiking: The closer between pre- and post-synaptic spikes, the larger the learning rate. The weight update direction is decided by which neuron spikes first: For the excitatory neuron, if the post-synaptic neuron n_j spikes later than n_i, the synapse will be strengthened; otherwise, it will be decayed; for the inhibitory neuron, vice versa. When every synaptic weight no longer changes or is set to 0/1, the learning process is finished. As an unsupervised method, STDP is mainly used as a feature extraction method. We cannot build a complete machine learning system only based on STDP. A classifier is usually required for practical recognition tasks. However, in our experiment, STDP method doesn't demonstrate enough efficiency of feature extraction. For example, we use the classic MNIST handwritten digit dataset [15] to test the performance with a support vector machine (SVM) [23] without a kernel, where two 50-dimension feature sets are extracted with STDP and principal component analysis (PCA). The PCA-SVM method achieves a recognition accuracy of 94% while the STDP-based method only reaches 91%. As PCA is usually the baseline for evaluating the performance of feature extraction, STDP does **NOT** demonstrate an efficient method for real-world cognitive applications or many other machine learning tasks.

11.3.2   Remote Supervision Method (ReSuMe)

Remote Supervision Method (ReSuMe) is a supervised learning method proposed in [24]. The algorithm introduces a supervised spike train for each synapse while training. The training process comes to an end if the post-synaptic spike train is the same as the supervised spike train. However, ReSuMe faces the difficulty on the pattern design of supervised spike trains and little guidance is offered on how to define the differences between different spike train. Although some papers [25] have attempted to build learning systems under ReSuMe learning algorithm, to the best of our knowledge, we have **NOT** seen any efficient way to solve a real-world task.

11.3.3   Neural Sampling Learning Scheme

The Neural Sampling learning scheme transforms the leaky Integrate-and-Fire (LIF) neuron into a nonlinear function (named Siegert function) [26] which represents the relationship between the input and output firing rate of a neuron, just as shown in Fig. 11.4. Moreover, Neftci demonstrates that nonlinear function, which is equivalent with LIF neuron, is satisfied with neural sampling conditions in [28] and can be approximated to sigmoid function under certain condition. Therefore, it can take advantage of contrastive divergence (CD), which is a classic algorithm exploited in restricted Boltzmann machine (RBM) to train the network. Moreover, the spiking RBM can be stacked into multi-layer to form the spiking deep belief network (DBN), which has demonstrated satisfying performance. In [26], Connor shows that a $784 \times 500 \times 500 \times 10$ spiking DBN achieves the recognition accuracy of 95.2% on MNIST dataset [15]. Recent research results show that it is unnecessary to introduce the Siegert approximation when transferring ANN to SNN, it is better for recognition accuracy if the ReLU neuron is introduced when training the original artificial network. The experiment results [29] show that spiking ConvNet achieves 99.1% accuracy on MNIST dataset when including ReLU neurons for original network training. The introduction of ReLU neuron makes it promising for high-performance large-scaled SNN model because of the better recognition accuracy for large-scaled ReLU-based artificial networks compared with Sigmoid-based (or tanh-based) ones.

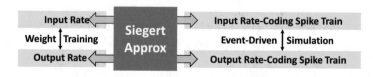

Fig. 11.4 Siegert approximation used in spiking neural network training [27]

In Sect. 11.4, we will make a hardware mapping of the spiking neural network which is trained under (1) STDP + three-layer ANN classier, (2) neural sampling learning method, to RRAM-based platform. The specific RRAM-based system is only used for forward (inference) process, while the training process done on the CPU platform will not be discussed in this work.

11.4   RRAM-Based Spiking Learning System

For an SNN system used for real-time classification applications, an offline training scheme is needed to decide the weights of the neural networks, i.e. coefficients in the crossbar matrix. To our best knowledge, there are two kinds of SNN training methods to build up classification systems: (1) unsupervised SNN training method, for example, Spike Timing Dependent Plasticity (STDP), is first introduced for extracting features; then the supervised classifier is introduced to finish the classification task. (2) First train an equivalent ANN using the gradient-based method, then transfer ANN to SNN and map SNN to the RRAM-based system for real-world applications. We design the two offline trained RRAM based SNN systems based on these two training methods [27, 30], and show them in the following subsections.

11.4.1   Unsupervised Feature Extraction + Supervised Classifier

As an unsupervised method, STDP is mainly used for feature extraction. We cannot build a complete classification system only based on STDP. A classifier is usually required for practical recognition tasks. Therefore, when mapping the system onto hardware, just as shown in Fig. 11.5, a five-layer neural network system is introduced: a two-layer spiking based neural network and a three-layer artificial neural network.

The first two layer SNN is trained using an unsupervised learning rule: spike timing dependent plasticity (STDP) [22], which updates the synaptic weights according to relative spiking time of pre- and post-synaptic neurons. The learning rate is decided by the time interval: the closer distance between pre- and post-synaptic spikes, the larger the learning rate. The weight updating direction is decided by which neuron spikes first: for the excitatory neuron, if the post-synaptic neuron spikes later, the synapse will be strengthened; otherwise, it will be decreased. When every synaptic weight no longer changes or is set to 0/1, the learning process is finished.

There is a converting module between the two layer SNN and 3-layer ANN to convert the spiking trains into the spike count vectors. Then the spike count vectors are sent into the following layers of the network (the 3-layer ANN). We use a 3-layer ANN as a classifier to process the features extracted from the input data by the

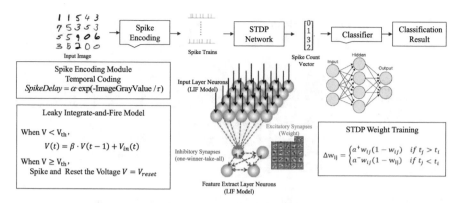

Fig. 11.5 System structure of unsupervised feature extraction + supervised classifier: 2-layer STDP based SNN + 3-layer ANN [27]

previous 2-layer SNN. We use the CMOS analog neuron in Sect. 11.2 for the LIF neuron; and the RRAM crossbar for synaptic computation in both 2-layer SNN (vector addition) and 3-layer ANN (matrix vector multiplication).

An experiment is made on MNIST digit recognition dataset to evaluate the performance of such system framework. The training algorithm is implemented on the CPU platform where LIF neurons are used in the first two layers and the sigmoid neurons are used in the last three layers. For the testing process (forward propagation of neural networks), we use circuit level simulation where the weight matrix is mapped to RRAM-based crossbar. Since the input images are 28×28 sized 256-level gray images, the first layer has 784 input channels. The five-layer spiking neural network system has five layers of neurons in all and the experiment result with the network size of "784×100SNN+100×50×10ANN" shows the recognition accuracy of 91.5% on CPU platform and 90% on RRAM-based crossbar model (circuit simulation result). The performance is a little worse than that of the three-layer ANN sized "784×100×10" with the recognition accuracy of 94.3% on CPU platform and 92% on RRAM-based crossbar model (circuit simulation result).

An interesting point comes from the energy consumption part, we find out that both ANN and SNN use the RRAM crossbar as the matrix vector multiplication part, ANN will consume more power than SNN with similar or even smaller neuron numbers. For example, the proposed "784×100SNN+100×50×10ANN" consumes 327.36 mW on RRAM while the power consumption increases to 2273.60 mW when we directly use "784×100×10ANN." The energy/power saving of SNN comparing ANN mainly comes from the different coding basis. The input voltage of SNN can be binary since it transforms the numerical information into the temporal domain, so there is no need for SNN to hold a large voltage range to represent multiple input states as implemented in ANN. ANN needs input voltages of 0.9 V, but SNN can work with much lower voltage supply (0.1 V). On the other hand, binary coding in SNN can avoid the usage of large number of AD/DA on the input and output interfaces, and the AD/DA consumes considerable large portion of power in the RRAM based NN systems [31].

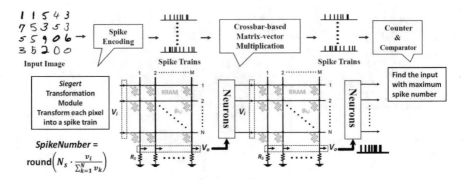

Fig. 11.6 System structure: transferring ANN to SNN—neural sampling method [30]

11.4.2 Transferring ANN to SNN: Neural Sampling Method

The neural sampling method provides a way to transfer ANN to SNN, thus offering a useful training scheme on classification tasks. A equivalent transformation is made between the nonlinear function (named Sigert function, which is similar to sigmoid function) of ANN and the Leaky Integrate-and-Fire (LIF) neuron of SNN. Therefore, it is possible to first train the ANN made up of the stacked Restricted Boltzmann Machine (RBM) structure using Contrastive Divergence (CD) method. In this way, a satisfying recognition accuracy of ANN can be first achieved. And then, the spike-based stacked RBM network with the same synaptic weight matrices can also be implemented for the classification tasks. The system structure is shown in Fig. 11.6 [30].

Since spike trains propagate in the spiking neural network, original input $x = [x_1, \ldots, x_N]$ should be mapped to spike trains $X(t) = [X_1(t), \ldots, X_N(t)]$ before running the test samples where $X_i(t)$ is a binary train with only two states 0/1. For the ith input channel, the spike train is made of N_t spike pulses with each pulse width T_0, which implies that the spike train lasts for the length of time $N_t \cdot T_0$. Suppose the spike number of all input channels during the given time $N_t \cdot T_0$ is N_s, then the spike count N_i of the ith channel is allocated as:

$$N_i = \sum_{k=0}^{N_t-1} X_i(kT_0) = \text{round}\left(N_s \cdot \frac{v_i}{\sum_{k=1}^{N} v_k}\right) \qquad (11.6)$$

which implies

$$\frac{N_i}{N_s} = \frac{v_i}{\sum_{i=1}^{N} v_i} \qquad (11.7)$$

Then the N_i spikes of the ith channel is randomly set on the N_t time intervals. For an ideal mapping, we would like to have $N_i << N_t$ to keep the spike sparsity on

Table 11.1 Important parameters of the SNN system

Network size	$784 \times 500 \times 500 \times 10$
Number of input spike (N_s)	2000
Number of pulse interval (N_t)	128
Input pulse voltage (V)	1 V
The pulse width (T_0)	1 ns

the time dimension. However, for the speed efficiency, we would like the running time $N_t \cdot T_0$ to be short. Here, T_0 is defined by the physical clock, i.e. the clock of the pulse generator, which implies that we can only optimize N_t directly. Here, we define the bit level of the input as

$$\log \left(\frac{N_t}{\text{mean}(N_i)} \right) \tag{11.8}$$

which evaluates the tradeoff between time efficiency and the accuracy performance.

We train the SNN with the size of $784 \times 500 \times 500 \times 10$. And the parameters are shown in Table 11.1. The experiment results show that the recognition accuracy of MNIST dataset is 95.4% on the CPU platform and 91.2% on the ideal RRAM-based hardware implementation. The recognition performance decreases about 4% because it is impossible to satisfy with $N_t \ll N_s$ on the RRAM platform.

We show the results for recognition under different bit level quantization of input signal and RRAM devices, together with RRAM process variation and input signal fluctuation. The simulation results in Fig. 11.7a show that an 8-bit RRAM device is able to realize a recognition accuracy of nearly 90%. The simulation results in Fig. 11.7b show that the input signal above 6-bit level achieves satisfying recognition accuracy (>85%). Based on the 8-bit RRAM result, different levels of signal fluctuation are added on the 8-bit input signal. The result shown in Fig. 11.7c demonstrates that the performance of accuracy just decreases 3% given 20% variation. Figure 11.7d shows that when RRAM device is made in 8-bit level with the 6-bit level input, the performance does not decrease under 20% process variation. The sparsity of the spike train leads to the system robustness, making it insensitive to the input fluctuation and process variation.

The power consumption of the system is mainly contributed by three parts: the crossbar, the comparator, and the $R_{\text{mem}}C_{\text{mem}}$ leaky path. The simulation results show that the power consumption is about 3.5 mW on average. However, it takes $N_t = 128$ cycles with the physical clock $T_0 = 1$ ns. Though input conversion from numeral values to spike trains leads to about 100× clock rate decrease, the system is able to **complete the recognition task in real time** (~1 μs/sample), thanks to the short latency of RRAM device.

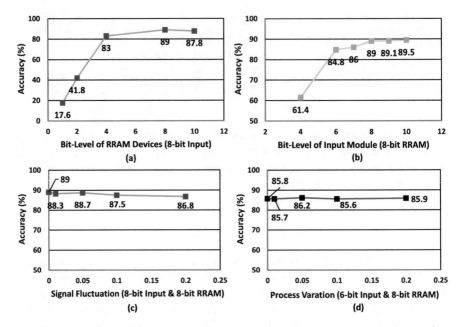

Fig. 11.7 Recognition accuracy under (**a**) different bit-level of RRAM devices, (**b**) different bit-level of input module, and (**c**) different degrees of input signal fluctuation, (**d**) different degrees of process variation of RRAM devices [30]

11.4.3 Discussion on How to Boost the Accuracy of SNN

The experiment results in the above subsections show that the recognition accuracy will decay after transferring an ANN to an SNN. However, due to the ultra-high integration density of the RRAM devices and the 0/1 based interfaces of SNN, SNN tends to consume much less circuit area and power compared with ANN. This result inspires us that we may integrate multiple SNNs with the same or even less circuit area and power consumption of ANN, and combine these SNNs together to boost the accuracy and robustness of the SNN system.

Previously, an ensemble method [31] is proposed to boosting the accuracy of RRAM-based ANN systems, named SAAB (Serial Array Adaptive Boosting), which is inspired by the AdaBoost method [32]. The basic idea of AdaBoost, which is also its major advantage, is to train a series of learners, such as ANNs or SNNs, sequentially, and every time we train a new learner, we try to "force" the new learner to pay more attention to the "hard" samples incorrectly classified by previous trained learners in the training set. The proposed technique can improve the accuracy of ANN by up to 13.05% on average and ensure the system performance under noisy conditions in approximate computation applications.

SAAB boost the computation accuracy at the cost of consuming more power and circuit area. As SNN usually consumes much less area and power compared with the

ANN, there is a chance to integrate multiple SNNs under the same circuit resource limitation of ANN. And these SNNs can be boosted together by the similar idea of SAAB. However, the inherent attributions of SNN systems should be considered when designing the boosting algorithm. According to our observation, there are two types of errors in the SNN-based classification tasks: (1) a traditional type: more than one neuron in the output layer spikes and the neuron spiking the most is not the target neuron; and (2) a special type of SNN: no neuron in the output layer spikes; It is interesting to observe that most of the wrong trials are the special type and it can be reduced slightly when increasing the input spike counts. We regard such samples as the difficult classifying cases. When seeking for the possibility to make up the performance loss after transferring ANN to SNN with a boosting-based method, this problem should be considered.

11.5 Conclusion

In this chapter, we first introduce the background knowledge of SNN and metal-oxide resistive switching random-access memory (RRAM). Then, we compare different training algorithms of SNN for real-world applications, and demonstrate that the Neural Sampling method is much more effective than other methods. We also explore the performance and energy efficiency by building the SNN-based energy efficient system for real time classification with RRAM devices. We implement different training algorithms of SNN, including Spiking Time Dependent Plasticity (STDP) and Neural Sampling method. Our RRAM-based SNN systems for these two training algorithms show good power efficiency and recognition performance on real-time classification tasks, e.g., the MNIST digit recognition. Finally, we discuss a possible direction to further improve the classification accuracy by boosting multiple SNNs.

However, there are still many challenges remaining in this spiking neural network structure. For example, the encoding mechanism from original data to spiking is not quite clear. It perhaps has a huge effect on system performance and power efficiency. Thus, how to design a proper encoding mechanism is one possible method of improving the performance of the system. In addition, the non-ideal circuit condition (e.g., the interconnection effect, the input variation) should be considered for future RRAM-based system design.

Acknowledgements This work was supported by 973 project 2013CB329000, National Science and Technology Major Project (2011ZX03003-003-01, 2013ZX03003013-003) and National Natural Science Foundation of China (Nos. 61373026, 61261160501, 61271269), and Tsinghua University Initiative Scientific Research Program. And we gratefully acknowledge the support from Prof. Shimeng Yu with the help of RRAM model.

References

1. L. Chang, Y.-K. Choi, D. Ha, P. Ranade, S. Xiong, J. Bokor, C. Hu, T.-J. King, Extremely scaled silicon nano-CMOS devices. Proc. IEEE **91**(11), 1860–1873 (2003)
2. W.A. Wulf, S.A. McKee, Hitting the memory wall: implications of the obvious. ACM SIGARCH Comput. Arch. News **23**(1), 20–24 (1995)
3. W. Maass, Networks of spiking neurons: the third generation of neural network models. Neural Netw. **10**(9), 1659–1671 (1997)
4. T. Masquelier, S.J. Thorpe, Unsupervised learning of visual features through spike timing dependent plasticity. PLoS Comput. Biol. **3**(2), e31 (2007)
5. D. Querlioz, W. Zhao, P. Dollfus, J. Klein, O. Bichler, C. Gamrat, Bioinspired networks with nanoscale memristive devices that combine the unsupervised and supervised learning approaches, in *2012 IEEE/ACM International Symposium on Nanoscale Architectures (NANOARCH)* (IEEE, 2012), pp. 203–210
6. P.A. Merolla, J.V. Arthur, R. Alvarezicaza, A.S. Cassidy, J. Sawada, F. Akopyan, B.L. Jackson, N. Imam, C. Guo, Y. Nakamura et al., A million spiking-neuron integrated circuit with a scalable communication network and interface, Science **345**(6197), 668–673 (2014)
7. B. Govoreanu, G.S. Kar, Y. Chen, V. Paraschiv, $10 \times 10\,nm^2$ HF/HFOx crossbar resistive ram with excellent performance, reliability and low-energy operation, *Electron Devices Meeting IEDM Technical Digest. International* (2011), pp. 31.6.1–31.6.4
8. B. Li, Y. Shan, M. Hu, Y. Wang, Memristor-based approximated computation, in *IEEE International Symposium on Low Power Electronics and Design* (2013), pp. 242–247
9. T. Tang, R. Luo, B. Li, H. Li, Energy efficient spiking neural network design with RRAM devices, in *International Symposium on Integrated Circuits* (2015), pp. 268–271
10. T. Tang, L. Xia, B. Li, R. Luo, Spiking neural network with RRAM: can we use it for real-world application? in *Design, Automation and Test in Europe* (2015), pp. 860–865
11. R. Wang, T.J. Hamilton, J. Tapson, A. Van Schaik, An FPGA design framework for large-scale spiking neural networks, in *Proceedings - IEEE International Symposium on Circuits and Systems* (2014), pp. 457–460
12. E. Painkras, L.A. Plana, J. Garside, S. Temple, Spinnaker: A 1-w 18-core system-on-chip for massively-parallel neural network simulation. IEEE J. Solid-State Circ. **48**(8), 1943–1953 (2013)
13. S.K. Esser, A. Andreopoulos, R. Appuswamy, P. Datta, D. Barch, A. Amir, J. Arthur, A. Cassidy, M. Flickner, P. Merolla, Cognitive computing systems: algorithms and applications for networks of neurosynaptic cores, in *The 2013 International Joint Conference on Neural Networks (IJCNN)* (2013), pp. 1–10
14. D. Kuzum, R.G. Jeyasingh, B. Lee, H.-S.P. Wong, Nanoelectronic programmable synapses based on phase change materials for brain-inspired computing. Nano Letters **12**(5), 2179–2186 (2011)
15. Y. Lecun, C. Cortes, The MNIST database of handwritten digits (1998)
16. E.M. Izhikevich, Which model to use for cortical spiking neurons? IEEE Trans. Neural Netw. **15**(5), 1063–1070 (2004)
17. G. Indiveri, A low-power adaptive integrate-and-fire neuron circuit, in *ISCAS (4)* (2003), pp. 820–823
18. S. Yu, B. Gao, Z. Fang, H. Yu, J. Kang, H.-S.P. Wong, Stochastic learning in oxide binary synaptic device for neuromorphic computing. Front. Neurosci. **7**,186 (2013)
19. M. Hu, H. Li, Q. Wu, G.S. Rose, Hardware realization of BSB recall function using memristor crossbar arrays, in *Proceedings of the 49th Annual Design Automation Conference* (ACM, 2012), pp. 498–503
20. P. Gu, B. Li, T. Tang, S. Yu, Y. Wang, H. Yang, Technological exploration of rram crossbar array for matrix-vector multiplication, in *ASP-DAC* (2015)
21. B. Li, Y. Wang, Y. Chen, H.H. Li, H. Yang, Ice: inline calibration for memristor crossbar-based computing engine, in *Proceedings of the conference on Design, Automation & Test in Europe* (European Design and Automation Association, 2014), p. 184

22. S. Song, K.D. Miller, L.F. Abbott, Competitive hebbian learning through spike-timing-dependent synaptic plasticity. Nat. Neurosci. **3**(9), 919–926 (2000)
23. C.-C. Chang, C.-J. Lin, Libsvm: a library for support vector machines. ACM Trans. Intell. Syst. Technol. (TIST) **2**(3), 27 (2011)
24. F. Ponulak, ReSuMe - new supervised learning method for spiking neural networks, Institute of Control and Information Engineering, Poznan University of Technology. Available online at: http://d1.cie.put.poznan.pl/~fp/research.html (2005)
25. J. Hu, H. Tang, K.C. Tan, H. Li, L. Shi, A spike-timing-based integrated model for pattern recognition. Neural Comput. **25**(2), 450–472 (2013)
26. P. O'Connor, D. Neil, S.-C. Liu, T. Delbruck, M. Pfeiffer, Real-time classification and sensor fusion with a spiking deep belief network. Neuromorphic Eng. Syst. Appl **61**, 1–10 (2015)
27. T. Tang, R. Luo, B. Li, H. Li, Y. Wang, H. Yang, Energy efficient spiking neural network design with RRAM devices, in *14th International Symposium on Integrated Circuits (ISIC)* (IEEE, 2014), pp. 268–271
28. E. Neftci, S. Das, B. Pedroni, K. Kreutz-Delgado, G. Cauwenberghs, Event-driven contrastive divergence for spiking neuromorphic systems (2013). Preprint, arXiv:1311.0966
29. P.U. Diehl, D. Neil, J. Binas, M. Cook, Fast-classifying, high-accuracy spiking deep networks through weight and threshold balancing, in *International Joint Conference on Neural Networks* (2015)
30. T. Tang, L. Xia, B. Li, R. Luo, Y. Wang, Y. Chen, H. Yang, Spiking neural network with RRAM: can we use it for real-world application? In *Proceedings of the 2015 Design, Automation & Test in Europe Conference & Exhibition* (EDA Consortium, 2015), pp. 860–865
31. B. Li, L. Xia, P. Gu, Y. Wang, H. Yang, Merging the interface: power, area and accuracy co-optimization for RRAM crossbar-based mixed-signal computing system, in *Design Automation Conference* (2015), pp. 1–6
32. Z.-H. Zhou, *Ensemble Methods: Foundations and Algorithms* (CRC Press, 2012)

Chapter 12
Efficient Neuromorphic Systems and Emerging Technologies: Prospects and Perspectives

Abhronil Sengupta, Aayush Ankit, and Kaushik Roy

12.1 Introduction

Deep Learning Networks (DLN) are inspired from the hierarchical organization of neurons and synapses in the human brain and are an important class of machine learning algorithms. Since its inception, it has been widely adopted in multifarious recognition tasks. Lately, DLNs have redefined the state of the art for many cognitive applications including computer vision [1], speech recognition [2], and natural language processing [2] across various application domains. For instance, Baidu's Deep Speech 2 recently demonstrated English and Mandarin language recognition at par to human capabilities. Google's DeepMind recently defeated AlphaGo world champion. Tesla is using deep learning powered vision, sonar and radar processing in their self-driving systems. Also, deep learning is being adopted for ever-more tasks, for example, face recognition by Facebook, data analytics by IBM, recommender systems by Amazon, and so on.

DLN performance (accuracy) is a strong function of the network scale. Hence, the network size has been commensurate with the target accuracy and the problem complexity. LeCun et al. used Convolutional Neural Networks (CNN) for handwritten digits classification using 1 million parameters in 1998 [3]. In 2012, Krizhevsky et al. used a CNN with 60 million parameters for object classification having 1000 classes [4]. Deepface used 120 million parameters to classify human faces [5]. In a nutshell, large DLN models are preferred by Machine Learning practitioners and this necessitates developing efficient systems to power DLNs.

Lately, conventional computing systems which are primarily based on von-Neumann computing model have been extensively used in cognitive applications

---
A. Sengupta • A. Ankit • K. Roy (✉)
School of Electrical and Computer Engineering, Purdue University, 465 Northwestern Ave, West Lafayette, IN 47907, USA
e-mail: asengup@purdue.edu; aankit@purdue.edu; kaushik@purdue.edu

© Springer International Publishing AG 2017

A. Chattopadhyay et al. (eds.), *Emerging Technology and Architecture for Big-data Analytics*, DOI 10.1007/978-3-319-54840-1_12

and have shown fascinating results across many domains. Majority of the current DLN implementations are done on Graphic Processing Units (GPUs) owing to their ability to deal with intensive parallel computations. For example, Tesla's DLNs uses in-vehicle GPU based supercomputers. Facebook is using GPUs to power its machine learning algorithms. It seems appealing that systems based on conventional computing styles like GPUs are addressing the computational needs of DLNs and hence seems to be an ideal choice for accelerating DLN for cognitive applications. The observation and the inference are correct but this simple treatment of facts veil the skyrocketed power and memory requirements driving the modern computing systems which is many orders higher than human brain.

There has been recent research to work around the memory bottlenecks in GPUs by using techniques such as data batching and virtualized DLNs [6]. However, the latency requirements can be unsuitable for real-time applications. Previous works have also used specialized hardware for DLN acceleration [7]. This work underscores the memory access associated power requirements and shows that the DRAM accesses account for significant power consumption. Consequently, the DLN performance on conventional computing systems is presently limited by the memory and power bottlenecks. Hence, with complex, bigger, and deeper networks that can achieve brain-like cognitive capabilities being developed, there is a dire need to develop brain-like energy-efficient architectures that can drive them.

These power and memory bottlenecks in the conventional computing systems are primarily a consequence of the mismatch between the conventional computing systems and computing patterns involved in these cognitive applications. MOS transistors, being on/off switches, have served as an ideal match to the abstractions of switching functions and Boolean logic, which (together with von Neumann architecture) form the underpinnings of modern computing. While current computing platforms are well suited to applications that involve arithmetic computations and storing and retrieving large amounts of data, they are known to be highly inefficient—requiring orders of magnitude more energy consumption—for performing tasks that humans routinely perform, such as visual recognition, semantic analysis, and reasoning. This inefficiency stems from the realization of neuron and synapse functionality by translating the underlying mathematical functions to Boolean logic gates and subsequently to transistors, resulting in hundreds of transistors to mimic a single neuron/synapses. In this chapter we will review typical von-Neumann computing based CMOS architectures used for DLN implementations and demonstrate the manner in which spintronic crossbar array based "in-memory" computing platforms can lead to more compact and energy-efficient implementations.

12.2 Neural Network Basics

DLNs are feed-forward networks where the computational units—the neurons are connected to neurons in other layers through programmable connections termed as synapses. Figure 12.1a shows an NN with one hidden layer. DLNs are usually

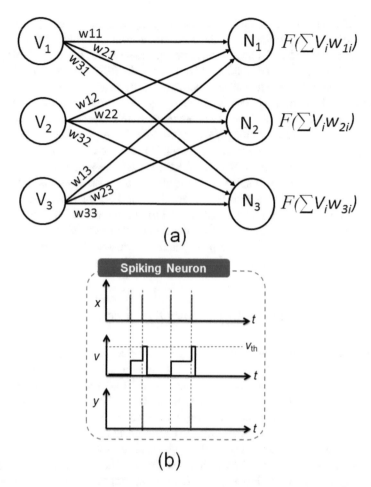

Fig. 12.1 (**a**) Feedforward NN consists of neurons in one layer connected to another layer through programmable synaptic weights, (**b**) IF spiking neuron computing model: x is the input spike train and v is the neuron membrane potential. The neuron generates an output spike y whenever v crosses the threshold voltage v_{th}

characterized by multiple such hidden layers. Depending on the type of connectivity, a DLN can either be fully connected (Multi-Layer Perceptron) or have sparse connectivity (Convolutional Neural Network). The net input a neuron receives is the weighted summation of its inputs. Thus, a neuron computation involves a weight fetch from a memory (SRAM) followed by a multiply-and-accumulate (MAC) operation for every input it receives. Once, all the inputs have been processed for the above-mentioned operations, a non-linear computation is done to compute the neuron's output activation potential. This output then acts as the input for the next layer neurons.

DLN architectures discussed above have achieved record-breaking results for many classification problems but the substantial computational cost of training and running deep networks have motivated the research for the more biologically plausible NN version called Spiking Neural Networks (SNN). In recent years, deep SNNs have become an increasingly active field of research which is primarily motivated by the extremely energy-efficient cognitive processing in human brain. Driven by brain-like event-driven computations, SNNs involve data processing in an input-triggered fashion. Unlike conventional ANNs where a vector is given at the input layer once and the corresponding output is produced after processing through several layers of the network, SNNs require the input to be encoded as a spike train. At a particular instant, each spike is propagated through the layers of the network while the neurons accumulate the spikes over time causing the output neuron to fire or spike. Thus, the spike information is used to communicate between the layers of the network. Figure 12.1b describes the functionality of an Integrate-Fire (IF) spiking neuron. Note that more bio-plausible models also include a leak-term in the membrane potential which causes it to decay in the absence of spikes.

12.3   General Purpose Computing Architecture

In this section we describe a general purpose architecture to implement SNNs. It involves an SRAM which stores the SNN weights and inputs and a computation core to perform the neuronal computations. Figure 12.2 shows the block diagram of the architecture and the logical dataflow between the constituent blocks. The SRAM memory stores the input data (image pixel values and weights) for the trained SNN. Efficient data movement is achieved by buffering the input data—image data and weight data in FIFOs. Image and weight data are read from SRAM memory and processed by an array of Neuron Units (NUs). The NUs keep accumulating weights depending on the input spikes, until all the inputs for a particular neuron are processed. After this, the NUs produce the output spikes which are written back to the SRAM.

Here we discuss the logical dataflow involved between the different components for a time-step in SNN computation. Input data read from the SRAM is stored into the Input FIFO to stream across the NU array as all the neurons in a layer share the same inputs. Corresponding to the input data, weight data are fetched from the SRAM and stored into the weight FIFOs for temporal reuse by the NUs. Each NU receives the weights from its dedicated weight FIFO. The input FIFO is flushed and new set of data are read from SRAM and put in the Input FIFO, after the input processing is finished. Similarly, the weight FIFO gets flushed, new weights read from SRAM and stored into weight FIFO, once the weight processing finishes. When all the computations for the neurons currently scheduled into the NUs are done, the next set of neurons are scheduled into the NUs, corresponding weights read from SRAM into their respective weight FIFOs. Each NU performs "accumulate-and-fire" operation. The NUs are connected in a serial fashion to allow data streaming from Input FIFO to the rightmost NU.

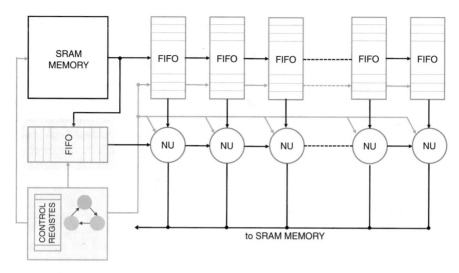

Fig. 12.2 General purpose computing architecture—logical dataflow and the constituent blocks

The neuron computations are done layer wise—read the inputs and weights from SRAM, compute all the outputs corresponding to the first layer, store back the outputs in SRAM and proceed to the next layer. Within a layer neurons are temporally scheduled in the NUs. First, the output computations for the first set of 'N' neurons are done, then the next set of 'N' neurons from the same layer are scheduled in the NU and this goes on until all the neurons in the current layer have been evaluated. Hence, we temporally map different layers of the neural network and different neurons within a layer to compute the entire neural network for a given input data.

For a typical fully connected network, the memory component of energy consumption (access + leakage) on this architecture is more than 50% on an average across various computing workloads. As discussed before, the general purpose computing frameworks have separate processing (core) and data storage components (SRAM). The volume of data to process has drastically increased with increasing DLN size. The DLN size has been ever-increasing to get better accuracy and solve more complex recognition tasks. Consequently, data movement between core and memory has been increasing and is becoming one of the most critical performance and energy bottlenecks in these computing systems. On one hand, this limits the research community to use sub-optimal architectures for solving the recognition problems and on the other hand this impedes their deployment on mobile computing platforms which are energy limited.

The size of the SRAM needed scales directly with the network size and the network size is a strong function of target accuracy and problem complexity. AlexNet [4], for instance, contains 5 convolutional layers with 2 fully connected layers and uses 2.3 million weights. The more recent VGG-16 [1] contains 16 convolution layers and 3 fully connected layers and uses 14.7 million weights. However, larger memories lead to increased power consumption due to increased memory access and leakage power and the memory component of energy (access + leakage) is typically higher than the computation component. Eventually, the memory-driven limitations affect the size of DLN that can be used. To work around these bottlenecks, Machine Learning practitioners must use less desirable DLN architectures (e.g., smaller number of layers and neurons). Additionally, memory imposed bandwidth also limits the number of NUs one can have in the NU-array for neuronal computations. Hence it is imperative to develop innovative architectures which addresses the rigid memory limitations of general purpose frameworks used towards DLN acceleration.

12.4 Underlying Device Physics

Before we present "In-Memory" computing architectures based on spintronic technologies that can potentially alleviate the memory-bandwidth limitations of conventional CMOS based neuromorphic architectures, let us discuss the underlying device physics of such emerging technologies that can provide a direct mapping to synaptic and neural functionalities.

Let us first illustrate the device structure and principle of operation of a Magnetic Tunnel Junction (MTJ: Fig. 12.3a) [8]. The MTJ consists of two ferromagnets (FMs) separated by a tunneling oxide barrier (MgO). The magnetization direction of one of the layers (denoted by "pinned" layer, PL) is magnetically hardened so that it serves as the reference layer. The magnetization of the "free" layer (FL), can be manipulated by an input charge current. The MTJ is characterized by two stable resistance states, namely the low-resistance parallel (P) configuration ("free" and "pinned" layer magnetizations are parallel) and the high-resistance anti-parallel (AP) configuration ("free" and "pinned" layer magnetizations are anti-parallel). The MTJ can be switched between the two stable states by charge current flow through the stack due to spin-transfer torque (STT) effect generated by charge current flowing through the "pinned" layer [9]. Recent experiments have shown that such an MTJ structure with in-plane magnetic anisotropy (IMA) can also be switched by a charge current flowing through a heavy-metal (HM) underlayer due to the injection of spins (whose polarization is transverse to the direction of both spin and charge current) at the FL-HM interface (assuming spin-Hall effect [10] to be the dominant underlying physical phenomenon). Such HM induced MTJ switching has been proven to be more energy efficient than STT induced switching [11, 12]. Further, it opens up the possibility of device structures that can exhibit decoupled "write" and "read" current paths, as will be explained in later sections. It is also worth noting here that at non-zero temperature, the magnetization dynamics of the MTJ is characterized by thermal noise, which can be accounted for by an additional thermal field. In the presence of thermal noise, the switching behavior of the MTJ due to the flow of a charge current through the "pinned" layer, during the "write" cycle, is stochastic in nature and the probability of switching increases with increase in the magnitude of input current [13].

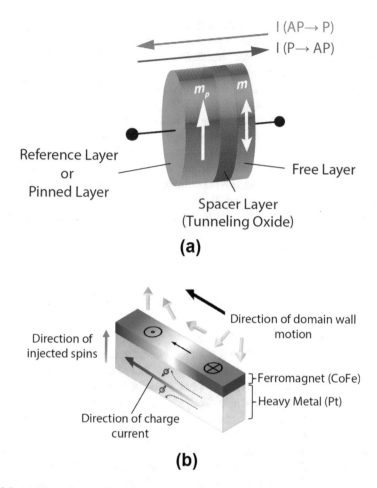

Fig. 12.3 (**a**) Magnetic tunnel junction consists of two nanomagnets separated by a spacer, (**b**) spin-orbit torque induced domain wall motion due to charge current flowing through a heavy-metal (HM) underlayer

In addition to monodomain magnets discussed above, we will also utilize multi-domain magnets having a domain wall separating oppositely polarized magnetic domains. Recent experiments on magnetic nanostrips of Pt/CoFe/MgO and Ta/CoFe/MgO have revealed high domain wall velocities due to charge current densities that are two orders of magnitude lower than that achievable by conventional spin-transfer torque (STT) [14, 15]. Additionally, domain wall motion was observed to be against the direction of electron flow (i.e., in the direction of current flow) in multilayer structures with Pt as the underlayer, thereby suggesting that current induced spin-orbit torque is the main mechanism of domain wall motion in such multilayer structures (with negligible contribution from conventional STT). In such magnetic heterostructures with high perpendicular magnetocrystalline anisotropy (PMA), spin orbit coupling and broken inversion symmetry leads to the stabilization of homochiral domain walls through the Dzyaloshinskii-Moriya exchange interaction (DMI). Such interfacial DMI at the FM-HM interface leads to the formation of a Néel domain wall with specific chirality. The DMI strength in such structures with HM underlayers has been observed to be sufficiently strong to impose a Néel wall configuration in FMs where conventional magnetostatics would have yielded a Bloch configuration [14, 15]. As shown in Fig. 12.3b, when an in-plane charge current is injected through the HM, a transverse spin-current is generated due to deflection of opposite spin-polarizations on the top and bottom surfaces of the HM due to spin-Hall effect. The accumulated spins at the FM-HM interface leads to DMI stabilized Néel domain wall motion [14, 15]. The direction of domain wall motion is in the direction of charge current flow and the final magnetization of the ferromagnet is given by the cross-product of the direction of injected spins at the FM-HM interface and the magnetization direction of the FM at the domain wall location.

12.5   Proposals for Spintronic Neuromimetic Devices

The building blocks of the brain (neurons and synapses), as well as artificial models thereof, are fundamentally different from the switching functions and Boolean logic gates that CMOS transistors naturally realize. The significant disparity between the brain and corresponding CMOS implementations results in area and power consumptions that are orders of magnitude higher than that involved in the brain. For example, almost 20 transistors would be required to implement the functionality of a single analog spiking neuron [16]. On the other hand, digital neurons would require area and power hungry multipliers and adders to compute the weighted summation of inputs. The situation is even worse for a synapse, where storing even a single-bit weight would require a 6-T/8-T SRAM cell [17]. To realize networks comprising billions of neurons and connectivity levels exceeding 10,000 synapses per neuron, nanoelectronic devices that can more naturally and efficiently mimic synaptic and neural functionalities are imperative.

Recent discoveries in spintronics have brought forward a set of device phenomena that can provide a direct mapping of neuronal and synaptic computations, laying the foundation for a quantum leap in the efficiency of neuromorphic computing. Consider the computational units in an ANN. Inputs to the neuron get multiplied by stored synaptic weights and are subsequently summed up and passed through a thresholding function. As shown in Fig. 12.4, the synaptic functionality can be implemented using a device structure composed of a Magnetic Tunnel Junction (MTJ) where the "free" layer is a domain wall motion based magnetic strip (DWM). A DWM is a ferromagnet with oppositely polarized magnetic domains separated by a transition region termed as domain wall (DW). In the device shown in Fig. 12.4, the position of the domain wall encodes the conductance of the MTJ, which represents the synaptic weight. Therefore, the read current represents the product of the input (read voltage) and synaptic weight (conductance).

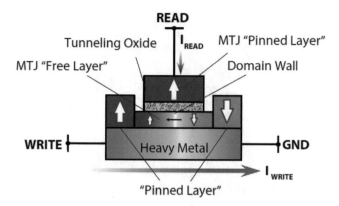

Fig. 12.4 Magnetic bilayer structure (DW-MTJ) as a spintronic synapse

As mentioned in the previous section, recent experiments on ferromagnet-heavy metal (HM) bilayers have provided a promising mechanism for efficient control of domain wall (DW) motion using current densities that can be ~100× lower than conventional spin-transfer torque driven DW motion [18]. Inspired by this development, the proposed device (shown in Fig. 12.4) is a magnetic heterostructure that also includes a Heavy Metal (HM) where a current flowing through the HM can be used for deterministic control of DW motion and hence, the MTJ conductance [18]. We will refer to this device as DW-MTJ for the rest of this chapter and demonstrate its application to realizing neural and synaptic units.

Let us next consider the neuron computation in ANNs, which involves summation of synaptic inputs and performing a thresholding operation on the result. Figure 12.5 [18] shows how the DW-MTJ device discussed above, together with a reference MTJ and a single transistor, can be used to realize this computation. The DW-MTJ together with the Reference MTJ (Fig. 12.5) forms a resistive divider network. Input synaptic current is fed into the HM layer, moves the domain wall, and changes the MTJ conductance, thereby causing a variation in the output current provided by the transistor, which represents the neuron output.

A more biologically realistic neural computing model in comparison to the artificial neuron is the spiking neuron model. Such a neuron receives input spikes and generates an output spike only when its membrane potential crosses a threshold. The neuron's membrane potential increases on the arrival of an input spike and leaks back to its resting potential in the absence of a spike. Interestingly, the magnetization dynamics of an MTJ offers a direct mapping to the functionality of such spiking neurons (Fig. 12.6). Incoming synaptic current flowing from the "free" to the "pinned" layer gets spin-polarized by the "pinned" layer and causes the "free" layer magnetization to be oriented parallel to the "pinned" layer. As shown in Fig. 12.6, when input spikes are transmitted to the MTJ, the magnetization or the conductance of the MTJ starts "integrating." However, upon removal of the stimulus the magnetization starts "leaking" back. The MTJ conductance increases

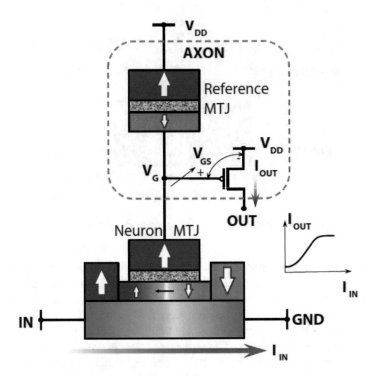

Fig. 12.5 Spintronic device (DW-MTJ) as an artificial neuron

by a specific amount for each spike and finally switches to the parallel (high-conductance) state at the end of the fifth input spike (Fig. 12.6) (analogous to the "spiking" of a biological neuron). Such a functionality can be exploited to build MTJ based spiking neurons. Further, thermal noise inherent in such devices can be exploited to perform probabilistic inference with stochastic spiking neurons [19, 20]. The MTJ switches probabilistically depending on the magnitude of the input synaptic current (Fig. 12.7). HM induced MTJ switching can significantly further improve the energy efficiency of this process.

12.6 Crossbar based "In-Memory" Computing Architecture

While the proposed spintronic devices realize the primitive functions required to implement individual neurons and synapses, realizing a multi-layer network where spin-neurons and synapses are cascaded requires hybrid circuits that involve spin devices and a few CMOS transistors. Furthermore, to support the broad range of neuromorphic applications, there is a need to design computing fabrics that can realize networks of varying sizes and topologies, and can perform both training/learning and evaluation. In doing so, it is critical to ensure that the intrinsic efficiency of spin neurons and synapses is preserved at the system level.

Fig. 12.6 MTJ as a leaky-integrate-fire spiking neuron. The magnetization or conductance integrates on each spike and starts leaking once the spike is removed (**a**) Input current stimulus and (**b**) MTJ conductance have been shown as a function of time

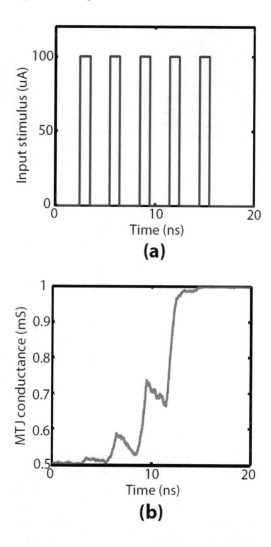

Note that the main computing kernel in a DLN is dot-product computation between the inputs and corresponding synaptic weights for each neuron followed by neuron processing. Figure 12.8 shows a possible design of such a spintronic computing kernel that consists of a crossbar array of spin-synapses driving spin-neurons. Input voltages applied across each row of the array result in the generation of a weighted synaptic current (weights are spin-synapse conductances), which is summed up and provided as input to each neuron along the column. CMOS transistors operate as axons (gate voltage is modulated by a resistive divider network consisting of a reference MTJ and the DW-MTJ). Note that the proposed processing unit can be cascaded (the drain of each output transistor drives a row of the next

Fig. 12.7 (**a**) Probabilistic
neural inference can be
performed by spin-Hall effect
induced MTJ switching in
presence of thermal noise, (**b**)
variation of MTJ switching
probability with magnitude of
input current for different
pulse width durations

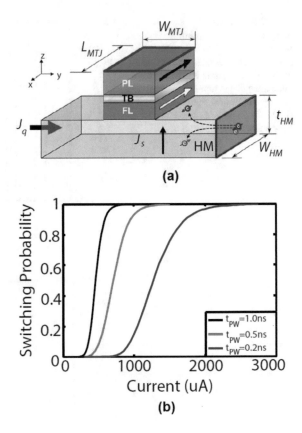

crossbar array). Hence, this unit can be used as a building block to construct scalable
neuromorphic architectures and assists in implementing the dot-product computing
kernel that is an essential component of such neuromimetic algorithms.

Let us analyze the spin-based ANN design shown in Fig. 12.8 and determine its
advantages over a CMOS based implementation. In order to provide an intuitive
insight to the energy efficiency of the proposed system, let us consider a "spin-
neuron" with "free layer" dimensions of 80 nm × 20 nm. Micromagnetic simulations
indicate that a current of ∼10.6 μA can displace the domain wall between the two
extreme edges within 2 ns leading to a maximum energy consumption of 0.1 fJ
(including energy consumption during neuron "reset" operation). This is almost two
orders of magnitude lower in comparison to analog (∼700 fJ) and digital (∼832.6 fJ)
CMOS neuron designs in 45 nm technology [18]. Additionally, the synaptic resistive
crossbar array can be operated at ultra-low voltages of ∼100 mV due to the low
current requirements of the spin-neurons. In contrast, the crossbar arrays have to be
operated at a much higher voltage (∼500 mV) to drive analog CMOS neurons. This
results in power (V^2/R) savings by a factor ∼25× per synapse and thereby helps in
reducing the overall power consumption of the neuromorphic system.

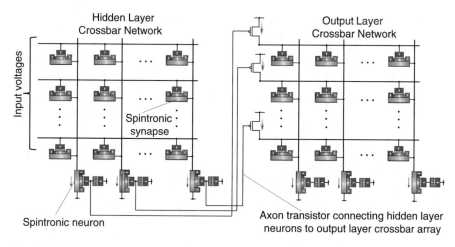

Fig. 12.8 Spintronic neuromorphic architecture (with DW-MTJ neurons and synapses) connecting different layers of the neural network

12.7   Conclusions

In this chapter, we provided a vision for "in-memory computing" architectures built on spintronic crossbars for neuromorphic computations. Crossbars alleviate the memory-bandwidth associated performance limitations in DLN acceleration. Additionally, it also removes the energy consumption associated with continuous data transfer between a separate memory and the computation core which is a dominant component of energy consumption in such data-intensive applications. Inner-product computing kernels based on spintronic crossbar arrays driving magneto-metallic spintronic neurons can pave the way for compact and energy-efficient neuromorphic architectures.

References

1. K. Simonyan, A. Zisserman, Very deep convolutional networks for large-scale image recognition (2014). arXiv preprint arXiv:1409.1556
2. T. Mikolov, M. Karafiát, L. Burget, J. Cernockỳ, S. Khudanpur, Recurrent neural network based language model.Interspeech **2**, 3 (2010)
3. Y. LeCun, L. Bottou, Y. Bengio, P. Haffner, Gradient-based learning applied to document recognition. Proc. IEEE **86**(11), 2278–2324 (1998)
4. A. Krizhevsky, I. Sutskever, G.E. Hinton, Imagenet classification with deep convolutional neural networks, in *Advances in Neural Information Processing Systems* (2012), pp. 1097–1105
5. Y. Taigman, M. Yang, M. Ranzato, L. Wolf, Deepface: closing the gap to human-level performance in face verification, in *Proceedings of the IEEE Conference on Computer Vision and Pattern Recognition* (2014), pp. 1701–1708

6. M. Rhu, N. Gimelshein, J. Clemons, A. Zulfiqar, S.W. Keckler, vDNN: Virtualized deep neural networks for scalable, memory-efficient neural network design (2016). arXiv preprint arXiv:1602.08124

7. Y. Chen, T. Luo, S. Liu, S. Zhang, L. He, J. Wang, L. Li, T. Chen, Z. Xu, N. Sun et al., Dadiannao: A machine-learning supercomputer, in *Proceedings of the 47th Annual IEEE/ACM International Symposium on Microarchitecture* (IEEE Computer Society, 2014), pp. 609–622

8. M. Julliere, Tunneling between ferromagnetic films. Phys. Lett. A **54**(3), 225–226 (1975)

9. J.C. Slonczewski, Conductance and exchange coupling of two ferromagnets separated by a tunneling barrier. Phys. Rev. B **39**(10), 6995 (1989)

10. J. Hirsch, Spin hall effect. Phys. Rev. Lett. **83**(9), 1834 (1999)

11. C.-F. Pai, L. Liu, Y. Li, H. Tseng, D. Ralph, R. Buhrman, Spin transfer torque devices utilizing the giant spin Hall effect of tungsten. Appl. Phys. Lett. **101**(12), 122404 (2012)

12. L. Liu, C.-F. Pai, Y. Li, H. Tseng, D. Ralph, R. Buhrman, Spin-torque switching with the giant spin Hall effect of tantalum. Science **336**(6081), 555–558 (2012)

13. A. Sengupta, S.H. Choday, Y. Kim, K. Roy, Spin orbit torque based electronic neuron. Appl. Phys. Lett. **106**(14), 143701 (2015)

14. S. Emori, U. Bauer, S.-M. Ahn, E. Martinez, G.S. Beach, Current-driven dynamics of chiral ferromagnetic domain walls. Nat. Mater. **12**(7), 611–616 (2013)

15. S. Emori, E. Martinez, K.-J. Lee, H.-W. Lee, U. Bauer, S.-M. Ahn, P. Agrawal, D.C. Bono, G.S. Beach, Spin Hall torque magnetometry of Dzyaloshinskii domain walls. Phys. Rev. B **90**(18), 184427 (2014)

16. G. Indiveri, A low-power adaptive integrate-and-fire neuron circuit, in *ISCAS (4)*. Citeseer (2003), pp. 820–823

17. P.A. Merolla, J.V. Arthur, R. Alvarez-Icaza, A.S. Cassidy, J. Sawada, F. Akopyan, B.L. Jackson, N. Imam, C. Guo, Y. Nakamura et al., A million spiking-neuron integrated circuit with a scalable communication network and interface. Science **345**(6197), 668–673 (2014)

18. A. Sengupta, Y. Shim, K. Roy, Proposal for an All-Spin Artificial Neural Network: emulating neural and synaptic functionalities through domain wall motion in ferromagnets, in *IEEE Transactions on Biomedical Circuits and Systems* (2016)

19. A. Sengupta, M. Parsa, B. Han, K. Roy, Probabilistic deep spiking neural systems enabled by magnetic tunnel junction. IEEE Trans. Electron Dev. **63**(7), 2963–2970 (2016)

20. A. Sengupta, P. Panda, P. Wijesinghe, Y. Kim, K. Roy, Magnetic tunnel junction mimics stochastic cortical spiking neurons. Sci. Rep. **6**, 30039 (2016)

Chapter 13
In-Memory Data Compression Using ReRAMs

Debjyoti Bhattacharjee and Anupam Chattopadhyay

The fast decline of Moore's law is paving the way for a new set of emerging technology devices that offer improved reliability, performance, endurance, and energy-efficiency. Resistive Random Access Memories (ReRAMs) have emerged as one of the most promising technologies for logic and memory applications [1]. ReRAMs are non-volatile, ultra compact memories with low leakage power and high endurance. Large passive crossbar arrays can be realized by means of devices such as a select device in series to a switch (1S1R) or a Complementary Resistive Switch (CRS), to prevent parasitic currents [2]. 1S1R-based devices offer non-destructive readout, unlike CRS-based devices in which readouts are destructive, which makes 1S1R devices suitable for implementation of logic.

Internet-of-things (IoT) is an umbrella term encompassing a wide range of applications and diverse devices, that generally share two common characteristics—connectivity and low energy requirements. Irrespective of the specific network topology, bit rates and communication standards, a considerable amount of energy budget of the IoT nodes is allocated for communication. The energy consumed by the communication sub-system is more or less directly proportional to the amount of data transmitted or received. Therefore, it is of paramount importance to compress the data before transmission.

LZ77 is a lossless compression technique, introduced by Abraham Lempel and Jacob Ziv [3]. LZ77 along with LZ78 forms the basis for multiple variations such as

D. Bhattacharjee (✉)
Hardware and Embedded Systems Laboratory, School of Computer Science and Engineering, Nanyang Technological University, Singapore 639798, Singapore
e-mail: debjyoti001@ntu.edu.sg

A. Chattopadhyay
School of Computer Science and Engineering, School of Physical and Mathematical Sciences, Nanyang Technological University, Block N4 Nanyang Avenue #02c-105, Singapore 639798, Singapore
e-mail: anupam@ntu.edu.sg

© Springer International Publishing AG 2017
A. Chattopadhyay et al. (eds.), *Emerging Technology and Architecture for Big-data Analytics*, DOI 10.1007/978-3-319-54840-1_13

LZW, LZSS, LZMA, and others. In addition, it forms the core of several ubiquitous compression schemes such as GIF, DEFLATE, etc. LZ77 was awarded as an IEEE Milestone in 2004.

This chapter is devoted to the introduction of an in-memory computing architecture using ReRAM crossbar arrays and how various functions can be realized using the architecture. We demonstrate a low-area implementation of LZ77 compression algorithm using the ReRAM based in-memory architecture. In Sect. 13.2, the **ReRAM** based **VLIW A**rchitecture for in-**M**emory com**P**uting (ReVAMP) architecture is introduced. In Sects. 13.2.1 and 13.2.2, realization of identity comparator and priority multiplexer is presented using ReVAMP. In Sect. 13.3, we present the details of compression using LZ77 algorithm on ReVAMP and the performance of the proposed implementation is analyzed. Section 13.5 presents a review of the existing works in the domain of in-memory computing using ReRAMs.

13.1 LZ77 Compression Algorithm

LZ77 is a lossless compression algorithm that forms the basis of multiple other compression algorithms [3]. In LZ77, compression is achieved by replacing repeated occurrences of data with reference to a single copy of that data that existed earlier in the uncompressed data stream. Such a match is encoded as a *length-distance* pair of numbers. This implies that the next *length* number of characters match the characters at *distance* characters behind it in the uncompressed scheme. The term *length* is also referred to as *offset*. The pseudo-code for LZ77 compression is shown in Algorithm 1.

LZ77 uses a sliding window data structure to find matches. The sliding window is divided into two parts, namely the Look-ahead buffer and the Dictionary buffer. The Dictionary buffer stores the most recent uncompressed data stream that is used to look for matches. The Look-ahead buffer contains the uncompressed data stream that is yet to be encoded. The larger the sliding window is, the encoder searches for finding longer matches, but it adds to the overhead of higher number of comparisons required for finding the longest prefix. Determining the longest prefix is the major

---

Algorithm 1 LZ77 compression algorithm pseudo-code

---

1 Fill Look-ahead buffer from input ;
2 while *Look-ahead buffer is not empty* **do**
3   Find longest prefix *p* of view starting in Look-ahead buffer;
4   *offset* := position of *p* in window;
5   *length* := number of characters in *p*;
6   *X* := first char after *p* in view;
7   Output (*offset*, *length*, *X*);
8   Add *length*+1 chars to the Look-ahead buffer;
9 end

Table 13.1 LZ77 compression of string *aacaacbcabaaac* with dictionary buffer size 8 and Look-ahead buffer size 6

Sl#	Dictionary								Look-ahead						Output
	8	7	6	5	4	3	2	1	0	1	2	3	4	5	
1									a	a	c	a	a	c	(0,0,a)
2								a	a	c	a	a	c	b	(1,1,c)
3						a	a	c	a	a	c	b	c	a	(3,3,b)
4		a	a	c	a	a	c	b	c	a	b	a	a	a	(5,2,b)
5	c	a	a	c	b	c	a	b	a	a	a	c			(7,2,a)
6	c	b	c	a	b	a	a	a	c						(8,1,$)

computation in the LZ77 algorithm. To determine individual match in characters in Look-ahead buffer and Dictionary buffer, *identity comparator* is needed. For determining the correct values of *offset*, *length* and *X*, *priority multiplexers* would be needed, with the priority based on the length of the match.

Example 1 To facilitate understanding of the algorithm, we present an example for encoding the string *aacaacbcabaaac*. Table 13.1 demonstrates the encoding of the string using a dictionary buffer size of 8 and a Look-ahead buffer size of 6. The encoded output is shown in the last column of the table, is of the form (*distance, length, next character X*). It should be noted that the distance is relative to the right edge of the dictionary buffer. The buffers operate on the principle of a sliding window, i.e. the data stream to be compressed is pushed left into the buffer. As noted in the algorithm, the shift is equal to the length of the match found in the dictionary, and a further position.

Initially, the dictionary buffer is empty and there are no matches, hence (0,0,a) is the output. The next character *a* in the Look-ahead buffer is a match with one character at distance 1 in the dictionary buffer, and hence the output is (1,1,*c*). At distance 3, three characters match the left most three characters of the Look-ahead buffer, and thus the output is (3,3,*b*). In the following step, two characters at distance 5 match the two left-most character of Look-ahead buffer, so the output is (5,2,*b*). In this step, two characters match and the output is (7,2,*a*). In the last step, the output is (8,1,$) as the last character c of the uncompressed string matches and to signify the end of string, $ symbol is used.

13.2 ReVAMP Architecture for In-Memory Computing

In this section, we explain the general purpose in-memory computing platform, ReVAMP, introduced in [4]. We also demonstrate how comparator and priority multiplexer can be realized using instructions of ReVAMP.

The ReVAMP architecture, presented in Fig. 13.1, utilizes two ReRAM crossbar memories with light weight peripheral circuitry. A ReRAM crossbar memory

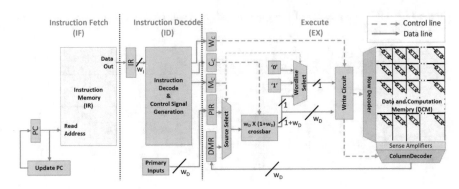

Fig. 13.1 ReVAMP architecture [4]

Fig. 13.2 ReVAMP
instruction set

$$\boxed{\text{Read}\ wl\ }$$

$$\boxed{\text{Apply}\ wl\ s\ ws\ wb\,(\text{v}\ \text{val}_{w_D-1})\ \dots\ (\text{v}\ \text{val}_0)}$$

consists of multiple 1S1R ReRAM devices [5], arranged in the form of a cross-bar [6]. Like conventional RAM arrays, ReRAM memories are accessed as w_D-bit wide words.

One of the memory arrays is used as instruction memory (IM). The IM is used as regular memory, with the program counter (PC) being used to access the next instruction. The other array is used as data storage and computation memory (DCM). In the DCM, in-memory computation using ReRAM devices takes place.

Each ReRAM device has two input terminals, namely the wordline wl and bitline bl. The internal resistive state Z of the ReRAM acts as a third input and stored bit. The next state of the device Z_n can be expressed as Boolean majority function with three inputs, with the bitline input inverted, as shown in the following equation.

$$Z_n = M_3(Z, wl, \overline{bl})$$

This forms the fundamental logic operation that can be realized using ReRAM devices. Using the intrinsic function Z_n, inversion operation can be realized. Since majority and inversion operation form a functionally complete set, any Boolean function can be realized using the Z_n.

The ReVAMP architecture has a three-stage pipeline with Instruction Fetch (IF), Instruction Decode (ID), and Execute (EX) stages. The ReVAMP architecture can be programmed using two instructions—*Read* and *Apply*, with the format shown in Fig. 13.2.

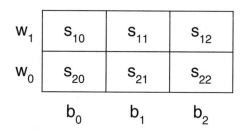

Fig. 13.3 A ReRAM crossbar array with two wordlines and three bitlines

Read instruction reads a specified word, wl from the DCM and stores it in the Data Memory Register (DMR). The read out word, available in the DMR, can be used as input by the next instruction.

The *Apply* instruction is used for computation in the DCM. The address wl specifies the word in the DCM that will be computed upon. A bit flag s chooses whether the inputs will be from primary input register (PIR) or DMR. Two-bit flag ws is used to select the wordline input—11 selects '1', 10 selects '0', 00 selects wb bit within the chosen data source for use as wordline input while 01 is an invalid value for ws. Pairs (v, val) are used to specify individual bitline inputs. Bit flag v indicates if the input is NOP or a valid input. Similar to wb, bits val specifies the bit within the chosen data source for use as bitline input.

We introduce the notations used for the implementation of the logic operations on ReRAM crossbars. Figure 13.3 shows a ReRAM array with two wordlines and three bitlines. Input w_1 and w_0 are the wordline inputs while b_2, b_1, and b_0 are the bitline inputs. The variable s_{ij} represents the internal states of device at wordline i and bitline j. Input '1', '0', and 0 represent V/2, $-$V/2, and GND, respectively. From the perspective of logic, inputs '1' and '0' represent Boolean logic 1 and 0, respectively, while input 0 represents no-operation. In a readout phase, the presence of a $5\,\mu A$ current is considered as Boolean logic 1 while absence of current is interpreted as Boolean logic 0.

Figure 13.4 shows how a 32-bit word can be loaded into the DCM using the ReVAMP instructions. The word $w_{31:0}$, available in the PIR, is loaded into the DCM by using an Apply instruction. This loads the words in inverted form in word i, as shown in Fig. 13.4a. In the next cycle, the inverted word is read out from word i using Read instruction, which stores it in the DMR. Another Apply instruction is used to write the word in non-inverted form to word j, by selecting writing the contents of the DMR as shown in Fig. 13.4c. The reader should understand the equivalence of the ReVAMP instructions and the representation of the crossbar operations, since these notations will be used interchangeably.

Figure 13.5 presents realization of basic Boolean functions. Figure 13.5a shows how an array can be reset to 0, irrespective of its contents. This is true because $M_3(x, 0, \neg 1) = 0$. To compute AND of two inputs, the first input is loaded into the array and then the negated second input is applied to the bitlines with '0' as wordline input, because $M_3(0, a, \neg(\neg b)) = a.b$. Similarly, OR of two inputs can be computed, by changing the wordline input to '1', since $M_3(0, a, \neg(\neg b)) = a + b$.

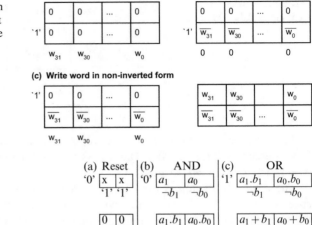

Fig. 13.4 Loading a word into DCM (**a**) Load word in inverted form, (**b**) Read out inverted word and (**c**) Write word in non-inverted form

Fig. 13.5 Realization of basic Boolean functions

This concludes the description of the ReVAMP architecture and basics of logic function realization using it. In the following subsections, we present the realization for identity comparator and priority multiplexer, which are required for LZ77 compression.

13.2.1 Comparator Design

An identity comparator compares the value of the inputs and generates a HIGH output only when both the inputs are identical, otherwise the output is LOW. An identity comparator for 4-bit values can be represented by the following equation:

$$c_4 = (a_0 \odot b_0).(a_1 \odot b_1).(a_2 \odot b_2).(a_3 \odot b_3) \tag{13.1}$$

$$a \odot b = \overline{a}.\overline{b} + \overline{\overline{a} + \overline{b}} \tag{13.2}$$

where a_i and b_i represent the ith bits of input data signals a and b, respectively. $\overline{a}$ represents the negated value of Boolean variable a. Operators ., +, and $\odot$ represent Boolean AND, OR, and XNOR operations, respectively. The XNOR operation can be expressed in terms of AND and OR operations as shown in (13.2).

Without loss of generality, we demonstrate the implementation of identity comparator using ReRAM arrays, for 4-bit inputs, as shown in Fig. 13.6. The grayed wordline represents the read out word.

Step 1: Word a is read out and an inverted copy of the word is created.
Step 2: Similar to step 1, another copy of $\overline{a}$ is created.
Step 3: In this step, $\overline{a_i}.\overline{b_i}$ is computed in the by reading out and applying b via the bitlines and '0' as wordline input, since $M_3(\overline{a}, 0, \overline{b_i}) = \overline{a_i}.\overline{b_i}$.

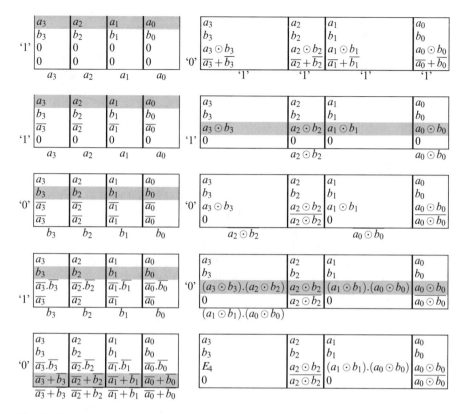

Fig. 13.6 Four bit identity comparator realization

Step 4: $\overline{a_i} + \overline{b_i}, 0 \le i \le 3$ is computed in the by reading out and applying b via the bitlines and '1' as wordline input, since $M_3(\overline{a_i}, 1, \overline{b_i}) = \overline{a_i} + \overline{b_i}$.

Step 5: The intermediate term $\overline{a_i} + \overline{b_i}$ is read out and ORed with corresponding $\overline{a_i} . \overline{b_i}$ to compute $a_i \odot b_i$. This completes completion of XNOR computation. It should be noted that as long as the number of bits in the word is less than w_D, the bitwise XNOR of the words a and b can be computed using fixed number of steps.

Step 6: The word holding the intermediate results $\overline{a_i} + \overline{b_i}$ is reset to 0.

Step 7: Now, the computed XNOR terms are combined together using an AND-reduction tree, as shown in Fig. 13.7. XNOR terms $a_2 \odot b_2$ and $a_0 \odot b_0$ are read out and stored in inverted forms.

Step 8: The inverted XNOR terms are read out and ANDed with the appropriate $a_i \odot b_i$ terms.

Step 9–10: The last two steps are similar to **Step 7–8** and compute the final identity comparator result E_4.

Fig. 13.7 AND-reduction
tree for identity comparator

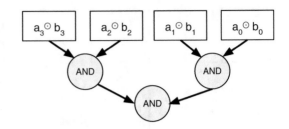

13.2.1.1 Analysis

Each step except Step 6 involves a read operation followed by a computation—
which implies a Read instruction followed by an Apply instruction and therefore
requires two cycles. For computation of the bit-wise XOR, ten cycles are required.
One cycle is required to reset the word holding the intermediate term. For the AND-
reduction tree, there are $\lceil \log_2 n \rceil$ levels, where n is the number of bits in the inputs.
Each level requires four cycles, therefore the reduction tree computation requires
$4\lceil \log_2 n \rceil$ cycles. Thus, an n-bit ($n \leq w_D$) identity comparator would require $11 +
4\lceil \log_2 n \rceil$ cycles to be realized on ReVAMP architecture.

13.2.2 Priority Multiplexer Design

A priority multiplexer selects from one of the n data signals, based on the n control
signals, which have a predefined priority. Basically, the priority multiplexer selects
input signal a_k, if control signal s_k is '1' and none of the other control signals with
priority more than s_k are '1'. If none of the select signals are '1', then the output
is invalid. A 4-bit priority multiplexer is represented by the truth table in Fig. 13.8b
and the following equations.

$$p_4 = s_3.a_3 + \overline{s_3}.s_2.a_2 + \overline{s_3}.\overline{s_2}.s_1.a_1 + \overline{s_3}.\overline{s_2}.\overline{s_1}.s_0.a_0 \qquad (13.3)$$

$$V = s_0 + s_1 + s_2 + s_3 \qquad (13.4)$$

where s_j and a_k represent the control and data signals, respectively. Priority of
control signal s_j is greater than s_k, if $j < k$. The output signal valid V is '1' when the
output is valid, otherwise it is low.

We demonstrate how priority multiplexers for 4, 3, 2, and 1 input can be realized
simultaneously using ReRAMs, with the overall delay being determined by the
delay of the 4-input priority multiplexer computation. Let a, b, c, and d by the data
inputs and s^1, s^2, s^3, and s^4 be the select signals to the four priority multiplexers,
respectively. The initial steps of the computation is shown in Fig. 13.9. In Step 1,
the select signals are read out and an inverted copy is written back. In Step 2, the s_i^t
is ANDed with the appropriate data signal. From Steps 3–5, the $\overline{s_i^t}$ terms are ANDed

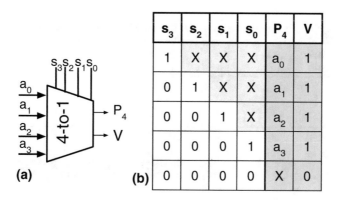

Fig. 13.8 4-Input priority multiplexer. (**a**) Block diagram. (**b**) Truth table

with the appropriate intermediate AND terms. In Step 7, the wordline storing the inverted select signals is reset. From Step 8 onwards, the final result of the priority multiplexer P_n is computed by using an OR-reduction tree (similar to Fig. 13.7). To compute the valid output V, another OR-reduction tree for the select signals would be required.

In general, two cycles are required to compute and write the inverted select signals. n steps are required to compute all the AND terms with each step involving a Read and Apply instruction. The reset operation requires one cycle. Finally, the OR-reduction tree requires $4\lceil \log_2 n \rceil$ cycles, similar to the AND-reduction tree. For the computation of the valid output signal, additional $4\lceil log_2 n \rceil$ cycles would be required. Therefore, an n-input priority multiplexer requires $3 + 2n + 8\lceil log_2 n \rceil$ cycles to complete execution. For the specific case of 4-input priority multiplexer, 27 cycles are required.

In the next section, we demonstrate how LZ77 compression can be realized using logic operations on ReRAM. We will be required to use the comparator and priority multiplexer designs introduced above for LZ77 compression.

13.3 LZ77 Compression Using ReVAMP

In this section, we present the implementation details of LZ77 on the ReVAMP architecture. We assume word length w_D of the ReVAMP architecture to be 32. For the LZ77 compression, we assume each character to be 8-bits, since ASCII text representation uses 8-bits. In addition, we consider the Dictionary buffer to hold 4-characters and Look-Ahead buffer to hold 5-character. Let the contents of the Dictionary buffer and the Look-ahead buffer be as shown in Fig. 13.10.

The key computation in LZ77 is finding the longest prefix p of view starting in Look-ahead buffer that is present in the dictionary buffer. Initially, the individual characters are compared in the Look-ahead buffer and Dictionary buffer—s_i^j is the

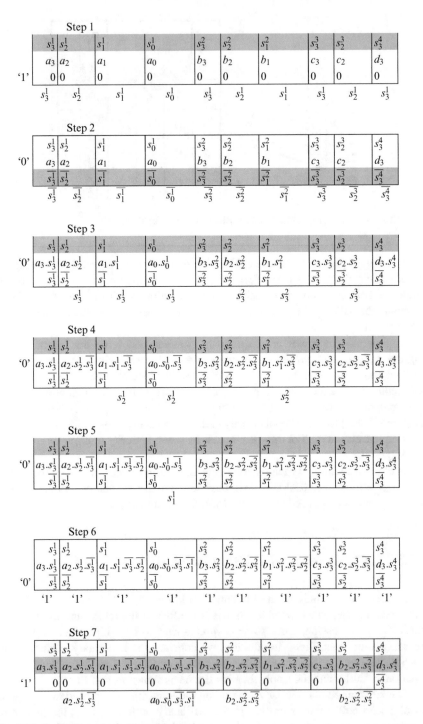

Fig. 13.9 Realization of priority multiplexer

Fig. 13.10 Dictionary and Look-ahead buffer

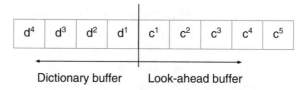

Dictionary buffer   Look-ahead buffer

Fig. 13.11 DCM layout

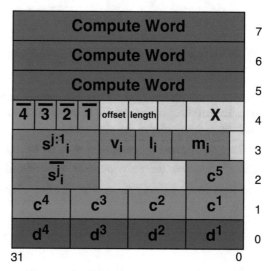

result of comparison of ith character in Look-ahead buffer and jth character in the Dictionary buffer. This is followed by determining what are the series of characters that match in the dictionary buffer—$s_i^{j:1}$ is 1, if characters from 1 to j positions in the Look-ahead buffer matches the characters from location i in Dictionary buffer. Using these results, the *offset* are determined. This is followed by determining the length of the priority multiplexers—l_i indicates that a prefix of length i is present. Using l_i, the outputs *length* and next character X is determined. Finally, the Dictionary buffer is shifted appropriately, depending on the length of the longest prefix.

The computation in the ReVAMP architecture takes place in the Data and Computation Memory (DCM). The layout of the DCM is shown in Fig. 13.11. Word 0 holds the contents of the Dictionary buffer while word 1 and the first 8-bits of word 2 act as Look-ahead buffer. In addition, word 2 holds results of character comparisons. Word 3 holds the select signals for the priority multiplexers, priority multiplexer outputs, and the valid bits. Word 3 holds constants 4, 3, 2, and 1 in inverted forms and will also store the *offset*, *length*, and next character X that is output by the algorithm. Finally, words 5–7 are used for computation.

To do so, we undertake the following sequence of operations. Compare the characters to determine if a prefix of length 1 is present. $s_i^1 = (d^i == c^1), 1 \geq i \geq 4$.

Similarly, comparisons are undertaken to determine if prefix of length 2, 3, and 4 are present. Each set of comparisons to determine a prefix of certain length t can be performed in parallel on the words 5–6 in the DCM, using the comparator realization

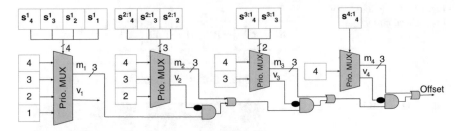

Fig. 13.12 Offset computation using Priority multiplexers

present in subsection 13.2.1. Once the comparison is complete, the s_i^t terms are read out and written in inverted form to the word 2. Then the words 5–6 are reset and the next set of comparisons are performed.

This is followed by computation of all the terms $\overline{s_i^{j:1}}$, $4 \geq \{i,j\} \geq 1$ in parallel. The $\overline{s_4^{4:1}}$ requires the most number of cycles to be computed, equal to 8 cycles. Finally, two cycles are need to read out the $\overline{s_i^{j:1}}$ terms and write it to word 3 in non-inverted form.

$$s_i^{j:1} = s_i^j . s_i^{j-1} \ldots s_i^2 . s_i^1 \qquad (13.5)$$

Using the $s_i^{j:1}$ terms, the offset can be determined by using a series of priority multiplexers as shown in Fig. 13.12 along with some additional computation for the AND and ORs. The priority multiplexers are realized in parallel, by the steps described in Sect. 13.2.2. Once the priority multiplexer computations are over, the AND and OR terms are computed sequentially to compute "*offset*." The computed offset is written to word 4. The computation words 5–7 are reset once again.

For computation of "*length*," a single priority multiplexer is used with select signals l_i, as defined below.

$$l_4 = s_4^{4:1} \qquad (13.6)$$

$$l_3 = s_4^{3:1} + s_3^{3:1} \qquad (13.7)$$

$$l_2 = s_4^{2:1} + s_3^{2:1} + s_2^{2:1} \qquad (13.8)$$

$$l_1 = s_4^{1:1} + s_3^{1:1} + s_2^{1:1} + s_1^{1:1} \qquad (13.9)$$

The select signals are computed in parallel with $\overline{l_i}$. This requires six cycles. Additional cycles are required to read out $\overline{l_i}$ and store l_i. This is followed by computation for the priority multiplexer to determine "length."

In order to compute the output character X, another priority multiplexer computation is used, with l_i as select signals as shown in Fig. 13.13, similar to that for computation of "length." If there is no-match in the Dictionary buffer, then the valid bit v is 0 and c^1 is the next character X. Using this, we can determine the correct next

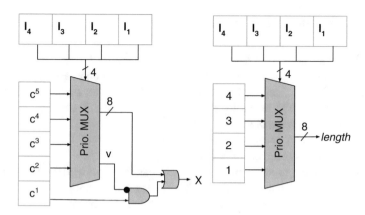

Fig. 13.13 Computation of output "length" and character X

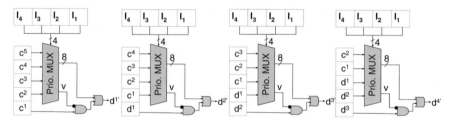

Fig. 13.14 Dictionary buffer update

character. This completes computation of the output (*offset, length, X*) for this round of the LZ77 algorithm. The computed output is readout using a Read instruction and available in DMR to be read out.

The dictionary buffer and the Look-ahead buffer need to be updated before the next iteration of the LZ77 can begin. For updating each character in the dictionary buffer, a priority multiplexer operation is used followed by an OR operation, with an inverted input. The priority multiplexers and the corresponding inputs and select signals are shown in Fig. 13.14. We should note that the valid bit v is computed once, since the select signals to the priority multiplexers are identical. Once the new character at a given position has been computed, the old character is reset in word 0 and the new character is written.

All the characters in the Dictionary buffer locations are reset to 0 and based on the length output, the contents of the Look-ahead buffer are loaded via the PIR and Apply instructions for the next iteration of LZ77 algorithm.

13.4   Performance Estimation

About 375 cycles are required for each iteration of the LZ77 algorithm. Updating the dictionary buffer requires 148 cycles while the initial comparisons along with computation of the $s_i^{j:1}$ terms require 92 cycles. To estimate the performance, we assume mature ReRAM technology with 1 ns access time, based on [7]. For the uncompressed text *aacaacbcabaaac* given in Example 1, seven iterations would be needed to compress it using the proposed implementation of LZ77 and 2.625 μs would be required to complete all the iterations.

The area of the proposed implementation can be measured in terms of the number of words required in DCM and IM. The proposed implementation requires seven words only, with each word 32-bit wide in DCM. Assuming the DCM to be addressed by 3-bits, each Apply instruction would require 201 bits and we assume that the Read instruction is padded with 0 s to make it of the same length as the Apply instruction. The proposed implementation requires ≈5.46 KB of memory for storing the instructions, considering 32-bit aligned memory access.

13.5   Related Works

The majority of the work related to in-memory computing related to ReRAMs can be broadly classified into three categories—dedicated circuit proposals, general purpose computing architectures using ReRAMs, and design automation tools for the architectures.

In [8], ReRAM cells were shown to be conditionally switchable sequential logic devices, thereby allowing logic-in-memory operations directly. Feasibility and performance of multiple logic-in-memory adder designs have been presented in the recent literature by means of memristive simulations [9–11]. Level-1 and Level-2 Binary Basic Linear Algebra Sub-routines (BiBLAS) were realized using ReRAM crossbas arrays [12, 13]. Neuromorphic computing has also been realized using ReRAMs [14–16]. Authors in [17] utilized the crossbar array as a Content-Addressable Memory (CAM) structure similar to those earlier proposed in [18] for realizing integer matrix multiplication.

A general approach to designing in-memory architecture for data-intensive applications was presented in [19]. Gaillardon et al. [20] introduced a light weight controller to enable general purpose computing using ReRAM arrays, using a bit-serial operation mode with a single instruction. The ReVAMP architecture [4] has two instructions and uses separate instruction and computation memories and allows bit-level parallel operations, thereby offering considerable speedup over PLiM computer [20].

Considerable amount of research has been undertaken for developing automation tools related to logic synthesis and technology mapping using memristors. In [9], the authors presented a basic methodology for computing Boolean functions using memristive devices. In [21], it has been shown that with two working memristors which realize material implication, any Boolean expression can be computed. In [22] and [23], logic synthesis solution for memristors that realize material implication has been proposed. In [24], a compiler for flow for generating RM_3 instructions, for the ReRAM based PLiM computer [20] for realization of Boolean functions has been presented. In [25], heuristics for logic synthesis of MIG for two variants of ReRAM has been proposed —one realizing material implication and the other realizing majority function. In [25], the authors used a naïve technology mapping with delay of $3k + c$ cycles, for an MIG with k levels and c number of levels with ingoing complemented edges. In [26], the authors demonstrated logic realization using memristive crossbar arrays using multi-bit adders and multipliers as case studies. In [27], the authors proposed a delay optimal technology mapping solution for memristive devices. Further, area-constrained technology mapping for ReRAM devices was presented in [28].

13.6 Summary

This chapter introduced the ReVAMP in-memory computing architecture that utilizes stateful logic operations on ReRAM crossbar arrays. Realizations of comparator and priority multiplexer was presented using the ReVAMP instructions. We presented implementation of LZ77 compression algorithm using the ReVAMP instructions and analyzed the performance in terms of number of cycles and area in terms of number of devices. Finally, we presented the landscape of research in the field of in-memory computing using memristors.

References

1. R. Waser, R. Dittmann, G. Staikov, K. Szot, Redox-based resistive switching memories–nanoionic mechanisms, prospects, and challenges. Adv. Mater. **21**, 2632–2663 (2009)
2. E. Linn, R. Rosezin, C. Kügeler, R. Waser, Complementary resistive switches for passive nanocrossbar memories. Nat. Mater. **9**(5), 403–406 (2010)
3. J. Ziv, A. Lempel, A universal algorithm for sequential data compression. IEEE Trans. Inf. Theory **23**(3), 337–343 (1977)
4. D. Bhattacharjee, R. Devadoss, A. Chattopadhyay, ReVAMP : ReRAM based VLIW Architecture for in-Memory comPuting, in *Design, Automation & Test in Europe Conference & Exhibition, DATE 2017* (2017)
5. A. Siemon, S. Menzel, A. Marchewka, Y. Nishi, R. Waser, E. Linn, Simulation of TaO$_x$-based complementary resistive switches by a physics-based memristive model, in Circuits and Systems *ISCAS* (2014, IEEE International Symposium on), pp. 1420–1423
6. E. Linn, R. Rosezin, S. Tappertzhofen, U. Böttger, R. Waser, Beyond von neumann-logic operations in passive crossbar arrays alongside memory operations. Nanotechnology **23**(30), 305205 (2012)
7. Emerging Research Devices (ERD) report, International Technology Roadmap for Semiconductors (ITRS) (2013)

8. J. Borghetti, G.S. Snider, P.J. Kuekes, J. Joshua Yang, D.R. Stewart, R. Stanley Williams, 'Memristive' switches enable 'stateful' logic operations via material implication. Nature **464**(7290), 873–876 (2010)

9. E. Lehtonen, M. Laiho, Stateful implication logic with memristors, in *Proceedings of the 2009 IEEE/ACM International Symposium on Nanoscale Architectures* (2009), pp. 33–36

10. S. Kvatinsky, G. Satat, N. Wald, E.G. Friedman, A. Kolodny, U.C. Weiser, Memristor-based material implication (imply) logic: design principles and methodologies. IEEE TVLSI **22**(10), 2054–2066 (2014)

11. A. Siemon, S. Menzel, R. Waser, E. Linn, A complementary resistive switch-based crossbar array adder. IEEE JETCAS **5**(1), 64–74 (2015)

12. D. Bhattacharjee, F. Merchant, A. Chattopadhyay, Enabling in-memory computation of binary blas using reram crossbar arrays, in *2016 IFIP/IEEE International Conference on Very Large Scale Integration (VLSI-SoC)*, September 2016, pp. 1–6

13. D. Bhattacharjee, A. Chattopadhyay, Efficient binary basic linear algebra operations on reram crossbar arrays, in *2017 30th International Conference on VLSI Design*, January 2017

14. D.B. Strukov, D.R. Stewart, J. Borghetti, X. Li, M. Pickett, G.M. Ribeiro, W. Robinett, G. Snider, J. P. Strachan, W. Wu, Q. Xia, J.J. Yang, R.S. Williams, Hybrid cmos/memristor circuits, in *ISCAS*, pp. 1967–1970 (2010)

15. K.-H. Kim, S. Gaba, D. Wheeler, J. M. Cruz-Albrecht, T. Hussain, N. Srinivasa, W. Lu, A functional hybrid memristor crossbar-array/cmos system for data storage and neuromorphic applications. Nano Letters **12**(1), 389–395 (2011)

16. M.P. Sah, H. Kim, L.O. Chua, Brains are made of memristors. IEEE Circuits Syst. Mag. **14**(1), 12–36 (2014)

17. L. Ni, Y. Wang, H. Yu, W. Yang, C. Weng, J. Zhao, An energy-efficient matrix multiplication accelerator by distributed in-memory computing on binary RRAM crossbar, in *ASP-DAC*, January 2016, pp. 280–285

18. F. Alibart, T. Sherwood, D.B. Strukov, Hybrid cmos/nanodevice circuits for high throughput pattern matching applications, in *Adaptive Hardware and Systems (AHS), 2011 NASA/ESA Conference on* (2011), pp. 279–286

19. S. Hamdioui, L. Xie, H. Anh Du Nguyen, M. Taouil, K. Bertels, H. Corporaal, H. Jiao, F. Catthoor, D. Wouters, L. Eike, J. van Lunteren, Memristor based computation-in-memory architecture for data-intensive applications, in *Proceedings of the 2015 Design, Automation & Test in Europe Conference & Exhibition, DATE 2015, Grenoble, March 9–13, 2015* (2015), pp. 1718–1725

20. P.-E. Gaillardon, L. Amaru, A. Siemon, E. Linn, R. Waser, A. Chattopadhyay, G. De Micheli, The Programmable Logic-in-Memory (PLiM) computer, in *DATE* (2016), pp. 427–432

21. J.H. Poikonen, E. Lehtonen, M. Laiho, On synthesis of Boolean expressions for memristive devices using sequential implication logic. IEEE TCAD **31**(7), 1129–1134 (2012)

22. A. Raghuvanshi, M. Perkowski, Logic synthesis and a generalized notation for memristor-realized material implication gates, in *ICCAD* (2014), pp. 470–477

23. A. Chattopadhyay, Z. Endre Rakosi, Combinational logic synthesis for material implication, in *IEEE/IFIP 19th International Conference on VLSI and System-on-Chip, VLSI-SoC 2011, Kowloon, Hong Kong* (2011), pp. 200–203

24. M. Soeken, S. Shririnzadeh, P.-E. Gaillardon, L. Amarú, R. Drechsler, G. De Micheli, An mig-based compiler for programmable logic-in-memory architectures, in *DAC* (2016)

25. S. Shirinzadeh, M. Soeken, P.-E. Gaillardon, R. Drechsler, Fast logic synthesis for RRAM-based in-memory computing using majority-inverter graphs, in *DATE* (2016)

26. L. Xie, H. Anh Du Nguyen, M. Taouil, K. Bertels, S. Hamdioui, Fast boolean logic mapped on memristor crossbar, in *33rd IEEE International Conference on Computer Design, ICCD 2015*, New York City, NY, October 18–21 (2015), pp. 335–342

27. D. Bhattacharjee, A. Chattopadhyay, Delay-optimal technology mapping for in-memory computing using reram devices, in *Proceedings of the 35th International Conference on Computer-Aided Design, ICCAD 2016*, Austin, TX, November 7–10 (2016), p. 119
28. D. Bhattacharjee, A. Easwaran, A. Chattopadhyay, Area-constrained technology mapping for in-memory computing using reram devices, in *Asia and South Pacific Design Automation Conference, ASP-DAC* (2017), pp. 1–6

Chapter 14
Big Data Management in Neural Implants: The Neuromorphic Approach

Arindam Basu, Chen Yi, and Yao Enyi

14.1 Introduction: Brain as a Source of Big Data

In the age of the Internet of Things (IoT) with millions of interconnected sensors spewing out data, we are facing a data deluge—there is a need for solutions to store and process this data. A unique set of IoT applications relates to the human body—in particular wearables and implantables to collect data from the human brain for neuroscientific research, prostheses or medical interventions [1–6]. The study of the human brain is one of the most important frontiers in science research today—there is a lot of emphasis on this with several billion dollar efforts worldwide to understand more about the brain [7, 8]. To get an idea about the scale of data generated by the brain, we first note from anatomy that the average adult human cortex has approximately 10^{11} neurons, widely regarded as the fundamental computational unit of the brain, with 10^{14} synapses or interconnections [9]. Assuming average cortical firing rates (a neural firing or discharge refers to a digital like pulse also called a spike or action potential) of 1–10 Hz [10], the human brain is generating at least 10^{11} spikes or events per second and about 10^{14} synaptic operations per second. Assigning an unique address or identifier to each neuron would need $b_{\mathrm{addr}} = \log_2(10^{11}) \approx 35$ bits—hence, the data rate generated by the brain is a whopping 3.5 Terabits/second. To put this in perspective, the exponential growth of data has put internet data in the exascale (10^{18} bits). One human brain can generate approximately the same amount of data in 10^6 s or 50 days! Of course, this is an extreme case and we are not aiming to store all the neural firings of a human brain over his or her lifetime (at least not at this moment) and neither do we currently possess the technology to access this data (but we are constantly striving

A. Basu (✉) • C. Yi • Y. Enyi
School of EEE, Nanyang Technological University, Singapore, Singapore
e-mail: arindam.basu@ntu.edu.sg

© Springer International Publishing AG 2017
A. Chattopadhyay et al. (eds.), *Emerging Technology and Architecture for Big-data Analytics*, DOI 10.1007/978-3-319-54840-1_14

Fig. 14.1 The brain as a source of big data: a single human brain generates data at a rate of 3.5 Terabits/second. The total data can reach exabyte scale within a year

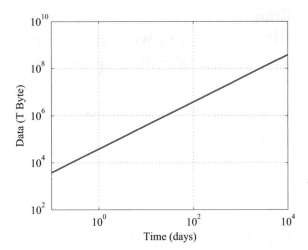

to record data from more neurons and this is one of the prime goals of the Brain initiative)—but this helps to give an idea about the scale of the problem. Figure 14.1 shows the rapid scaling of data generated from a single human brain over time.

Just like any other application related to big-data, the problems of storage and manipulation exist in this data generated by the brain. However, an added problem stems in this case from the strict power dissipation requirement of electronics implanted within the brain to collect the data. Any electronics in contact with the cortical tissue cannot generate heat larger than $80\,mW/cm^2$ [11, 12] to avoid damaging the neural tissue (temperature rise less than $1\,°C$). Instead of implants, another option is to collect data non-invasively through EEG from the scalp—however, EEG provides a highly filtered (both spatially and temporally) picture of the brain activity and is not informative enough for activities with many degrees of freedom such as upper limb prostheses [13, 14]. Therefore, in the rest of this chapter, we only consider the case of neural recording from implanted electrodes that can provide enough information for dexterous motor control.

14.2 The Nature of Neural Data

The signals recorded by neural implants are obtained typically through microelectrode arrays such as the Utah or Michigan arrays [15–17]. The neural signals can be broadly divided into two categories—(1) Local Field Potentials (LFP) that are $1–10\,mV$ in amplitude occupying a bandwidth of $1–100\,Hz$ produced by combined activity of groups of neurons and (2) neural spikes or action potentials which are much smaller ($10–100\,\mu V$ in amplitude) but occupy a much larger bandwidth of $\approx 0.2–5\,kHz$. While both signals have useful information [18, 19], most of the studies on neural prosthetics that require fine motor manipulation typically use

Fig. 14.2 A neural spike recorded from the pre-frontal cortex of a rat. Neural spikes typically have a small amplitude ≈ 10–100 μV while occupying a large bandwidth of ≈ 0.2–5 kHz

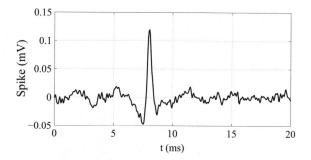

neural spikes [20–23]. In this chapter, we will therefore focus on neural recording systems for sensing and transmitting neural spikes. Unlike LFP signals where the amplitude is informative, it is believed that spikes are like digital signals [24] where the amplitude is non-informative but the timing and firing rate of spikes are important. An example of a spike recorded from pre-frontal cortex of a rat is shown in Fig. 14.2.

14.3 System Architectures for Neural Spike Recording Systems: Neuromorphic Compression Schemes

The different blocks comprising a typical neural recording system are shown in Fig. 14.3a. In a typical system, the neural signal is amplified by a low-noise amplifier (LNA) [25–29], followed by an optional variable gain amplifier (VGA) and finally an analog-digital converter (ADC) [29–32] before being transmitted wirelessly. We can estimate the data rate for such a system under some mild assumptions. Denoting the number of recording channels as N_{chan}, ADC sampling rate and bit resolution as f_{ADC} and b_{ADC}, respectively, the data rate R_{typ} of a typical neural implant is given by:

$$R_{\text{typ}} = N_{\text{chan}} \times f_{\text{ADC}} \times b_{\text{ADC}} \qquad (14.1)$$

As an example, for moderate values of $N_{\text{chan}} = 100, f_{\text{ADC}} = 20\,\text{kHz}$, and $b_{\text{ADC}} = 10$ bits, we get $R_{\text{typ}} = 20\,\text{Mbps}$—a huge data rate that will drain out an implant's battery in a matter of hours given typical power requirements of ≈ 50–1000 pJ/bit for wireless transmitters [33–36]. Hence, it is imperative to compress the data and reduce the concomitant power dissipation so that the neural recording system can be scaled in future to thousands or millions of channels. One possible way to do this is to take inspiration from the brain—in the absence of the implant, the brain would have processed the thousands of neural spikes recorded by the implant and given a refined command to the next region. Similarly, we can also use electronics to perform this signal processing on the implant, thus reducing the bandwidth of data to be transmitted. Figure 14.3 shows three different modes of compression based on

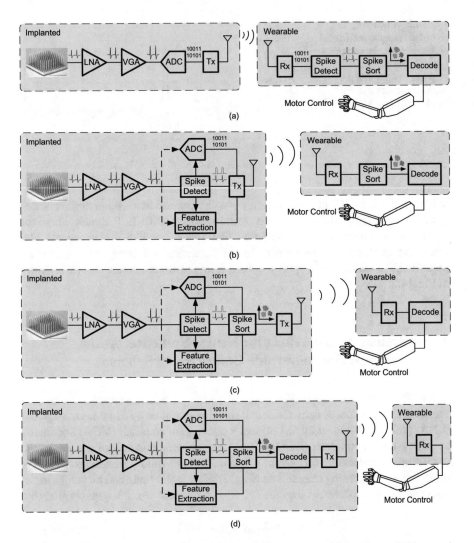

Fig. 14.3 Block diagram of a typical neural recording system which senses, digitizes and wirelessly transmits the neural data. As an alternative to sending raw data, different neuromorphic schemes may be used as shown to achieve different rates of compression. (**a**) Typical. (**b**) Mode 1. (**c**) Mode 2. (**d**) Mode 3

the amount of signal processing kept on the implant. There is a trade-off in this case between amount of extra area and energy expended on signal processing in-implant versus the energy saved in reduced transmission. Clearly, it is not beneficial if the added circuits for signal processing burn as much energy as the energy saved in reduced data rate!

One way to perform the processing at very low energy/area overheads is to use neuro-inspired analog circuits, sometimes also called 'neuromorphic' circuits following Carver Mead's seminal paper [37]. Mead and others [38] have shown that analog circuits require less energy and area than digital counterparts when processing signals at a low resolution, typically $\leq$ 8 bits. The brain also uses a similar principle by computing using analog quantities such as charge, currents and ionic concentrations and this is cited as one of the reasons for its power efficiency. This is hence well suited for processing noisy sensory signals where precision is limited by input signal to noise ratios. In the rest of the chapter, we will explore several such schemes to compress neural recording data by extracting information from it.

14.3.1 Compression Mode 1: Spike Detection

The first scheme is inspired by a communication protocol used in neuromorphic chips. Several neuromorphic sensors and neural networks have been designed using brain-inspired analog processing principles [39–44] while noise robust digital pulses are used for communication [45, 46]. Since digital communication is much faster ($\sim$ 10 Gbps) than the average firing rate of a neuron ($\sim$ 10 Hz), the firing information of multiple neurons can be multiplexed on the same serial bus where the identity of the source neuron is encoded in a simultaneously transmitted digital address. This protocol is referred to as Address Event Representation (AER) and allows neuromorphic spiking chips to communicate data from N neurons using only $\log_2(N)$ wires.

The AER scheme can be adopted for neural implants as well since in many cases, we are interested in only knowing the occurrence of spikes. In that case, circuits are needed to distinguish spikes from background noise—these are called spike detectors. Figure 14.3b denotes this scheme as Mode 1 with three possible variants. The earliest instance of such detectors is based on simple thresholding circuits[24] where it is assumed that the amplitude of the spike is larger than background noise by a certain amount. A feedback loop is used to track the baseline noise level and the spike detection threshold is set to a multiple of this value. However, this method was found to produce high false positives in noisy conditions and hence an improved detection method using a non-linear energy operator (NEO) has been proposed. The NEO operator is defined as:

$$\text{NEO}(V) = \left(\frac{dV}{dt}\right)^2 - \frac{d^2V}{dt^2} \cdot V \qquad (14.2)$$

Several analog implementations of the NEO scheme have been reported [47–50] and an example of spike detection waveforms from the implementation in [47] is shown in Fig. 14.4. We refer to this method as Mode 1-A.

The spike detection method discards all information about the amplitude and shape of the neural spike—this information may, however, be useful at a later stage to decide the identity of the source neuron. Hence, two other variants of

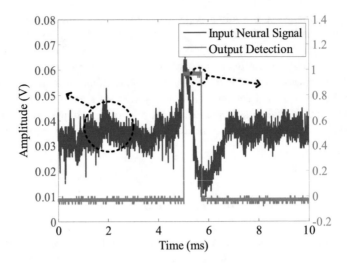

Fig. 14.4 Input noisy neural signal and corresponding digital spike detection output from the implementation in [47]. Only the detection result can be transmitted thus eliminating background data

the previously mentioned detection scheme have been commonly used. In some cases [51, 52], the authors use a regular spike detector to trigger the capture of a pre-defined number of samples of the neural spike signal so that all the features of the wave shape are retained for future extraction. We refer to this method as Mode 1-B. The other prevalent approach is to extract the relevant features (such as maximum, minimum, temporal width, derivative extrema) from the neural spike waveform when triggered by the spike detector [36, 48, 53–56]. Only these features are now digitized and transmitted providing a good trade-off between data reduction and signal information retention. We refer to this as Mode 1-C.

We can now derive the data rates R_{1-A}, R_{1-B}, and R_{1-C} required by each of the compression schemes. Denoting the number of biological neurons recorded by the sensor as N_{neu} (different from N_{chan}), firing rates of each neuron as f_{bio} we can write the equations as:

$$R_{1-A} = N_{neu} \times f_{bio} \times \lceil \log_2(N_{chan}) \rceil \tag{14.3}$$

$$R_{1-B} = N_{neu} \times f_{bio} \times f_{ADC} \times b_{ADC} \times t_{spk} \tag{14.4}$$

$$R_{1-C} = N_{neu} \times f_{bio} \times N_f \times b_{ADC} \tag{14.5}$$

where t_{spk} denotes the time span of the neural signal per spike transmitted in Mode 1-B, N_f denotes the number of features extracted in Mode 1-C and other variables have same meaning as defined earlier. We can estimate the degree of compression by assuming some nominal values of the parameters: $N_{neu} = 200$, $f_{bio} = 10$ Hz, $t_{spk} = 3$ ms, $N_f = 4$, $N_{chan} = 100$, $f_{ADC} = 20$ kHz and $b_{ADC} = 10$ bits. Then the three data rates become $R_{1-A} = 14$ kbps, $R_{1-B} = 120$ kbps and $R_{1-C} = 80$ kbps. Compared to the typical data rate, these modes offer a compression between ≈ 100–$1000\times$.

14.3.2 Compression Mode 2: Spike Sorting

The next possible scenario for compression is to use the features of the spike waveform to separate or classify each different wave shape into its own category representing a different source neuron. This method of assigning each distinct neural spike shape recorded on the same channel one unique identifier is called 'spike sorting' [57, 58]. Each category, in which spikes have similar shape, is believed to be generated by one neuron. The reasoning behind spike sorting is that the shape of spikes generated by neurons and recorded by an electrode is stereotypical, determined by the morphology of the dendritic trees of the neuron and the transmission pathway to the electrode. It is therefore believed that the shape of spikes from different neurons are distinct from each other and does not change over time, or at least over a significant amount of time. Though some work has demonstrated spike sorting may not be necessary for robust decoding performance [59, 60], the majority of work today still uses spike sorting to squeeze out as much information as possible from the neural recording implant.

Some authors have integrated a spike sorting classifier on the implant [61, 62]. While there are some implementations that have used supervised methods similar to template matching [63], most other approaches [64, 65] use unsupervised clustering techniques due to the advantage of not needing explicit training sessions. Figure 14.5 depicts the typical steps involved in spike sorting. After sorting, only the distinct identifier of the source neuron needs to be sent resulting in huge compression. We can estimate this data rate in Mode 2 as:

$$R_2 = N_{neu} \times f_{bio} \times \lceil \log_2(N_{neu}) \rceil \tag{14.6}$$

where the symbols have the same meaning defined earlier. Using the same values of the parameters used in the earlier Sect. 14.3.1, we can estimate the data rate for this mode to be $R_2 = 16$ kbps equivalent to a compression of $\approx 1000\times$ compared to a typical case.

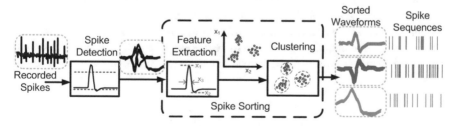

Fig. 14.5 The steps involved in spike sorting include feature extraction followed by unsupervised clustering to separate the neural spikes into distinct categories according to their shape

14.3.3   Compression Mode 3: Intention Decoding

The final and most advanced mode of compression is attained when the last stage of signal processing—decoding intentions from the recorded multi-channel spike train—is also integrated in the implant. This is shown in Fig. 14.3c as Mode 3. In this chapter, we focus on systems for motor prosthesis only—hence, in this case, intentions refer to 'motor' intentions or desire to move a limb. The fundamental of current decoding algorithms can be referred back to the work done by Georgopoulos and his colleagues [66, 67]. It is revealed in the experiment that the activity intensity of some neurons in the motor cortex is tuned to be a sinusoidal function of the movement direction of the arm with respect to a preferred direction where the activity reaches its maximum. They therefore proposed to represent each neuron by a vector indicating its preferred direction. The population vectors can be obtained by linear combination of all preferred vectors in the group weighted by the firing rate in the short time period of tens of millisecond, leading to a prediction on the velocity of upcoming arm movement [68].

Current state-of-the-art decoding algorithms for mapping population activity into motor intention can be categorized into two broad subgroups: inferential decoders [69–71] and classifiers [1, 20, 72]. However, most of these algorithms are run using bulky computers with wires connecting to the patient which impairs free movement and are a risk for infection. Recently, some approaches have been proposed for custom, low-power, compact hardware implementations of decoding algorithms [73–75] of which only one has shown measured results from a low-power integrated circuit [76] to decode motor intentions for dexterous finger movement as done in [20]. In the rest of the chapter, we elaborate on the details of this design, show the decoding performance and estimate achievable data compression using this scheme.

14.3.3.1   Algorithm: Extreme Learning Machine

The machine learning algorithm used in this work is the Extreme Learning Machine (ELM) [77, 78]. It is a two-layer neural network (Fig. 14.6) where the first layer of weights from inputs to hidden neurons (w_{ij} denotes weight from i-th input to j-th hidden neuron) are fixed and random. Only the weights in the second layer from the hidden neurons to output neurons need to be trained. Using β_{ki} to denote the weight from the i-th hidden neuron to the k-th output neuron, we can express the k-th output o_k as:

$$o_k = \sum_i^L \beta_{ki} g(\mathbf{w}_i, \mathbf{x}, b_i) = \sum_i^L \beta_{ki} h_i = \mathbf{h}^T \boldsymbol{\beta}_k$$

$$\mathbf{w}_i, \mathbf{x} \in \Re^D; \beta_{ki}, b_i \in \Re; \mathbf{h}, \boldsymbol{\beta}_k \in \Re^L \tag{14.7}$$

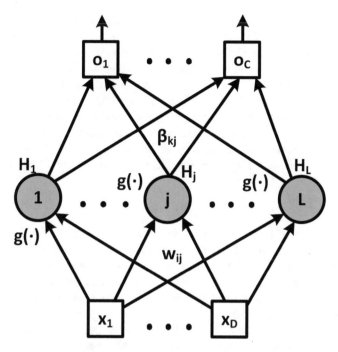

Fig. 14.6 Extreme Learning Machine (ELM) is a two-layer neural network where the weights of the first layer are random and fixed. Only second layer weights are tuned according to the task

where **x** denotes the D-dimensional input vector, **h** is the L-dimensional output of the hidden layer, $g()$ is the non-linear activation function of the hidden layer and b_i denotes the bias of the i-th hidden layer neuron. One of the commonly used activation functions is the additive node where $h_i = g(w_i^T x + b_i)$ and $g : \Re \to \Re$ is any non-linear function with finitely many discontinuities. While the outputs o_k can be directly used for regression, for classification, we assign the input sample to the class belonging to the output neuron with the highest value.

The second layer weights can be obtained by a direct solution instead of typically used iterative methods such as back propagation for multi-layer neural networks—hence, the training time for ELM based systems is much smaller. The output weights for each of the C classes can be optimized separately by using the same hidden layer values. Suppose there are p samples and let H denote the $p \times L$ hidden layer matrix where each row stores the output of the hidden neurons for one sample. Further, let $T_k \in \Re^p$ denote the target or desired values for the k-th hidden neuron. Then, the ideal weights $\hat{\boldsymbol{\beta}}_k$ for the k-th hidden neuron is obtained as solution of the following optimization problem [78]:

$$\hat{\boldsymbol{\beta}}_k = \underset{\beta_k}{\arg\min} \, \|H\boldsymbol{\beta}_k - T_k\|_2 + \gamma \|\boldsymbol{\beta}_k\|_2 \qquad (14.8)$$

where the second term in the equation is needed for regularization and γ is optimized on the validation set as a hyper-parameter. Closed form solutions to the value of $\boldsymbol{\beta}_k$ can be obtained in two different ways for the cases where the number of training samples is less or more than the number of hidden neurons [78].

To apply this neural network to neural decoding, the authors use an approach similar to [20] where the Artificial Neural Network is replaced by an ELM. The ELM decodes the onset time as well as the type of movement from the asynchronous neural spikes every $T_s = 20$ ms. First, instantaneous firing rate $r_i(t_k)$ at time t_k of each biological neuron is computed by counting the number of spikes in a time window $T_w = 100$ ms. Then, the input feature vector to the ELM at time t_k is defined by:

$$x(t_k) = [r_1(t_k), r_2(t_k) \ldots r_D(t_k)] \tag{14.9}$$

The total number of output neurons C in this case is equal to $M + 1$ where there are M movement types and one extra neuron is used to classify the onset time of movement. For training, the last output for onset time is trained on the entire dataset while the others are trained only on neural data during movement. Also, the last neuron is trained to solve a regression problem where the target function is trapezoidal—it gradually rises from 0 to 1 to mimic the gradually increasing activity of biological neuron ensembles. To reduce false positives in detecting movement onset, further processing is done on this 'primary' output by voting across the decision for several consecutive time samples [76] to produce the post-processed output. Another special signal processing feature of the IC is ability to include time delayed versions of neuronal activity as additional inputs to the ELM, i.e the number of inputs D to the ELM may be larger than the number of biological neurons N. This feature, referred to as Time-delay based dimension increase (TDBDI), is especially useful for chronic implants where the signal quality from many probes degrades with time due to scarring and fibrotic encapsulation.

The main reason for choosing the ELM algorithm is that most of the multiplications to be done in this architecture are the $D \times L$ random scalings in the first stage which can be done in very low energy and area using analog neuromorphic circuits. The mismatch induced errors [79] in analog circuits is not a problem in this case but can be part of the random coefficients. To get high accuracy, the trainable weights of the second stage can be implemented using digital circuits. However, this does not degrade system level energy efficiency as long as $D >> C$ which ensures that the number of multiplications in second stage are much less than that in the first stage. The circuit implementation of this algorithm is shown next.

14.3.3.2 Chip Architecture

The system architecture for the neuromorphic ELM chip is shown in Fig. 14.7. Since biological firing rate are sparse, the AER protocol described in Sect. 14.3.1 is used to send the neural spikes to a desired channel based on the address or identity of the source neuron. Then, the input handling circuits (IHC) compute an average

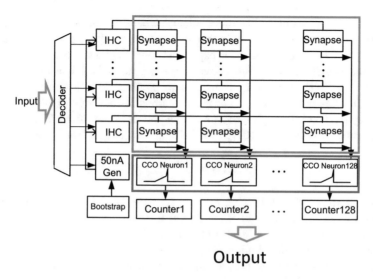

Fig. 14.7 Overall architecture of the ELM based decoder IC has a decoder to pass input spikes to desired channel, input handling circuits (IHC) to calculate average firing rate of spikes as a feature, a synapse array to create the random weighting of inputs needed in stage 1 of ELM and an array of hidden neurons

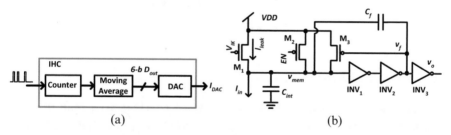

Fig. 14.8 (**a**) The IHC block comprises a counter and a moving average circuit to compute average firing rate in digital domain. The DAC then converts the digital number to an analog current. (**b**) The neuron is made of a current controlled oscillator (CCO) that clocks a counter (not shown)

firing rate using digital circuits in two steps (Fig. 14.8a). First, a counter estimates instantaneous firing rates by counting the number of spikes in a time interval T_s. Then a moving average circuit finds average firing rate in a time window T_w. This digital number is then converted to an analog current I_{DAC} using a digital to analog converter (DAC) so that following steps can be implemented in the analog domain. The major task of multiplication by a random number is performed by the synapse— a current mirror comprising identical minimum sized transistors. Ideally, without statistical variations, the current mirror would produce same output as its input. However, due to mismatch and sub-threshold operation of the transistors, the output current from a mirror is given by:

$$I_{out} = e^{\Delta V_T/U_T} I_{in} \tag{14.10}$$

where ΔV_T denotes threshold voltage mismatch between the two mirror transistors and U_T denotes thermal voltage. In this architecture, the diode connected transistor for every row is shared while the synapse just consists of a single mirror transistor. Hence, the weight of the synapse connecting i-th input to the j-th neuron is given by $w_{ij} = e^{\Delta V_{T.ij}/U_T}$. The sum of these currents are obtained by just wiring the drains of the mirror transistors together. Finally, this current is converted to the hidden layer output by passing it through a neuron circuit shown in Fig. 14.8b. The neuron is a current controlled oscillator (CCO) whose frequency of oscillation is given by:

$$f_{\text{CCO}} = \frac{I_{\text{in}} - I_{\text{leak}}}{C_f \times VDD} \tag{14.11}$$

This equation is valid as long as $I_{\text{in}} \ll I_{\text{rst}}$ where I_{rst} denotes the reset current flowing through transistor M3 when turned fully on. The current I_{leak} serves the function of the bias term b_i in Eq. (14.7). Similar to the weights w_{ij}, these also follow a log-normal distribution. The digital pulses from the CCO are used to clock a counter which is enabled along with the neuron for T_{en} seconds. Also, the counter can be stopped at a digitally programmable count value $h_{\max}$ which provides a saturating nonlinearity. Hence, the hidden layer output after the counter can be expressed as:

$$h = f_{\text{CCO}}T_{\text{en}} \text{ if } f_{\text{CCO}}T_{\text{en}} < h_{\max}$$

$$= h_{\max} \text{ otherwise.} \tag{14.12}$$

14.3.3.3   Measurement Results

The chip described above was fabricated in $0.35\,\mu$m CMOS process. With 128 input channels and 128 hidden neurons, the die size of this chip was $4.95 \times 4.95 mm^2$. An example of the mismatch is shown in the variability in measured tuning curves of the hidden neurons (Fig. 14.9) when the input spike frequency of only one

Fig. 14.9 Measured transfer curves of the 128 hidden layer neurons on the chip obtained by sweeping the input spike frequency of one of the channels. The variation of the curves is due to statistical variations in the chip

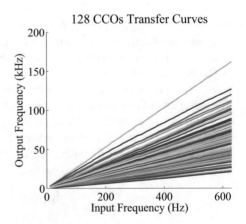

128 CCOs Transfer Curves

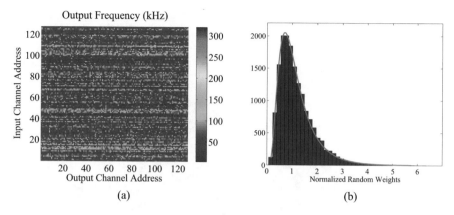

Fig. 14.10 (**a**) A map of the threshold variation across the 128×128 synaptic current mirror transistors on one of the dies. (**b**) The weights due to mismatch fit a log-normal distribution as expected

channel is varied. A more detailed characterization of the mismatch across the entire synaptic array is shown in Fig. 14.10a. This figure is obtained by giving a fixed input frequency to each channel one by one and recording the hidden neuron firing frequency. These weights are fit to a log-normal distribution in Fig. 14.10b implying an underlying gaussian distribution of ΔV_T. Across eight different dies, the mean of the gaussian distribution varies from -0.1 to $0.57 \, \text{mV}$ and the standard deviation varies from 16.2 to $17.6 \, \text{mV}$.

The authors in [76] have applied the IC for decoding flexion and extension of fingers and wrist from neural activity recorded from the M1 region of a non-human primate. The experiment with the monkey is described in detail in [20]. In brief, monkeys are trained to move individual fingers and wrist based on visual input while simultaneously, a single-unit recording device implanted in the motor cortex is used to record the brain activity. This data contains information about the monkey's motor intention and is used for the decoding. The entire data set has experiments performed on three monkeys. This pre-recorded data was fed into the IC and the hardware performance has been benchmarked with software decoding results reported in [20].

Figure 14.11 shows an example of the decoding being performed—three different trials are shown. The bottom part of the figure shows neural spikes obtained after sorting from $N = 40$ M1 neurons. The middle panel shows the onset detection while the top panel shows predicted movement type. The authors reported that the decoding accuracy increases to $\approx 96\%$, at par with software results, for a hidden layer size of $L = 60$ neurons. It is also important to see how the decoding accuracy degrades when less number of biological M1 neurons are available for recording. This is shown in Fig. 14.12 for 8 different samples of the IC. It can be seen that using delayed samples to increase dimension (TDBDI) helps in boosting decoding accuracy for all samples. The result is specially significant when the number of M1 neurons is small. This clearly shows the benefit of TDBDI for chronic implants. For this IC, the authors report a power dissipation of $414 \, \text{nW}$ for the case of $D = 40$

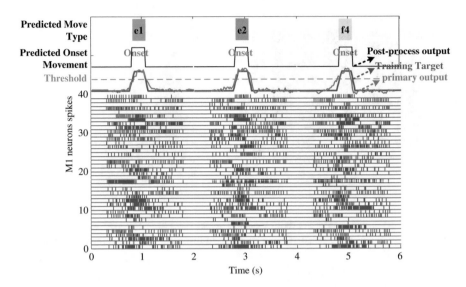

Fig. 14.11 Example of a neural decoding trial where the chip uses $L = 60$ hidden layer neurons to decode the onset time and type of movement from $N = 40$ biological neurons recorded from the M1 region of a non-human primate. 12 types of movement are considered here—flexion and extension of five fingers and wrist

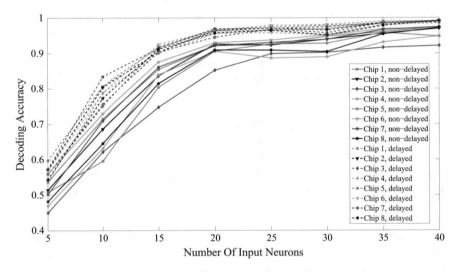

Fig. 14.12 Using the time delayed samples for extra information helps in increasing decoding accuracy especially when the number of biological M1 neurons is small. The results are verified from 8 chips

and $L = 60$ resulting in an ultra-low energy per operation of 3.45 pJ/MAC where MAC refers to multiply and accumulate. This is much smaller than recently reported digital multiplier which requires 16–70 pJ/MAC [80–82].

We can now estimate the amount of data compression achievable in this mode of operation with an integrated neural decoder. In the beginning of a session, this system needs to transmit the raw data rate of R_{typ} or R_1 or R_2. This data is used for training. Once trained however, the data rate R_3 to be transmitted is given by:

$$R_3 = f_{\text{deco}} \times \lceil \log_2(C) \rceil \qquad (14.13)$$

where C is the number of classes of movement and f_{deco} is the rate of classification. As an example, for the case described earlier with $f_{\text{deco}} = 50\,\text{Hz}$ and $C = 13$, $R_3 = 200\,\text{bps}$ with a compression factor of 10^5 over R_{typ} showing the huge potential of compression obtainable this way.

14.4   Conclusion and Discussions

Implantable brain machine interfaces are an emerging area of research which can be used by patients with motor disabilities to interact naturally with prosthetics or devices such as wheelchairs. More broadly, neural implants can be used to treat other neural diseases such as Parkinson's, epilepsy or depression. In this chapter, we showed the issue of scaling neural implants to thousands of channels in the future stems from increasing wireless transmission rates of the order of 200 Mbps. It was also shown that it is possible to achieve variable rates of compression from 10–10^5 by incorporating more processing steps into the implanted chip as opposed to leaving it to the receiver module outside the body. To make this viable, the processing has to be done in ultra low power so that the power budget of the implant is not exceeded.

Neuromorphic or neuro-inspired analog circuits provide a viable alternative for reducing power dissipation beyond what is achievable from current digital circuits. In this chapter, we presented an extensive survey of the different levels of compression that are achievable when integrating spike detection, sorting or intention decoding within the neural implant. The most promising scheme for the future large scale implants—intention decoding—is described in great detail starting from the algorithm to chip architecture and details of sub-circuits. In the long term, we envision that as brain sensing technologies mature so that thousands of neurons can be simultaneously probed, integrated machine learners for intention decoding will become a common feature for managing the 'big data' originating from neural implants. However, to allow chronic or long-term recording using such devices, some challenges still need to be overcome. One of the major issues in long-term recordings is parameter drift such as change of probe impedance due to scarring or gliosis. Though the current solution has a feature of TDBDI to counter this, there is no automatic detection strategy of when to apply this and to which channels. This is a topic that deserves more attention in future. Also, the current method of training the machine learner used a trial structure where the time of movement was known— in real life operation, there will not be any such precise temporal markers and the training algorithm has to be modified to suit this. One promising possibility is reinforcement learning based training [83] but more work is needed in this direction.

Lastly, the current training paradigm used data from a monkey performing actual movements. To move to a prosthetic control using imagined movements only, there will be an aspect of visual feedback that will alter the neural data recorded by the chip—a phenomenon referred to as 'closed-loop' decoder training. In this case, we have to retrain the machine learner iteratively over several closed-loop experimental trials and convergence of such training for ELM based decoders is an open avenue for research.

Acknowledgements The authors acknowledge funding support from NTU and MOE, Mediatek for supporting chip design and Prof. Nitish Thakor for providing neural data from primate experiments.

References

1. J. Wessberg, C. Stambaugh, J. Kralik, P. Beck, M. Laubach, J. Chapin, J. Kim, J. Biggs, M. Srinivasan, M. Nicolelis, Real-time prediction of hand trajectory by ensembles of cortical neurons in primates. Nature **408**, 361–365 (2000)
2. L. Hochberg, M. Serruya, G. Friehs, J. Mukand, M. Saleh, A. Caplan, A. Branner, D. Chen, R. Penn, J. Donoghue, Neuronal ensemble control of prosthetic devices by a human with tetraplegia. Nature **442**, 164–171 (2006)
3. L. Hochberg, D. Bacher, B. Jarosiewicz, N. Masse, J. Simeral, J. Vogel, S. Haddain, J. Liu, S. Cash, P. der Smagt, J. Donoghue, Reach and grasp by people with tetraplegia using a neurally controlled robotic arm. Nature **485**, 372–375 (2012)
4. M. Lebedev, M. Nicolelis, Toward a whole-body neuroprosthetic. Prog. Brain Res. **194**, 47–60 (2011)
5. M. Lebedev, M. Nicolelis, Brain-machine interfaces: past, present and future. Trends Neurosci. **29**(9), 536–546 (2006)
6. A. Kübler, B. Kotchoubey, J. Kaiser, J. Wolpaw, N. Birbaumer, Brain-computer communication: unlocking the locked in, American Psychological Association, Washington, DC, 2001
7. Human brain project official website https://www.humanbrainproject.eu/
8. The BRAIN initiative, NIH website http://www.nih.gov/science/brain/
9. K. Micheva, B. Busse, N. Weileer, N. O'Rourke, S. Simith, Single-synapse analysis of a diverse synapse population: proteomic imaging methods and markers. Neuron **68**, 639–653 (2010)
10. G. Buzsaki, K. Mizuseki, The log-dynamic brain: how skewed distributions affect network operations. Nat. Rev. Neurosci. **15**, 264–78 (2014)
11. T.M. Seese, H. Harasaki, G.M. Saidel, C. Davies, Characterization of tissue morphology, angiogenesis, and temperature in the adaptive response of muscle tissue in chronic heating. Lab. Invest. **78**, 1553–1562 (1998)
12. S. Kim, R. Normann, R. Harrison, F. Solzbacher, Preliminary study of the thermal impact of a microelectrode array implanted in the brain, in *Proceedings of IEEE Engineering in Medicine and Biology Conference* (2006), pp. 2986–2989
13. A. Usakli, Improvement of EEG signal acquisition: an electrical aspect for state of the art of front end. Comput. Intell. Neurosci. **2010**, 630649 (2010)
14. P. Konrad, T. Shanks, Implantable brain computer interface: challenges to neurotechnology translation. Neurobiol. Dis. **38**, 369–375 (2010)
15. A. Hoogerwerf, K. Wise, A three-dimentional microelectrode array for chronic neural recording. IEEE Trans. Biomed. Eng. **41**(12), 1136–1146 (1994)
16. C.T. Nordhausen, E.M. Maynard, R.A. Normann, Single unit recording capabilities of a 100 microelectrode array. Brain Res. **726**, 129–140 (1996)
17. A.L. Owens, T.J. Denison, H. Versnel, M. Rebbert, M. Peckerar, S.A. Shamma, Multi-electrode array for measuring evoked potentials from the surface of ferret primary auditory cortex. J. Neurosci. Methods **58**, 209–220 (1995)

18. V. Aggarwal, M. Mollazadeh, A.G. Davidson, M.H. Schieber, N.V. Thakor,   State-based decoding of hand and finger kinematics using neuronal ensemble and LFP activity during dexterous reach-to-grasp movements. J. Neurophysiol. **109**(12), 3067–3081 (2013)
19. K. Rupp, M. Schieber, N.V. Thakor, Local field potentials mitigate decline in motor decoding performance caused by loss of spiking units, in *Annual International Conference of the IEEE Engineering in Medicine and Biology Society (EMBC)*, Chicago, Aug 2014, pp. 1298–1301
20. V. Aggarwal, S. Acharya, F. Tenore, H. Shin, R. Etienne-Cummings, M. Schieber, N. Thakor, Asynchronous decoding of dexterous finger movements using M1 neurons. IEEE Trans. Neural Syst. Rehabil. Eng. **16**, 3–14 (2008)
21. S. Acharya, F. Tenore, V. Aggarwal, R. Etienne-Cummings, M. Schieber, and N. Thakor, Decoding individuated finger movements using volume-constrained neuronal ensembles in the M1 hand area. IEEE Trans. Neural Syst. Rehabil. Eng. **16**, 15–23 (2008)
22. M. Velliste, S. Perel, M. Spalding, A. Whitford, A. Schwartz, Cortical control of a prosthetic arm for self-feeding. Nature **453**, 1098–1101 (2008)
23. V. Gilja, P. Nuyujukian, C.A. Chestek, J.P. Cunningham, B.M. Yu, J.M. Fan, M.M. Churchland, M.T. Kaufman, J.C. Kao, S.I. Ryu, K.V. Shenoy, A high-performance neural prosthesis enabled by control algorithm design. Nat. Neurosci. **15**, 1752–1757 (2012)
24. R. Harrison, The design of integrated circuits to observe brain activity. Proc. IEEE **96**(7), 1203–1216 (2008)
25. Y. Chen, A. Basu, L. Liu, X. Zou, R. Rajkumar, G.S. Dawe, M. Je, A digitally assisted, signal folding neural recording amplifier. IEEE Trans. Biomed. Circuits Syst. **8**(4), 528–542 (2014)
26. R. Harrison, A low-power integrated circuit for adaptive detection of action potentials in noisy signals, in *Proceeding of the 25th Annual International Conference of the IEEE EMBS* (2003)
27. W. Wattanapanitch, M. Fee, R. Sarpeshkar, An energy-efficient micropower neural recording amplifier. IEEE Trans. Biomed. Circuits Syst. **1**(2), 136–147 (2007)
28. R. Ginosar, Y. Perelman, Analog frontend for multichannel neuronal recording system with spike and LFP separation. J. Neurosci. Methods **153**, 21–26 (2006)
29. F. Shahrokhi, K. Abdelhalim, D. Serletis, P.L. Carlen, R. Genov, The 128-channel fully differential digital integrated neural recording and stimulation interface. IEEE Trans. Biomed. Circuits Syst. **4**(3), 149–161 (2010)
30. M. Mollazadeh, K. Murari, G. Cauwenberghs, N. Thakor, Micropower CMOS integrated low-noise amplification, filtering, and digitization of multimodal neuropotentials. IEEE Trans. Biomed. Circuits Syst. **3**, 1–10 (2009)
31. W. Wattanapanitch, R. Sarpeshkar, A low-power 32-channel digitally programmable neural recording integrated circuit. IEEE Trans. Biomed. Circuits Syst. **5**, 592–602 (2011)
32. R.R. Harrison, P.T. Watkins, R.J. Kier, R.O. Lovejoy, D.J. Black, B. Greger, F. Solzbacher, A low-power integrated circuits for a wireless 100-electrode neural recording system. IEEE J. Solid State Circuits **42**(1), 123–133 (2007)
33. M. Yin, D.A. Borton, J. Aceros, W.R. Patterson, A.V. Nurmikko, A 100-channel hermetically sealed implantable device for chronic wireless neurosensing applications. IEEE Trans. Biomed. Circuits Syst. **7**(2), 115–128 (2013)
34. J. Tan, W.S. Liu, C.H. Heng, Y. Lian, A 2.4 GHz ULP reconfigurable asymmetric transceiver for single-chip wireless neural recording IC. IEEE Trans. Biomed. Circuits Syst. **8**(4), 497–509 (2014)
35. S.X. Diao, Y.J. Zheng, Y. Gao, S.J. Cheng, X.J. Yuan, M.Y. Je, A 50-Mb/s CMOS QPSK/O-QPSK transmitter employing injection locking for direct modulation. IEEE Trans. Microwave Theory Tech. **60**(1), 120–130 (2012)
36. M. Chae, Z. Yang, M. Yuce, L. Hoang, W. Liu, A 128-channel 6 mW wireless neural recording IC with spike feature extraction and UWB transmitter. IEEE Trans. Neural Syst. Rehabil. Eng. **17**(4), 312–321 (2009)
37. C. Mead, Neuromorphic electronic systems. IEEE Proc. **78**(10), 1629–1636 (1990)
38. R. Sarpeshkar, Efficient precise computation with noisy components: extrapolating from an electronic cochlea to the brain. PhD thesis, California Institute of Technology, Pasadena, CA (1997)

39. P. Lichtsteiner, C. Posch, T. Delbruck, A 128 × 128 120dB 15us latency asynchronous temporal contrast vision sensor. IEEE J. Solid State Circuits 43(2), 566–576 (2008)
40. V. Chan, S.-C. Liu, A. van Schaik, AER EAR: a matched silicon cochlea pair with address event representation interface. IEEE Trans. Circuits Syst. I 54(1), 48–59 (2007)
41. G. Indiveri, E. Chicca, R. Douglas, A VLSI array of low-power spiking neurons and bistable synapses with spike-timing dependent plasticity. IEEE Trans. Neural Netw. 17(1), 211–221 (2006)
42. G. Indiveri, E. Chicca, R.J. Douglas, Artificial cognitive systems: from VLSI networks of spiking neurons to neuromorphic cognition. Cogn. Comput. 1, 119–127 (2009)
43. S. Brink, S. Nease, P. Hasler, S. Ramakrishnan, R. Wunderlich, A. Basu, B. Degnan, A learning-enabled neuron array IC based upon transistor channel models of biological phenomenon. IEEE Trans. Biomed. Circuits Syst. 7(1), 71–81 (2013)
44. B.V Benjamin, P. Gao, E. McQuinn, S. Choudhary, A.R. Chandrasekaran, J.-M. Bussat, R. Alvarez-Icaza, J.V. Arthur, P.A. Merolla, K. Boahen, Neurogrid: a mixed-analog-digital multichip system for large-scale neural simulations. Proc. IEEE 102(5), 699–716 (2014)
45. K. Boahen, Point-to-point connectivity between neuromorphic chips using address events. IEEE Trans. Circuits Syst. II 47(5), 416–434 (2000)
46. S. Furber, F. Galluppi, S. Temple, L. Plana, The SpiNNaker project. Proc. IEEE 102(5), 652–665 (2014)
47. Y. Enyi, C. Yi, A. Basu, A 0.7 V, 40 nW compact, current-mode neural spike detector in 65 nm CMOS. IEEE Trans. Biomed. Circuits Syst. 10(2), 309–318 (2016)
48. J. Holleman, A. Mishra, C. Diorio, B. Otis, A micro-power neural spike detector and feature extractor in .13 μm CMOS, in *Proceedings of the IEEE Custom Integrated Circuits Conference*, Sept 2008, pp. 333–336
49. E. Koutsos, S.E. Paraskevopoulou, T.G. Constandinou, A 1.5 uW NEO-based spike detector with adaptive-threshold for calibration-free multichannel neural interfaces, in *Proceedings of the International Symposium on Circuits and Systems*, May 2013, pp. 1922–1925
50. Y.-G. Li, Q. Ma, M.R. Haider, Y. Massoud, Ultra-low-power high sensitivity spike detectors based on modified nonlinear energy operator, in *Proceedings of the International Symposium on Circuits and Systems*, May 2013, pp. 137–140
51. L. Liu, L. Yao, X. Zou, W.L. Goh, M. Je, Neural recording front-end IC using action potential detection and analog buffer with digital delay for data compression, in *Annual International Conference of the IEEE Engineering in Medicine and Biology Society (EMBC)*, Osaka, July 2013, pp. 747–750
52. Y. Perelman, R. Ginosar, An integrated system for multichannel neuronal recording with spike/LFP separation, integrated A/D conversion and threshold detection. IEEE Trans. Biomed. Eng. 54(1), 130–137 (2007)
53. R.H. Olsson, K.D. Wise, A three-dimensional neural recording microsystem with implantable data compression circuitry. IEEE J. Solid State Circuits 40(12), 2796–2804 (2016)
54. T. Horiuchi, T. Swindell, D. Sander, P. Abshire, A low-power CMOS neural amplifier with amplitude measurement for spike sorting, in *Proceedings of the 2004 International Symposium on Circuits and Systems*, vol. 4 (2004), pp. 23–26
55. T. Horiuchi, D. Tucker, K. Boyle, P. Abshire, Spike discrimination using amplitude measurements with a low-power CMOS neural amplifier, in *IEEE International Symposium on Circuits and Systems (ISCAS)* (2007)
56. A. Bhaduri, E. Yao, A. Basu, Pulse-based feature extraction for hardware-efficient neural recording systems, in *International Symposium on Circuits and Systems (ISCAS)*, Montreal, May 2016
57. R.Q. Quiroga, Spike sorting. Scholarpedia 2(12), 3583 (2007)
58. R.Q. Quiroga, Z. Nadasdy, Y. Ben-Shaul, Unsupervised spike detection and sorting with wavelets and superparamagnetic clustering. Neural Comput. 16(8), 1661–1687 (2004)
59. E. Stark, M. Abeles, Predicting movement from multiunit activity. J. Neurosci. 27, 8387–8394 (2007)
60. V. Ventura, Spike train decoding without spike sorting. Neural Comput. 20, 923–963 (2008)

61. S. Gibson, J. Judy, D. Markovic, Spike sorting: the first step in decoding the brain. IEEE Signal Process. Mag. **29**, 124–143 (2012)
62. V. Karkare, S. Gibson, C. Yang, H. Chen, D. Markovic, A 75 uW, 16-channel neural spike-sorting processor with unsupervised clustering, in *IEEE Symposium on VLSI Circuits Digest of Technical Papers* (2011)
63. A. Patil, S. Shen, E. Yao, A. Basu, Random projection for spike sorting: decoding neural signals the neural network way, in *Biomedical Circuits and Systems (BioCAS)*, Atlanta, Oct (2015)
64. V. Karkare, S. Gibson, D. Marković, A 130-W, 64-channel neural spike-sorting DSP chip. IEEE J. Solid State Circuits **46**(5), 1214–1222 (2011)
65. V. Karkare, S. Gibson, D. Markovic, A 75-μW, 16-channel neural spike-sorting processor with unsupervised clustering. IEEE J. Solid State Circuits **48**(9), 2230–2238 (2013)
66. A. Georgopoulos, J. Kalaska, R. Caminiti, J. Massey, On the relations between the direction of two-dimensional arm movements and cell discharge in primate motorcortex. J. Neurosci. **2**, 1527–1537 (1982)
67. A. Georgopoulos, J. Kalaska, R. Caminiti, J. Massey, Spatial coding of movement: a hypothesis concerning the coding of movement direction by motor cortical populations. Exp. Brain Res. Suppl. **7**, 327–336 (1983)
68. A. Georgopoulos, A. Schwartz, R. Kettner, Neuronal population coding of movement direction. Science **233**, 1357–1440 (1986)
69. W. Wu, M. Black, D. Mumford, Y. Gao, E. Bienenstock, J. Donoghue, Modeling and decoding motor cortical activity using a switching Kalman filter. IEEE Trans. Biomed. Eng. **51**, 933–942 (2004)
70. W. Wu, Y. Gao, E. Bienenstock, J. Donoghue, M. Black, Bayesian population decoding of motor cortical activity using a Kalman filter. Neural Comput. **18**, 80–118 (2006)
71. A. Brockwell, A. Rojas, R. Kass, Recursive bayesian decoding of motor cortical signals by particle filtering. J. Neurophysiol. **91**, 1899–1907 (2004)
72. S. Lin, J. Si, A. Schwartz, Self-organization of firing activities in monkey's motor cortex: trajectory computation from spike signals. Neural Comput. **9**, 607–621 (1997)
73. B. Rapoport, W. Wattanapanitch, H. Penagos, S. Musallam, R. Andersen, R. Sarpeshkar, A biomimetic adaptive algorithm and low-power architecture for implantable neural decoders, in *31st Annual International Conference of the IEEE EMBS* (2009)
74. B. Rapoport, L. Turicchian, W. Wattanapanitch, T. Davidson, R. Sarpeshkar, Efficient universal computing architectures for decoding neural activity. PLoS ONE **7**, e42492 (2012)
75. J. Dethier, V. Gilja, P. Nuyujukian, S.A. Elassaad, K.V. Shenoy, K. Boahen, Spiking neural network decoder for brain-machine interfaces, in *5th International IEEE/EMBS Conference on Neural Engineering (NER), 2011* (2011)
76. C. Yi, Y. Enyi, A. Basu, A 128 channel extreme learning machine based neural decoder for brain machine interfaces. IEEE Trans. Biomed. Circuits Syst. **10**(3), 679–692 (2016)
77. G.B. Huang, Q.Y. Zhu, C.K. Siew, Extreme learning machines: theory and applications. Neurocomputing **70**, 489–501 (2006)
78. G.-B. Huang, H. Zhou, X. Ding, R. Zhang, Extreme learning machine for regression and multiclass classification. IEEE Trans. Syst. Man Cybern. B Cybern. **42**(2), 513–529 (2012)
79. P.R. Kinget, Device mismatch and tradeoffs in the design of analog circuits. IEEE J. Solid State Circuits **40**(6), 1212–1224 (2005)
80. Y. He, C.H. Chang, A new redundant binary booth encoding for fast 2n-bit multiplier design. IEEE Trans. Circuits Syst. I **56**(6), 1192–1201 (2009)
81. K.S. Chong, B.H. Gwee, J.S. Chang, A micropower low-voltage multiplier with reduced spurious switching. IEEE Trans. VLSI **13**(2), 255–265 (2005)
82. M. La Guia de Solaz, R. Conway, Razor based programmable truncated multiply and accumulate, energy-reduction for efficient digital signal processing. IEEE Trans. VLSI **23**(1), 189–193 (2015)
83. J. DiGiovanna, B. Mahmoudi, J. Fortes, J. Principe, J. Sanchez Co-adaptive brain machine interface via reinforcement learning. IEEE Trans. Biomed. Eng. **54**(64), 56–61 (2009)

Chapter 15
Data Analytics in Quantum Paradigm: An Introduction

Arpita Maitra, Subhamoy Maitra, and Asim K. Pal

15.1 Introduction

The basic model of classical computers was initially visualized by Alan Turing, Von Neumann, and several other researchers in the 1930s and the decade after that. However the model of computers, that Turing or Neumann studied, are limited by classical physics and thus termed as classical computers. Till the end of nineteenth century, most scientists believed that Newtonian laws governing the motion of material bodies and Maxwell's theory of electromagnetism are the fundamental areas of physics. However, the discovery of X-rays and electrons towards the end of that century finally helped the physicists to understand quantum mechanics around 1925. The limitation of classical mechanics could be understood clearly after that. In 1982, Richard Feynman presented the seminal idea of a universal quantum simulator or more informally, a quantum computer.

Informally speaking, a quantum system of more than one particles can be explained by a Hilbert space whose dimension is exponentially large in the number of particles. Thus, one naturally expects that a quantum system can efficiently solve a problem that may require exponential time on a classical computer. During the 1980s, the initial works by Deutsch-Jozsa [12] and Grover [17] could explain quantum algorithms that are exponentially faster than the classical ones. Most importantly, in 1994, Shor discovered that in quantum paradigm, factorization and discrete log problems can be efficiently solved [37]. This result had a major impact in classical cryptography. This is because, there are lot of public key cryptosystems

A. Maitra • A.K. Pal
Indian Institute of Management Calcutta, Kolkata, India
e-mail: arpita76b@gmail.com; asim@iimcal.ac.in

S. Maitra (✉)
Indian Statistical Institute, Kolkata, India
e-mail: subho@isical.ac.in

© Springer International Publishing AG 2017
A. Chattopadhyay et al. (eds.), *Emerging Technology and Architecture for Big-data Analytics*, DOI 10.1007/978-3-319-54840-1_15

that are based on these two tools. The internet communication as a whole, including the online banking system, depends on the security of these. Thus, in the field of public key cryptography, this warranted for cryptographic primitives that can resist attacks even with the existence of quantum computers. While commercial quantum computers are still elusive, the recent developments in the area of experimental physics are gaining huge momentum as evident from the award of Nobel prize for Physics in 2012 to Wineland and Haroche for "ground-breaking experimental methods that enable measuring and manipulation of individual quantum systems," a study on the particle of light, the photon. The Nobel prize in Physics, 2016, is awarded to Thouless, Haldane, and Kosterlitz for "theoretical discoveries of topological phase transitions and topological phases of matter." These results might have importance towards actual implementation of a quantum computer. Thus it shows that this domain of research is indeed one of the top priorities in international scientific community.

Data analytics is the technology of investigating raw data towards obtaining valid conclusions regarding relevant information. Such techniques are exploited by organizations to identify better business decisions towards verifying or disproving the models they study. As these algorithms, in many cases, require high complexity, it would always be interesting to investigate whether one can have more efficient solutions in the quantum domain. Consider the example of a share market. There we require huge computation in short time, need to communicate those data quickly among different parties, and at the same time the data security has to be considered with priority. While the data communication and security issues may be handled as a part where much competition might not be involved, each of the companies will be interested to have a better forecast than the other. Towards a better forecast, which is the main purpose of data analytics, one requires to have huge statistical calculations, which finally boils down to arithmetic, algebraic, combinatorial, and symbolic computations. Thus, the main question here is whether we can have better computational facilities in quantum paradigm. This is the focus of this material. At the same time, we also touch a few issues in communication and security domain that are relevant in data analytics and where the quantum paradigm has efficient tools to offer.

Before proceeding further, let us present brief introductory materials. For detailed technical understanding, one may refer to [29].

15.1.1 Basics of a Qubit and the Algebra

As a bit (0 or 1) is the basic element of a classical computer, the quantum bit (called the qubit) is the fundamental element in the quantum paradigm, whose physical counterpart is a photon. A qubit is represented as

$$\alpha|0\rangle + \beta|1\rangle,$$

where $\alpha, \beta \in \mathbb{C}$ (i.e., complex numbers), and $|\alpha|^2 + |\beta|^2 = 1$. If one measures the qubit in $\{|0\rangle, |1\rangle\}$ basis, then $|0\rangle$ is observed with probability $|\alpha|^2$, and $|1\rangle$ with $|\beta|^2$. The original state gets destroyed after the observation and collapse to the observed state.

That is, the qubits $|0\rangle, |1\rangle$ are the quantum counterparts of the classical bits 0, 1. The qubit $|0\rangle$ can be represented as $\begin{bmatrix} 1 \\ 0 \end{bmatrix}$ and $|1\rangle$ can be represented as $\begin{bmatrix} 0 \\ 1 \end{bmatrix}$. The superposition of $|0\rangle, |1\rangle$, i.e., $\alpha|0\rangle + \beta|1\rangle$ can be written as $\alpha \begin{bmatrix} 1 \\ 0 \end{bmatrix} + \beta \begin{bmatrix} 0 \\ 1 \end{bmatrix} = \begin{bmatrix} \alpha \\ \beta \end{bmatrix}$, where $\alpha, \beta \in \mathbb{C}, |\alpha|^2 + |\beta|^2 = 1$.

Based on this definition, one may theoretically pack infinite amount of information in a single qubit. However, it is not clear how to extract such information. Further in actual implementation of quantum circuits, it might not be possible to perfectly create a qubit for any α, β. Nevertheless, it is clear that a single qubit may contain huge information compared to a bit.

The basic algebra relating to more than one qubits can be interpreted as tensor products. Thus, consider tensor product of two qubits as

$$(\alpha_1|0\rangle + \beta_1|1\rangle) \otimes (\alpha_2|0\rangle + \beta_2|1\rangle) = \begin{bmatrix} \alpha_1 \\ \beta_1 \end{bmatrix} \otimes \begin{bmatrix} \alpha_2 \\ \beta_2 \end{bmatrix} = \begin{bmatrix} \alpha_1 \begin{bmatrix} \alpha_2 \\ \beta_2 \end{bmatrix} \\ \beta_1 \begin{bmatrix} \alpha_2 \\ \beta_2 \end{bmatrix} \end{bmatrix} = \begin{bmatrix} \alpha_1\alpha_2 \\ \alpha_1\beta_2 \\ \beta_1\alpha_2 \\ \beta_1\beta_2 \end{bmatrix}$$

$$= \alpha_1\alpha_2 \begin{bmatrix} 1 \\ 0 \\ 0 \\ 0 \end{bmatrix} + \alpha_1\beta_2 \begin{bmatrix} 0 \\ 1 \\ 0 \\ 0 \end{bmatrix} + \beta_1\alpha_2 \begin{bmatrix} 0 \\ 0 \\ 1 \\ 0 \end{bmatrix} + \beta_1\beta_2 \begin{bmatrix} 0 \\ 0 \\ 0 \\ 1 \end{bmatrix}$$

$= \alpha_1\alpha_2|00\rangle + \alpha_1\beta_2|01\rangle + \beta_1\alpha_2|10\rangle + \beta_1\beta_2|11\rangle$. That is,
$(\alpha_1|0\rangle + \beta_1|1\rangle) \otimes (\alpha_2|0\rangle + \beta_2|1\rangle) = \alpha_1\alpha_2|00\rangle + \alpha_1\beta_2|01\rangle + \beta_1\alpha_2|10\rangle + \beta_1\beta_2|11\rangle$.

However, any 2-qubit state may not always be decomposed as above. Consider the state $\gamma_1|00\rangle + \gamma_2|11\rangle$ with $\gamma_1 \neq 0, \gamma_2 \neq 0$. This can never be written as $(\alpha_1|0\rangle + \beta_1|1\rangle) \otimes (\alpha_2|0\rangle + \beta_2|1\rangle)$. This phenomenon is described as entanglement. An example of maximally entangled state is $\frac{|00\rangle + |11\rangle}{\sqrt{2}}$, which is an example of Bell states or EPR pairs. We will later explain how to produce such entangled states and why they are important in quantum information.

15.1.2 Quantum Gates

Now let us briefly describe the quantum gates. Such gates are basic primitives in building a quantum computer. A quantum gate can be considered as a reversible circuit having n qubits as inputs and n qubits as outputs. Mathematically, they can be seen as $2^n \times 2^n$ unitary matrices where the elements are complex numbers. Let us first present a few examples of single input single output quantum gates. In matrix

Quantum input	Quantum gate	Quantum output
$\alpha\lvert 0\rangle + \beta\lvert 1\rangle$	X	$\beta\lvert 0\rangle + \alpha\lvert 1\rangle$
$\alpha\lvert 0\rangle + \beta\lvert 1\rangle$	Z	$\alpha\lvert 0\rangle - \beta\lvert 1\rangle$
$\alpha\lvert 0\rangle + \beta\lvert 1\rangle$	H	$\alpha\frac{\lvert 0\rangle + \lvert 1\rangle}{\sqrt{2}} + \beta\frac{\lvert 0\rangle - \lvert 1\rangle}{\sqrt{2}}$

form, the gate operations are as follows.

- X gate: $\begin{bmatrix} 0 & 1 \\ 1 & 0 \end{bmatrix}\begin{bmatrix} \alpha \\ \beta \end{bmatrix} = \begin{bmatrix} \beta \\ \alpha \end{bmatrix}$;

- Z gate: $\begin{bmatrix} 1 & 0 \\ 0 & -1 \end{bmatrix}\begin{bmatrix} \alpha \\ \beta \end{bmatrix} = \begin{bmatrix} \alpha \\ -\beta \end{bmatrix}$;

- H gate: $\begin{bmatrix} \frac{1}{\sqrt{2}} & \frac{1}{\sqrt{2}} \\ \frac{1}{\sqrt{2}} & -\frac{1}{\sqrt{2}} \end{bmatrix}\begin{bmatrix} \alpha \\ \beta \end{bmatrix} = \begin{bmatrix} \frac{\alpha+\beta}{\sqrt{2}} \\ \frac{\alpha-\beta}{\sqrt{2}} \end{bmatrix}$.

Note that $\frac{\alpha+\beta}{\sqrt{2}}\lvert 0\rangle + \frac{\alpha-\beta}{\sqrt{2}}\lvert 1\rangle = \alpha\frac{\lvert 0\rangle + \lvert 1\rangle}{\sqrt{2}} + \beta\frac{\lvert 0\rangle - \lvert 1\rangle}{\sqrt{2}}$.

The 2-input 2-output quantum gates can be seen as 4×4 unitary matrices. An example is the CNOT gate which works as follows: $\lvert 00\rangle \rightarrow \lvert 00\rangle$, $\lvert 01\rangle \rightarrow \lvert 01\rangle$, $\lvert 10\rangle \rightarrow \lvert 11\rangle$, $\lvert 11\rangle \rightarrow \lvert 10\rangle$. The related matrix is $\begin{bmatrix} 1 & 0 & 0 & 0 \\ 0 & 1 & 0 & 0 \\ 0 & 0 & 0 & 1 \\ 0 & 0 & 1 & 0 \end{bmatrix}$.

As an application of these gates, let us describe the circuit in Fig. 15.1 to create certain entangled states as follows: $\lvert \beta_{00}\rangle = \frac{\lvert 00\rangle + \lvert 11\rangle}{\sqrt{2}}$, $\lvert \beta_{01}\rangle = \frac{\lvert 01\rangle + \lvert 10\rangle}{\sqrt{2}}$, $\lvert \beta_{10}\rangle = \frac{\lvert 00\rangle - \lvert 11\rangle}{\sqrt{2}}$, and $\lvert \beta_{11}\rangle = \frac{\lvert 01\rangle - \lvert 10\rangle}{\sqrt{2}}$.

15.1.3 No Cloning

While it is very easy to copy an unknown classical bit (i.e., either 0 or 1), it is now well known that it is not possible to copy an unknown qubit. This result is known as the "no cloning theorem" and was initially noted in [13, 43]. It has a huge implications in quantum computing, quantum information, quantum cryptography, and related fields.

Fig. 15.1 Quantum circuit for creating entangled state

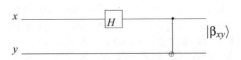

The basic outline of the proof is as follows. Consider a quantum slot machine with two slots labeled A and B. Here A is the data slot set in a pure unknown quantum state $|\psi\rangle$ whereas B is target slot set in a pure state $|s\rangle$ where A will be copied. Let there exist a unitary operator which does the copying procedure. Mathematically, it is written as $U(|\psi\rangle|s\rangle) = |\psi\rangle|\psi\rangle$. Note that, U being a unitary operator, $UU^\dagger = I$, where $(U^\dagger)_{ij} = \overline{U}_{ji}$, transpose of the matrix and scalar complex conjugate for each element. Let this copying procedure work for two particular pure states, $|\psi\rangle$ and $|\phi\rangle$. Then we have

$$U(|\psi\rangle|s\rangle) = |\psi\rangle|\psi\rangle, U(|\phi\rangle|s\rangle) = |\phi\rangle|\phi\rangle.$$

From the inner product: $\langle s|\langle\psi|U^\dagger U|\phi\rangle|s\rangle = \langle\psi|\langle\psi||\phi\rangle|\phi\rangle$. This implies $\langle\psi|\phi\rangle = (\langle\psi|\phi\rangle)^2$.

Note that $x = x^2$ has only two solutions: $x = 0$ and $x = 1$. Thus we get either $|\psi\rangle = |\phi\rangle$ or inner product of them equals to zero, i.e., $|\psi\rangle$ and $|\phi\rangle$ are orthogonal to each other. This implies that a cloning device can only clone orthogonal states. Therefore a general quantum cloning device is impossible. For example, given that the unknown state is one of $|0\rangle$, $\frac{|0\rangle+|1\rangle}{\sqrt{2}}$, two nonorthogonal states, it is not possible to clone the state without knowing which one it is.

This provides certain advantages as well as disadvantages. The advantages are in the domain of quantum cryptography, where by the laws of physics copying an unknown qubit is not possible. However, in terms of copying or saving unknown quantum data, this is actually a potential disadvantage. At the same time, it should be clearly explained that given a known quantum state, it is always possible to copy it. This is because, for a known quantum state, we know how to create it deterministically and thus it is possible to reproduce it with the same circuit.

For explaining with an example, one may refer to Fig. 15.2. If an unknown qubit $|\mu\rangle$ is either $|0\rangle$ or $|1\rangle$, then it will be copied perfectly without creating any disturbance to $|\mu\rangle$. However, if $|\mu\rangle = \frac{|0\rangle+|1\rangle}{\sqrt{2}}$, say, then at the output we will get entangled state $\frac{|00\rangle+|11\rangle}{\sqrt{2}}$. Thus copying is not successful here.

This concept can also be applied towards distinguishing quantum states. Given two orthogonal states $\{|\psi\rangle, |\psi_\perp\rangle\}$, it is possible to distinguish them with certainty. For example, the pair of states

$$\{|0\rangle, |1\rangle\}\};$$

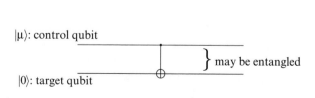

$|\mu\rangle$: control qubit

$\}$ may be entangled

$|0\rangle$: target qubit

Fig. 15.2 Explanation of no cloning with a simple circuit

$$\left\{ \frac{1}{\sqrt{2}}(|0\rangle + |1\rangle), \frac{1}{\sqrt{2}}(|0\rangle - |1\rangle) \right\} ;$$

$$\left\{ \frac{1}{\sqrt{2}}(|0\rangle + i|1\rangle), \frac{1}{\sqrt{2}}(|0\rangle - i|1\rangle) \right\}$$

are orthogonal and can be certainly distinguished.

However, two non-orthogonal quantum states, this is not possible. For example, given the two states are $|0\rangle$, $\frac{|0\rangle+|1\rangle}{\sqrt{2}}$, which are nonorthogonal, it is not possible to exactly identify each one with certainty. These ideas are back-bone to the famous BB84 Quantum Key Distribution (QKD) protocol [4].

15.2 A Brief Overview of Advantages in Quantum Paradigm

Next we like to briefly mention a couple of areas where the frameworks based on quantum physics provide advantageous situations over the classical domain. We will consider one example each in the domain of communication as well as computation.

15.2.1 Teleportation

Teleportation is one of the important ideas that shows the strength of quantum model over the classical model [5]. Given a sharing of a pair of entangled states by the two parties at distant locations, one just needs to send two classical bits of information to send an unknown quantum state (this may contain information corresponding to infinitely many bits) from one side to another side (Fig. 15.3).

As an example take $|\beta_{xy}\rangle = |\beta_{00}\rangle$, G the CNOT gate, i.e., $00 \rightarrow 00\rangle$, $|01\rangle \rightarrow |01\rangle$, $|10\rangle \rightarrow |11\rangle$, $|11\rangle \rightarrow |10\rangle$. Further consider $A = H, B = X^{M_2}, C = Z^{M_1}$.

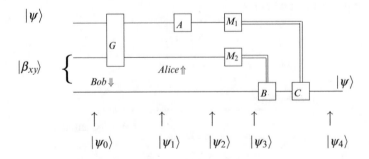

Fig. 15.3 Quantum circuit for Teleporting a qubit

This will provide the basic teleportation circuit. As a simple extension, one can use any $|\beta_{xy}\rangle$, G as CNOT and $A = H, B = X^{M_2 \oplus x}, C = Z^{M_1 \oplus y}$. The step by step explanation for teleportation is as follows.

- $|\psi_0\rangle = |\psi\rangle|\beta_{00}\rangle = (\alpha|0\rangle + \beta|1\rangle)\frac{(|00\rangle+|11\rangle)}{\sqrt{2}}$
- $|\psi_1\rangle = \alpha|0\rangle\frac{(|00\rangle+|11\rangle)}{\sqrt{2}} + \beta|1\rangle\frac{(|10\rangle+|01\rangle)}{\sqrt{2}}$
- $|\psi_2\rangle = \alpha\frac{|0\rangle+|1\rangle}{\sqrt{2}}\frac{(|00\rangle+|11\rangle)}{\sqrt{2}} + \beta\frac{|0\rangle-|1\rangle}{\sqrt{2}}\frac{(|10\rangle+|01\rangle)}{\sqrt{2}} = \frac{1}{2}(|00\rangle(\alpha|0\rangle + \beta|1\rangle) + |01\rangle(\beta|0\rangle + \alpha|1\rangle) + |10\rangle(\alpha|0\rangle - \beta|1\rangle) - |11\rangle(\beta|0\rangle - \alpha|1\rangle))$
- Observe 00, nothing to do. Observe 01, apply X. Observe 10, apply Z. Observe 11, apply both X, Z.

The importance of this technique in data analytics is that if two different places may share entangled particles, then it is possible to send a huge amount of information (in fact theoretically infinite) by just communicating two classical bits. Again, one important issue to be noted is that, even if we manage to transport a qubit, in case it is unknown, it might not be possible to extract the relevant information from that.

15.2.2 Deutsch-Jozsa Algorithm

Deutsch-Jozsa algorithm [12] is possibly the first clear example that demonstrates quantum parallelism over the standard classical model. Take a Boolean function $f : \{0, 1\}^n \rightarrow \{0, 1\}$. A function f is constant if $f(x) = c$ for all $x \in \{0, 1\}^n$, $c \in \{0, 1\}$. Further f is called balanced if $f(x) = 0$ for 2^{n-1} inputs and $f(x) = 1$ for the rest of 2^{n-1} inputs. Given the function f as a black box, which is either constant or balanced, we need an algorithm, that can answer which one this is. It is clear that a classical algorithm needs to check the function for at least $2^{n-1} + 1$ inputs in worst case to come to a decision. Quantum algorithm can solve this with only one input. Note that given a classical circuit f, there is a quantum circuit of comparable efficiency which computes the transformation U_f that takes input like $|x, y\rangle$ and produces output like $|x, y \oplus f(x)\rangle$ (Fig. 15.4).

The step by step operations of the technique can be described as follows.

- $|\psi_0\rangle = |0\rangle^{\otimes n}|1\rangle$

Fig. 15.4 Quantum circuit to implement Deutsch-Jozsa algorithm

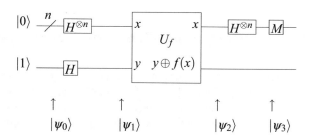

- $|\psi_1\rangle = \sum_{x\in\{0,1\}^n} \frac{|x\rangle}{\sqrt{2^n}} \left[\frac{|0\rangle - |1\rangle}{\sqrt{2}} \right]$
- $|\psi_2\rangle = \sum_{x\in\{0,1\}^n} \frac{(-1)^{f(x)}|x\rangle}{\sqrt{2^n}} \left[\frac{|0\rangle - |1\rangle}{\sqrt{2}} \right]$
- $|\psi_3\rangle = \sum_{z\in\{0,1\}^n} \sum_{x\in\{0,1\}^n} \frac{(-1)^{x\cdot z\oplus f(x)}|z\rangle}{2^n} \left[\frac{|0\rangle - |1\rangle}{\sqrt{2}} \right]$
- Measurement: all zero state implies that the function is constant, otherwise it is balanced.

The importance of explaining this algorithm in the context of data analytics is that it is often important to distinguish between two objects very efficiently. The example of Deutsch-Jozsa algorithm [12] demonstrates that it is significantly efficient compared to the classical domain.

At this point we like to present two important aspects of Deutsch-Jozsa algorithm [12] in terms of data analytics and machine learning. First of all, one must note that we can obtain the equal superposition of all 2^n many n-bit states just by using n many Hadamard gates. For this, note the first part of $|\psi_1\rangle$ which is $\sum_{x\in\{0,1\}^n} \frac{|x\rangle}{\sqrt{2^n}}$. This provides an exponential advantage in quantum domain as in the classical domain we cannot access all the 2^n many n-bit patterns efficiently. The second point is related to machine learning. As we have discussed, we may have the circuit of f available as a black-box and we like to learn several properties of the function efficiently. In this direction, Walsh transform is an important tool. What we obtain as the output of the Deutsch-Jozsa algorithm just before measurement is $|\psi_3\rangle$ and the first part of this is $\sum_{z\in\{0,1\}^n} \sum_{x\in\{0,1\}^n} \frac{(-1)^{x\cdot z\oplus f(x)}|z\rangle}{2^n}$. Note that, the Walsh spectrum of the Boolean function f at a point z is defined as $W_f(z) = \sum_{x\in\{0,1\}^n}(-1)^{x\cdot z\oplus f(x)}$. That is, $\sum_{z\in\{0,1\}^n} \sum_{x\in\{0,1\}^n} \frac{(-1)^{x\cdot z\oplus f(x)}|z\rangle}{2^n} = \sum_{z\in\{0,1\}^n} \frac{W_f(z)}{2^n}|z\rangle$. This means that using such an algorithm, we can efficiently obtain a transform domain spectrum of the function, which is not achievable in classical domain.

Testing several properties of Boolean functions in classical as well as quantum paradigm is an interesting area of research in property testing [6], which are in turn useful in learning theory. There are several interesting properties of Boolean functions, mostly in the area of coding theory and cryptology, that need to be tested efficiently. However, in many of the cases, the efficient algorithms are elusive. The Deutsch-Jozsa Algorithm [12] is the first step in this area in quantum computational model. In a larger view, the details of various quantum algorithms can be obtained from [32].

15.3 Preliminaries of Quantum Cryptography

In any commercial environment, confidentiality of data is one of the most important issues. Due to Shor's result [37] on efficient factorization as well as solving discrete logarithm in quantum domain, classical public key cryptography will be completely broken in case a quantum computer can actually be built. One must note that many of the commercial security systems, including banking, are based on algorithms whose security are promised by hardness of factorization or discrete log problems. In this regard, we present a few basic issues in classical and quantum cryptography that must be explained in any data centric environment.

The main challenge in cryptology in early seventies was how to decide on a secret information between two parties over a public channel. The solution to this has been proposed by Diffie and Hellman in 1976 [14]. The protocol is as follows.

- In public domain, the information about a suitable group G is made available. For example, one can consider $G = (\mathbb{Z}_p^*, \cdot)$ where the elements are $\{1, \ldots, p-1\}$ and the multiplication is modulo p. The prime p should be very large, say of the order of 1024 bits.
- Given the generator g (which is again known in public domain) and another element h, it is hard (using a classical computer) to obtain i such that $h = g^i$. This is well known as Discrete Logarithm Problem (DLP).
- Thus, it is believed that in the classical paradigm, it is not easy to obtain g^{ab} using g^a, g^b only (which are available to the adversary from the public channel) without any knowledge of a, b. Here g^{ab} is used as the secret key for further secured communication. That is, this secret key is the output of the key distribution algorithm which will be secretly shared by the participating parties after communication over a public channel.

Now let us describe the famous RSA cryptosystem [35]. The RSA cryptosystem has been invented by Rivest, Shamir, and Adleman in 1977 and this is undoubtedly the most popular public key cryptosystem which is used in various electronic commerce protocols. The security of this cryptosystem relies on the difficulty of factoring a number into its two constituent primes. In practice, the prime factors of interest will be several hundred bits long. A modulus $N = p \times q$ of 1024 bits, for example, would be common. Let us now briefly describe the scheme.

Key Generation Algorithm

- Choose primes p, q (generally same bit size, $q < p < 2q$)
- Construct modulus $N = pq$, and $\phi(N) = (p-1)(q-1)$
- Set e, d such that $d = e^{-1} \bmod \phi(N)$
- Public key: (N, e) and Private key: d

Encryption Algorithm: $C = M^e \bmod N$
Decryption Algorithm: $M = C^d \bmod N$

The RSA cryptosystem relies on the efficiency of the following:

- finding two large primes p, q, and computing $N = pq$;
- computing $d = e^{-1} \bmod \phi(N)$ given $N = pq$ and e;
- computing modular exponentiations $M^e \bmod N$ and $C^d \bmod N$.

While it is very clear that if one can factor the modulus N, then RSA can be immediately broken, the other two security problems are the following.

- To compute $d = e^{-1} \mod \phi(N)$ given N, e.
- To compute $M = C^{1/e} \mod N$ given N, e, C [RSA Problem].

Naturally, in classical domain, there is no efficient algorithm to solve the above two problems.

Till date, there is no efficient algorithm to solve DLP or RSA in classical domain. However, in the famous work by Shor [37], it has been shown that both these problems can be solved efficiently in quantum paradigm. This opens a new area called post-quantum cryptography [31], where the cryptosystems are studied considering that the adversary can attack the systems using quantum computers. There are certain classical public key cryptosystems, for example, lattice based and code based schemes for which no efficient quantum attack is known. However, understanding these algorithms requires advanced background in mathematics and computer science. Further, the commercial implementation of these schemes is not as efficient as RSA.

On the other hand, Bennett and Brassard provided the idea of Quantum Key Distribution [4] (QKD) where the physical laws are exploited towards the security proof. This idea is quite elegant and easy to understand. More interestingly, while the commercial quantum computers are still elusive, several QKD schemes have already been implemented for commercial purposes [33, 34]. We now describe this idea in more detail.

15.3.1   Quantum Key Distribution and the BB84 Protocol

Based on the above discussion, it is clear that the community needs a key distribution scheme that can resist a quantum adversary. The famous BB84 [4] protocol provides a secure quantum key distribution scheme which is secure under certain assumptions. The scheme has received huge attention in the research community as evident from its citation; it has also been implemented in commercial domain as well.

Bennett and Brassard (the BB of BB84) initiated the seminal idea of QKD in 1979 based on the pioneering concept proposed by Wiesner in 1970. Both these ideas have been published much later, i.e., the idea of Wiesner in 1983 [41] and that of Bennet and Brassard in [3, 4]. The work published in 1984 [4] received more prominence and that is why the 84 of BB84 comes. Interested readers may have a look at [9] for a detailed history in this area. Informally speaking, the security of BB84 protocol comes from no-cloning theorem and indistinguishability of non-orthogonal quantum states. The basic steps of BB84 QKD may be described as follows.

- One needs to transmit 0 or 1 securely.
- For this, one may consider the bases

$$\{|0\rangle, |1\rangle\};$$

$$\left\{ \frac{1}{\sqrt{2}}(|0\rangle + |1\rangle), \frac{1}{\sqrt{2}}(|0\rangle - |1\rangle) \right\}.$$

- Choosing any one of the above bases, one may encode 0 to one qubit and 1 to the other qubit in that basis.
- If only a single basis is used, then the attacker can measure in that basis to obtain the information and reproduce.
- Thus Alice needs to encode randomly with more than one bases.
- Bob will also measure in random basis.
- Basis will match in a proportion of cases and from that the secret key will be prepared.

This is the brief idea to obtain a secret key between two parties over an insecure public channel using the BB84 [4] protocol. After obtaining the secret key, one may use a symmetric key cryptosystem (for example, a stream cipher or a block cipher, see [38] for details) for further communication in encrypted mode. One may refer to [22] for state-of-the-art results of quantum cryptanalysis on symmetric ciphers, though it is still not as havoc as it had been on classical public key schemes.

15.3.2 Secure Multi-Party Computation

Let us now consider another important aspect of cryptology that might be relevant in data analytics. Take the example of an Automated Teller Machine (ATM) for money transaction. This is a classic example of secure two or multi-party computation. Due to such transactions and several other application domains which are related to secure data handling, Secure Multi-Party Computation (SMC) has become a very important research topic in data intensive areas. In a standard model of SMC, n number of parties wish to compute a function $f(x_1, x_2, \ldots, x_n)$ of their respective inputs $x_1, x_2, \ldots, x_n$, keeping the inputs secret from each other. Such computations have wide applications in online auction, negotiation, electronic voting, etc. Yao's millionaire's problem [44] is considered as one of the initial attempts in the domain of SMC. Later, this has been studied extensively in classical domain (see [18] and the references therein). The security of classical SMC usually comes from some computational assumptions such as hardness of factorization of a large number.

In quantum domain, Lo [24] showed the impossibility for secure computation in certain two-party scenario. For example, "one out of two parties secure computation" means that only one out of two parties is allowed to know the output. As a corollary to this result [24], it had been shown that one out of two oblivious transfer is impossible in quantum paradigm. It has been claimed in [23] that given an implementation of oblivious transfer, it is possible to securely evaluate any polynomial time computable function without any additional primitive in classical domain. However, it seems that such a secure two-party computation might not work in quantum domain. Hence, in case of two-party quantum computation, some additional assumptions, such as the semi-honest third party, etc., have been introduced to obtain the secure private comparison [40].

In [45], Yao had shown that any secure quantum bit commitment scheme can be used to implement secure quantum oblivious transfer. However, Mayers [27] and Lo et al [25] independently proved the insecurity of quantum bit commitment. Very recently some relativistic protocols [26] have been proposed in the domain of quantum SMC. Unfortunately, these techniques are still not very promising for practical implementations. Thus, considering quantum adversaries, it might not be possible to achieve SMC and in turn collaborative multi-party computation in distributed environments without compromising the security.

15.4   Data Analytics: A Critical View of Quantum Paradigm

Given the background of certain developments in quantum paradigm over the classical world, now let us get into some specific issues of data analytics. The first point is, if we consider use of one qubit just as storing one bit of data, then that would be a significant loss in terms of exploiting the much larger (theoretically infinite) space of a qubit. On the other hand, for analysis of classical data, we may require to consider new implementation of data storage that might add additional overhead as data need to be presented in quantum platform. For example, consider the Deutsch-Jozsa [12] algorithm. To apply this algorithm, we cannot use an n-input 1-output Boolean function, but we require a form where the same function can be realized as a function with equal number of input and output bits. Further the same circuit must be implemented with quantum circuits so that the superposition of qubits can be handled. These are the overheads that need to be considered.

Next let us come to the issue of structured and unstructured data. In classical domain, if a data set with N elements are not sorted, then in worst case, we require $O(N)$ search complexity to find a specific data. In quantum domain, the seminal Grover's algorithm [17] shows that this is possible in only $O(\sqrt{N})$ effort. For a huge unsorted data set, this is indeed a significant gain. However, in any efficient database, the individual data elements are stored in a well-structured manner so that one can identify a specific record in $O(\log N)$ time. This is exponentially small in comparison with both $O(N)$ and $O(\sqrt{N})$ and thus, in such a scenario, quantum computers may not be of significant advantage.

15.4.1   Related Quantum Algorithms

To achieve any kind of data analysis, we require several small primitives. Let us first consider finding minimum or maximum from an unsorted list. Similar ideas as in [17] can be applied to obtain minimum or maximum value from an unsorted list of size N in $O(\sqrt{N})$ time as explained in [15] and [2], respectively. The work [20] considers in detail quantum searching in ordered list and sorting. However, in such a scenario where ordered lists are maintained, quantum algorithms do not provide very significant improvements. Matrix related operations are necessary elements in any kind of data analytics. Given $n \times n$ matrices, A, B, C, the matrix product verification problem is to decide whether $A \times B = C$. While the classical domain algorithms must require $\Omega(n^2)$ time, we have $O(n^{\frac{5}{3}})$ algorithm in quantum domain [10]. Such algorithms heavily use results related to quantum walks [39]. In a related direction, solution of a system of linear equations had naturally received serious attention in quantum domain and there are interesting speed-up in several cases. Further these results [19] have applications towards solving linear differential equations, least square techniques and in general, in the domain of machine learning. One may refer to [32] for a detailed description of quantum algorithms and then compare their complexities with the classical counterparts.

While there are certain improvements in specific areas, the situation is not always hopeful and a nice reference in this regard is [1], where Aaronson says

> "Having spent half my life in quantum computing research, I still find it miraculous that the laws of quantum physics let us solve any classical problems exponentially faster than today's computers seem able to solve them. So maybe it shouldn't surprise us that, in machine learning like anywhere else, Nature will still make us work for those speedups."

One may also have a look at [8, 21] for very recent state-of-the-art discussions on quantum supremacy. While most of the explanations do not provide a great recommendation towards advantages of quantum machine learning, for some initial understanding of this area from a positive viewpoint, one may refer to [42].

15.4.2   Database

The next relevant question is if we have significant development in the area of quantum database. In this direction there are some initial concept papers such as [36]. This work presents a novel database abstraction that allows to defer the finalization of choices in transactions until an entity forces the choices by observation in quantum terminology. Following the quantum mechanical idea, here a transaction is in a quantum state, i.e., it could be one of many possible states or might be a superposition. This is naturally undecided and unknown until observed by some kind of measurement. Such an abstraction enables late binding of values read from the database. The authors claimed that this helps in obtaining more transactions to succeed in a situation with high contention. This scenario might be useful for applications where the transactions compete for physical resources represented by data items in the database, such as booking seats in an airline or buying shares. However, these are more at the conceptual level, where actual implementation related details cannot be exactly estimated.

Let us now look at what happens when we are interested in a series of computations which are possibly the most occurring phenomenon in practice. Consider two scenarios, one from a static data set (structured) and another from a dynamic data set where arbitrary search, addition, modification, and alteration are allowed. In static case, the database is generally maintained in such a manner so that the search efforts are always logarithmic. Now consider a little more complex scenario, where the database grows or shrinks arbitrarily and the search and other write operations are allowed in arbitrary sequence. Even in case of such dynamic updations, we always try to maintain some well-known balanced tree structures. Hence, in both the scenarios, we do not have any clear advantage in quantum domain.

15.4.3   Text Mining

Text mining is an integral part of data analytics given the popularity of social media. Consider a scenario involving text mining problem, which uses a bag of words and unsupervised or semi-supervised clustering technique. In the simplest situation, let there be N words in a given corpus (dictionary). Say, the topics are to be extracted in an unsupervised manner from a set of n stories or documents. Each document contains a set of words. Each topic can be seen as a distribution over the set of words in the corpus and also a document can be considered to be a distribution over the set of (unknown) k topics, where the value of k is determined at the beginning depending on the granularity of the topics required. A simple (or innermost iteration) requires going through the documents one by one, allocating the words in the document to topics, while simultaneously modifying the probability distributions of topics in the documents and words in the topics. Now consider just one iteration only. There are two main steps: (1) to create the dictionary (in this case, say the dictionary is fixed, cannot be modified), and (2) we can study one document at a time. For each document, we can allocate each word to a topic and topics to stories following the distributions. It is obvious to see that in classical computation the fixed dictionary is best to be organized as a sorted array. Once this is done, the search efforts are logarithmic in classical domain and we should not get any immediate improvement in the quantum counterpart. In this regard, we also need to refer to topic modeling. Given a corpus of words, topic modeling is more static in nature. However, with time the database of the corpus has to go through changes due to both additions and deletions. The corpus size will generally increase, along with rapid increase in number of stories to be analyzed. Further, with more and more computing capabilities, finer topics and sub-topics will have to be retrieved. Here big data analysis may play an important role and related algorithms should be evaluated in quantum paradigm.

Let us now refer to certain statistical analysis [7] in this domain on a classical model. The idea of Latent Dirichlet allocation (LDA) is described here. This is based on a generative probabilistic model for collections of discrete data, for example, text. LDA is a three-level hierarchical Bayesian model. Each item of a collection is considered as a finite mixture over an underlying set of topics. These techniques can be used in text classification. However, it is not very clear how these complex ideas can be lifted in quantum domain. In a follow-up work [16], this has been extended where the authors present a Markov chain Monte Carlo algorithm for inference (for quantum speed-up for Monte Carlo methods, one may refer to [28]). This algorithm is applied to analyze abstracts from scientific journals using Bayesian model selection to identify the number of topics. Text mining is one of the most important topics in the domain of analytics and thus this kind of scenarios need to be explored in quantum domain. One may refer to [11] where several ideas of quantum Markov chains are discussed from a different information-theoretic viewpoint and it is not very clear how long it will take to connect ideas from machine learning domain and the paradigm of quantum information to obtain meaningful commercial results.

15.5 Conclusion: Google, PageRank, and Quantum Advantage

In this review, we have taken an approach to present certain introductory issues in quantum paradigm and then explained how they relate to basics of data analytics. We described several aspects in the domain of computation, communication, and security and pointed out why the computational part should receive prime attention. In the quantum computational model, we have enumerated several significant improvements over the classical counterpart, but the two main concerns that remain are as follows.

- Can we fabricate a commercially viable quantum computer?
- (Even if we have a quantum computer) Can we have significant improvements in computational complexity for algorithms related to data analytics?

Let us now conclude with a very practical and well-known problem in the domain of data analytics that received a significant attention. This should help the reader to form his/her own opinion regarding the impact of quantum computation on a significant problem. The problem is related to PageRank. PageRank is an algorithm used by Google Search to rank the websites through their search engine results. It is a method of quantifying the importance of the web pages, i.e., PageRank may be viewed as a metric proposed by Google's owners Larry Page and Sergey Brin. According to Google:

> "PageRank works by counting the number and quality of links to a page to determine a rough estimate of how important the website is. The underlying assumption is that more important websites are likely to receive more links from other websites."

Informally speaking, the PageRank algorithm heuristically provides a probability distribution. This is used to represent the likelihood that an entity, randomly clicking on web links, will arrive at any particular page. It is very natural that this kind of technique will require huge amount of computational resources and further there will be continuous efforts in upgrading such strategies. Some parts of such effort might involve a lot of "rough" heuristics where exact quantification in such a complex environment might be very hard. In [30], it has been outlined that a quantum version of Google's famous search algorithm may be significantly faster. However, till date it is not clearly understood how such quantum algorithms may behave on a huge network. We have to wait and watch to experience how the quantum algorithms will evolve to solve the complex problems of data analytics in the coming days.

Acknowledgements Arpita Maitra is supported by the project "Information Security and Quantum Cryptography: Study of Secure Multi-Party Computation in Quantum Domain and Applications" at IIM Calcutta.

Subhamoy Maitra is supported by the project "Cryptography & Cryptanalysis: How far can we bridge the gap between Classical and Quantum Paradigm," awarded by the Scientific Research Council of the Department of Atomic Energy (DAE-SRC), the Board of Research in Nuclear Sciences (BRNS).

Asim K. Pal is supported by the projects "Sentiment analysis: An approach with data mining, computational intelligence and longitudinal analysis with Applications to finance and marketing" as well as "Information Security and Quantum Cryptography: Study of Secure Multi-Party Computation in Quantum Domain and Applications" at IIM Calcutta.

References

1. S. Aaronson, Quantum machine learning algorithms: read the fine print preprint (2015). Available at http://www.scottaaronson.com/papers/qml.pdf
2. A. Ahuja, S. Kapoor, A quantum algorithm for finding the maximum (1999). Available at https://arxiv.org/abs/quant-ph/9911082
3. C.H. Bennett, G. Brassard, Quantum cryptography and its application to provably secure key expansion, public-key distribution, and coin-tossing, in *Proceedings of IEEE International Symposium on Information Theory*, St-Jovite, p. 91, Sept 1983
4. C.H. Bennett, G. Brassard, Quantum Cryptography: public key distribution and coin tossing, in *Proceedings of the IEEE International Conference on Computers, Systems, and Signal Processing*, Bangalore (IEEE, New York, 1984), pp. 175–179
5. C.H. Bennett, G. Brassard, C. Crepeau, R. Jozsa, A. Peres, W.K. Wootters, Teleporting an unknown quantum state via dual classical and Einstein-Podolsky-Rosen channels. Phys. Rev. Lett. **70**, 1895–1899 (1993)
6. E. Bernstein, U. Vazirani, Quantum complexity theory, in *Proceedings of the 25th Annual ACM Symposium on Theory of Computing* (ACM Press, New York, 1993), pp. 11–20
7. D.M. Blei, A.Y. Ng, M.I. Jordan, Latent Dirichlet allocation. J. Mach. Learn. Res. **3**, 993–1022 (2003)
8. S. Boixo, S.V. Isakov, V.N. Smelyanskiy, R. Babbush, N. Ding, Z. Jiang, J.M. Martinis, H. Neven, Characterizing quantum supremacy in near-term devices. https://arxiv.org/abs/1608.00263, Aug 3 (2016)
9. G. Brassard, Brief history of quantum cryptography: a personal perspective, in *Proceedings of IEEE Information Theory Workshop on Theory and Practice in Information Theoretic Security*, Awaji Island, Oct 2005, pp. 19–23. [quant-ph/0604072]

10. H. Buhrman, R. Spalek, Quantum verification of matrix products, in *Proceedings of the 17th ACM-SIAM Symposium on Discrete Algorithms*, pp. 880–889 (2006). arXiv:quant-ph/0409035

11. N. Datta, M.M. Wilde. Quantum Markov chains, sufficiency of quantum channels, and Renyi information measures. J. Phys. A **48**(50), 505301 (2015). Available at https://arxiv.org/abs/1501.05636

12. D. Deutsch, R. Jozsa, Rapid solution of problems by quantum computation. Proc. R. Soc. Lond. A **439**, 553–558 (1992)

13. D. Dieks, Communication by EPR devices. Phys. Lett. A **92**(6), 271–272 (1982)

14. W. Diffie, M.E. Hellman, New directions in cryptography. IEEE Trans. Inf. Theory **22**, 644–654 (1976)

15. C. Durr, P. Hoyer, A quantum algorithm for finding the minimum (1996). Available at https://arxiv.org/abs/quant-ph/9607014

16. T.L. Griffiths, M. Steyvers, Finding scientific topics. Proc. Natl. Acad. Sci. U.S.A. **101** suppl. 1, 5228–5235 (2004). Available at www.pnas.org/cgi/doi/10.1073/pnas.0307752101

17. L. Grover, A fast quantum mechanical algorithm for database search, in *Proceedings of 28th Annual Symposium on the Theory of Computing (STOC)*, pp 212–219, May 1996. Available at http://xxx.lanl.gov/abs/quant-ph/9605043

18. S.D. Gordon, C. Hazay, J. Katz, Y. Lindell, Complete fairness in secure two-party computation, in *Proceedings of the 40-th Annual ACM symposium on Theory of Computing (STOC)* (ACM Press, New York, 2008), pp. 413–422

19. A.W. Harrow, A. Hassidim, S. Lloyd, Quantum algorithm for linear systems of equations. Phys. Rev. Lett. **103**(15), 150502 (2009). Available at https://arxiv.org/abs/0811.3171

20. P. Hoyer, J. Neerbek, Y. Shi, Quantum complexities of ordered searching, sorting, and element distinctness (2001). Available at https://arxiv.org/abs/quant-ph/0102078

21. https://rjlipton.wordpress.com/2016/04/22/quantum-supremacy-and-complexity/. April 22, 2016

22. M. Kaplan, G. Leurent, A. Leverrier, M. Naya-Plasencia, Breaking symmetric cryptosystems using quantum period finding, in *CRYPTO (2)*. Lecture Notes in Computer Science, vol. 9815, (Springer, New York, 2016), pp. 207–237

23. J. Killan, Founding cryptography on oblivious transfer, in *Proceedings of the 20th Annual ACM Symposium on the Theory of Computation (STOC)* (1988)

24. H.-K. Lo, Insecurity of quantum secure computations. Phys. Rev. A **56**, 1154–1162 (1997)

25. H.-K. Lo, H.F. Chau, Is quantum bit commitment really possible? Phys. Rev. Lett. **78**, 3410 (1997)

26. T. Lunghi, J. Kaniewski, F. Bussieres, R. Houlmann, M. Tomamichel, S. Werner, H. Zbinden, Practical relativistic bit commitment. Phys. Rev. Lett. **115**, 030502 (2015)

27. D. Mayers. Unconditionally secure quantum bit commitment is impossible. Phys. Rev. Lett. **78**, 3414 (1997)

28. A. Montanaro, Quantum speedup of Monte Carlo methods. Proc. R. Soc. A **471**, 20150301 (2015). Available at http://dx.doi.org/10.1098/rspa.2015.0301

29. M.A. Nielsen, I.L. Chuang, *Quantum Computation and Quantum Information* (Cambridge University Press, Cambridge, 2010)

30. G.D. Paparo, M.A. Martin-Delgado, Google in a quantum network. Sci. Rep. **2**, 444 (2012). Available at https://arxiv.org/abs/1112.2079

31. Post-quantum cryptography. http://pqcrypto.org/

32. Quantum algorithm zoo. http://math.nist.gov/quantum/zoo/

33. Quantum key distribution equipment. ID Quantique (IDQ). http://www.idquantique.com/

34. Quantum key distribution system (Q-Box). MagiQ Technologies Inc. http://www.magiqtech.com

35. R.L. Rivest, A. Shamir, L. Adleman, A method for obtaining digital signatures and public key cryptosystems. Commun. ACM **21**, 120–126 (1978)

36. S. Roy, L. Kot, C. Koch. Quantum databases, *The 6th Biennial Conference on Innovative Data Systems Research (CIDR)* (2013)

37. P.W. Shor, Algorithms for quantum computation: discrete logarithms and factoring, in *Foundations of Computer Science (FOCS) 1994* (IEEE Computer Society Press, New York, 1994), pp. 124–134
38. D. Stinson, *Cryptography Theory and Practice*, 3rd edn. (Chapman & Hall/CRC, Boca Raton, 2005)
39. M. Szegedy, Quantum speed-up of Markov chain based algorithms, in *Proceedings of the 45th IEEE Symposium on Foundations of Computer Science*, pp. 32–41 (2004)
40. H.Y. Tseng, J. Lin, T. Hwang, New quantum private comparison protocol using EPR pairs. Quantum Inf. Process. **11**, 373–384 (2012)
41. S. Wiesner, Conjugate coding. Manuscript 1970, subsequently published in SIGACT News 15:1, pp.78–88 (1983)
42. P. Wittek, Quantum machine learning: what quantum computing means to data mining. http://peterwittek.com/book.html (2014)
43. W.K. Wootters, W.H. Zurek, A single quantum cannot be cloned. Nature **299**, 802–803 (1982)
44. A.C. Yao, Protocols for secure computations, *23rd Annual Symposium on Foundations of Computer Science (FOCS)*, pp. 160–164 (1982)
45. A.C. Yao, Security of quantum protocols against coherent measurements, in *Proceedings of 26th Annual ACM Symposium on the Theory of Computing (STOC)*, vol. 67 (1995)